Marine Molecular Biotechnology

Subseries of Progress in Molecular and Subcellular Biology
Series Editor: Werner E. G. Müller

Progress in Molecular and Subcellular Biology **39**
Series Editors: W.E.G. Müller (Managing Editor),
Ph. Jeanteur, Y. Kuchino, A. Macieira-Coelho, R. E. Rhoads

Volumes published in the Series

Progress in Molecular and Subcellular Biology

Volume 25
Signaling Through the Cell Matrix
A. Macieira-Coelho (Ed.)

Volume 26
Signaling Pathways for Translation: Insulin and Nutrients
R.E. Rhoads (Ed.)

Volume 27
Signaling Pathways for Translation: Stress, Calcium, and Rapamycin
R.E. Rhoads (Ed.)

Volume 28
Small Stress Proteins
A.-P. Arrigo and W.E.G. Müller (Eds.)

Volume 29
Protein Degradation in Health and Disease
M. Reboud-Ravaux (Ed.)

Volume 30
Biology of Aging
A. Macieira-Coelho

Volume 31
Regulation of Alternative Splicing
Ph. Jeanteur (Ed.)

Volume 32
Guidance Cues in the Developing Brain
I. Kostovic (Ed.)

Volume 33
Silicon Biomineralization
W.E.G. Müller (Ed.)

Volume 34
Invertebrate Cytokines and the Phylogeny of Immunity
A. Beschin and W.E.G. Müller (Eds.)

Volume 35
RNA Trafficking and Nuclear Structure Dynamics
Ph. Jeanteur (Ed.)

Volume 36
Viruses and Apoptosis
C. Alonso (Ed.)

Volume 38
Epigenetics and Chromatin
Ph. Jeanteur (Ed.)

Subseries Marine Molecular Biotechnology

Volume 37
Sponges (Porifera)
W.E.G. Müller (Ed.)

Volume 39
Echinodermata
V. Matranga (Ed.)

Valeria Matranga (Ed.)

Echinodermata

With 75 Figures, 11 in Color

Springer

Dr. Valeria Matranga
Istituto die Biomedicina e
Immunologia Molecolare „Alberto Monroy"
Via U. La Malfa 153
90146 Palermo
Italy

ISSN 1611-6119
ISBN 3-540-24402-6 Springer-Verlag Berlin Heidelberg New York

Library of Congress Control Number: 2004117278

Springer-Verlag is a part of Springer Science+Business Media
springeronline.com

Printed in Germany

Production and typesetting: Friedmut Kröner, Heidelberg, Germany
Cover design: *design & production* GmbH, Heidelberg, Germany

Printed on acid free paper 39/3150 YK 5 4 3 2 1 0

Preface to the Series

Recent developments in the applied field of natural products are impressive, and the speed of progress appears to be almost self-accelerating. The results emerging make it obvious that nature provides chemicals, secondary metabolites, of astonishing complexity. It is generally accepted that these natural products offer new potential for human therapy and biopolymer science. The major disciplines which have contributed, and increasingly contribute, to progress in the successful exploitation of this natural richness include molecular biology and cell biology, flanked by chemistry. The organisms of choice, useful for such exploitation, live in the marine environment. They have the longest evolutionary history during which they could develop strategies to fight successfully against invading organisms and to form large multicellular plants and animals in aqueous medium. The first multicellular organisms, the plants, appeared already 1000 million years ago (MYA), then the fungi emerged and, finally, animals developed (800 MYA).

Focusing on marine animals, the evolutionary oldest phyla, the Porifera, the Cnidaria and the Bryozoa, as sessile filter feeders, are exposed not only to a huge variety of commensal, but also toxic microorganisms, bacteria and fungi. In order to overcome these threats, they developed a panel of defense systems, for example, their immune system, which is closely related to those existing in higher metazoans, the Protostomia and Deuterostomia. In addition, due to this characteristic, they became outstandingly successful during evolution: they developed a chemical defense system which enabled them to fight in a specific manner against invaders. These chemicals are of low molecular weight and of non-proteinaceous nature. Due to the chemical complexity and the presence of asymmetrical atom centers in these compounds, a high diversity of compounds became theoretically possible. In a natural selective process, during evolution, only those compounds could survive which caused the most potent bioactivity and provided the most powerful protection for the host in which they were synthesized. This means that during evolution nature continuously modified the basic structures and their derivatives for optimal function. In principle, the approach used in combinatorial chemistry is the same, but turned out to be painful and only in few cases successful. In consequence, it is advisable to copy and exploit nature for these strategies to select for bioactive drugs. Besides the mentioned metazoan phyla, other animal phyla, such as the higher evolved animals, the mollusks or tunicates, or certain algal groups, also produce compounds for their chemical defense which are of interest scientifically and for potential application.

There is, however, one drawback. Usually, the amount of starting material used as a source for the extraction of most bioactive compounds found in marine organisms is minute and, hence, not sufficient for their further application in biomedicine. Furthermore, the constraints of the conventions for the protection of nature limit the commercial exploitation of novel compounds, since only a small number of organisms can be collected from the biotope. Consequently, exploitation must be sustainable, i.e., it should not endanger the equilibrium of the biota in a given ecosystem. However, the protection of biodiversity in nature, in general, and those organisms living in the marine environment, in particular, holds an inherent opportunity if this activity is based on genetic approaches. From the research on molecular biodiversity, benefits for human society emerge which are of obvious commercial value; the transfer of basic scientific achievements to applicable products is the task and the subject of *Marine Molecular Biotechnology*. This discipline uses modern molecular and cell biological techniques for the sustainable production of bioactive compounds and for the improvement of fermentation technologies in bioreactors.

Hence, marine molecular biotechnology is the discipline which strives to define and solve the problems regarding the sustainable exploitation of nature for human health and welfare, through the cooperation between scientists working in marine biology/molecular biology/microbiology and chemistry. Such collaboration is now going on successfully in several laboratories. It is the aim of this new subset of thematically connected volumes within our series "Progress in Molecular and Subcellular Biology" to provide an actual forum for the exchange of ideas and expertise between colleagues working in this exciting field of "Marine Molecular Biotechnology". It also aims to disseminate the results to those researchers who are interested in the recent achievements in this area or are just curious to learn how science can help to exploit nature in a sustainable manner for human prosperity.

Werner E.G. Müller

Preface

When, nearly 20 years ago, I was participating in the organization of a workshop held in Palermo entitled "Regulation of Transcription in Sea Urchin Embryos" (see Fig. 1), under the expert direction of Prof. G. Giudice, at the time Director of both the University and CNR Institutes for Developmental Biology, I was not aware of the extent to which echinoderms (sea urchins) contribute to the advancement of science. It was spring 1986 and, among other well-known and those who would one day be well-known in the field, Tim Hunt (R. Timothy Hunt) presented his lecture on the "Role of maternal mRNA in the regulation of cell division in early cell cycles of the sea urchin embryos". I am sure that at the moment most, if not all, of the audience found the way in which a messenger RNA could promote the synthesis of a protein which was going up and down during cell cycles very bizarre. Many years later, his efforts, together with those of Paul M. Nurse and Leland H. Hartwell, were appreciated worldwide. They were awarded the Nobel Prize in Physiology and Medicine 2001 for "their discoveries of key regulators of the cell cycle". In fact, their work led to the discovery of cyclins, proteins widespread in the animal kingdom, that oscillate throughout the cell cycle, binding and activating cyclin-dependent kinases. Needless to say that nowadays everybody knows what cyclins are and how studies on their regulation are inspiring new therapies against human cancer. However, very few people remember that this all stemmed from a lesson from the sea urchin, just one example of how precious simple marine organisms can be in teaching scientists how to improve human health.

With this in mind, it was with great pleasure that I accepted the kind invitation of Prof. Müller to assemble a book that, by examining the recent productive research in support of marine biotechnology, would encourage further studies on the sustainable exploitation of biologically active compounds from echinoderms. Echinodermata, a phylum that appeared back in the Precambrian age, accounting for more than 7,000 living species, belongs to the branch of the animal kingdom known as the deuterostomes, a group that also includes man. Since they are phylogenetically more related to chordates than to other invertebrate groups, it is not surprising that echinoderms possess regulatory mechanisms rather similar to those of vertebrates. This explains the flourishing interest of scientists in the fields of developmental biology, molecular genomics and molecular evolution.

However, as many of the contributors to this book have pointed out, despite the incredible amount of research accomplished during the past centuries using echinoderms as a model organism in the fields of zoology, ecology, embryology, cell biology, molecular biology, etc., modest efforts of the echinoderm scientific community have been directed towards studies on the development of the sustainable production of bioactive compounds from echinoderms and their application in biomedicine.

This book illustrates the progress made in the exploitation of natural products from marine invertebrates (echinoderms) at a specialized high level. Studies describe: (1) the discovery of new potential therapeutic tools for human health; (2) the introduction of new biomolecular and biocellular sensors for the detection of environmental contamination; (3) the development of new composite materials for biomedical applications; and (4) the improvements and limitations of aquacultural techniques and farming. Past and recent findings are reported in the format of reviews on general topics or detailed explanations where the description of a recently defined method is needed.

Environmental and health issues are addressed by the 2004–2010 Action Plan of the European Community which focuses on: (1) an integrated environmental and health monitoring system; (2) the standardization of methods of analysis; and (3) common sampling and sample preparation procedures. Taking these points into consideration, I decided to invite only European scientists, with only one exception, to contribute their work and thoughts, most of them being involved in EU R&D projects which encourage and support multidisciplinary research with the aim of ensuring the rapid transfer of technology.

For all the above-mentioned reasons, I am particularly grateful to the *Marine Molecular Biotechnology* Series Editor, Prof. Werner Müller, who gave me the opportunity to select those studies which I think will be able to promote discussion in this rapidly growing field and open new routes for research on innovative bioactive compounds to be used in environmental and medical research.

Many thanks are extended to all the contributors to this book; I am confident that the high scientific value of their reviews on past and current findings will serve as a forum of ideas for the exploitation of echinoderm as a bioresource and will promote the development of studies in the new exciting field of marine molecular biotechnology.

I am also very grateful to the invaluable and friendly collaboration and assistance of all the actual members of the group (see Fig. 2) whose daily hard work and professional skills were part of the success of this enterprise. Special warm thanks go to my colleagues Francesca Zito and Rosa Bonaventura for their continuous and generous collaboration, support and assistance throughout all the difficulties related to the editorial work.

Fig. 1. Original sketch of a sea urchin, engaged in "transcribing", drawn by G. Giudice and used for the cover of the scientific program of the workshop on "Regulation of Transcription in Sea Urchin Embryos", held in Palermo in 1986

Finally, I express my gratitude to my husband Benedetto and my daughter Laura for their continuous intellectual and practical support. In spite of the time it took me away from them, their intelligent and open-minded attitude was invaluable in this scientific endeavour.

In closing, I would like to recall the words of Prof. Monroy, who used to say: "Science is one of the best ways for mutual understanding. Its language is universal and pays no attention to national barriers. This is to say that science is one of the best approaches to peace".

Valeria Matranga

Foreword 1

As Director of the Institute of Biomedicine and Molecular Immunology "Alberto Monroy", I am pleased to introduce this book on the marine molecular biotechnology of echinoderms, edited by Dr. Valeria Matranga, with the precious help of her team, who dedicated much time and effort. I know and appreciate the dedication of Valeria Matranga to her work on cell biology, especially her research on cell–environment interactions and on the validation of the role played by cellular and molecular biology in the detection and assessment of environmental pollution. I am confident that this book will have a great impact on our understanding of biological markers and their different functions in the cell. In addition, the advancements described here will be extremely useful for the evaluation of environmental risk. Moreover, this research field ultimately addresses issues on the protection of human health, since we can now monitor potentially dangerous changes in the sea environment along the coast where degradation products from the earth accumulate.

The topics addressed in the book, *Echinodermata*, will be of interest to scientists in various fields:

- Cell biologists will find it useful to study cell responses to stress factors;
- Marine biologists will be able to evaluate the damage to cells and organisms caused by coastal water pollution, as well as the role of monitoring such phenomena;
- Researchers dedicated to Nutritional Sciences will appreciate the "warning messages" provided by the environmental monitoring of seawater;
- Environmental scientists will take advantage of an additional way to fight seawater pollution, which could possibly be useful in the prevention and adequate correction of problems originating from ground pollution.

Finally, I wish Valeria Matranga and her team continuing success in their engrossing and promising research field.

Istituto di Biomedicina e Immunologia Molecolare (IBIM) "Alberto Monroy"
Palermo, Italy

Professor Giovanni Bonsignore
(Director)

Foreword 2

Biology owes a great deal to Echinodermata, especially in the field of development. Sea urchins have actually represented one of the main and one of the first biological materials on which the history of developmental biology was built.

Fertilization and parthenogenesis were, in fact, first clearly and correctly described in sea urchins. Oskar Hertwig, working at the Zoological Station of Naples, wrote in 1875: "I was lucky enough, by studying the egg of *Toxopneustes lividus*, to find an object in which phases and intimate phenomena of fertilization were clearly visible, that is: following fertilization, few minutes after semen addition, around the sperm head, lodged in the cortex, a series of rays was formed". The description of a series of important observations followed, which allowed him to conclude: "And therefore I was able to formulate the low: the fertilization rests on the union of two sexually different cells".

Again, with regard to sea urchins, Edmund B. Wilson (1906) wrote: "Fertilization accordingly consists of two distinct phenomena: first the introduction into the egg of the paternal hereditary characteristics potentially contained in some unknown manner in the substance of the sperm nucleus or of the chromosomes in which resolves itself. Second in the introduction into the egg of a centrosome which gives rise to the mechanisms by means of which the egg divides and the hereditary substance is distributed to the resulting cells". Thus, the concept of the centrosome is also due to sea urchins.

In 1895 it was again Hertwig who described parthenogenesis in sea urchins, while Jacque Loeb in 1909 showed that it is possible to induce parthenogenesis in sea urchins also by chemical treatments such as hypertonic seawater. This caused public surprise and also some concern, such that some people even advised women to avoid bathing in seawater owing to the danger of parthenogenesis!

Furthermore, genetics, in spite of a long-lasting metamorphosis, owes some crucial initial experiments to sea urchins. It suffices here to recall the experiments of Theodor Bovery, who in 1889 studied sea urchin hybrids, again at the Zoological Station of Naples, and concluded that chromosomes are qualitatively responsible for genetic character transmission, after observing the fate of hybrid merogones in which some chromosomes were selectively lacking.

Although not being the elective material for genetic studies, sea urchins represented and still represent a very important model for studies of molecular biology and molecular genomics. The story started with the demonstration by Alberto Monroy that, following fertilization, there is an activation of protein synthesis in the sea urchin egg. This demonstration was achieved by Eizo Nakano and Alberto Monroy in Palermo by preloading the amino acid pool of the unfertilized sea urchin egg with radioactive amino acids (which was carried out by injecting the labeled amino acids into the coelomic cavity of the adult female), and showing that immediately following fertilization, but not before, there was a quick transfer of the amino acids from the pool to the proteins. This injection of amino acids into the adult sea urchin was depicted by a cartoonist (myself, at that time one of Monroy's students; Fig. 1).

The story of molecular genetics has continued in sea urchins, especially in the USA, where many research groups flourished, especially that of Eric Davidson at Caltech, of which I will only recall here the regulatory gene networks described in sea urchins. It has continued also in Europe, e.g. in Palermo with Monroy's former students, including myself and younger coworkers, both at the University and at the National Research Council of Palermo. It has also continued at the Zoological Station of Naples and elsewhere, e.g. in Villefranche sur Mer, just to quote an example. Other groups are studying molecular genomics of sea urchins in Japan.

It should be noted here that the hypothesis of a role for DNA methylation in differentiation was first proposed in sea urchin embryos in 1958 by Edoardo Scarano and by colleagues still working in Naples.

It is also worth recalling the contribution of sea urchins to the field of cell interaction. Curt Herbst, in 1891, working at the Zoological Station of Naples, found that lowering the calcium concentration of seawater brought about a loosening of contacts between sea urchin blastomeres, which remained inside the fertilization envelope and continued to develop, although with a "kranklike" aspect. This observation allowed Hans Driesch to easily separate the first two blastomeres and to show that each gave rise to an entire, albeit smaller,

Fig. 1. Original sketch by G. Giudice representing a furious sea urchin chasing Prof. Eizo Nakano, armed with a syringe, and Prof. Alberto Monroy, first in the row, at the time of their joint experiments in the mid-1950s.

pluteus. Many years later, in 1961, I succeeded in dissociating sea urchin blastulae into single cells, essentially by removing calcium. These cells were able to reaggregate and to develop into pluteus-like structures, which represented the first example of entire larvae of any kind of embryos reconstituted from dissociated cells. These studies were followed by those of some colleagues of mine, e.g. Letizia Vittorelli and Valeria Matranga, and by others, including Yukio Yokota, Hans Noll and David McClay.

The theory of morphogen gradients, so popular in developmental biology, also originated in the study of sea urchins, following the beautiful microtransplantation experiments done by Sven Hörstadius in 1928 and by the intelligent speculations of John Runnström.

Finally, the beginning of so-called chemical embryology can be attributed to studies on sea urchins: it was in fact Otto Waburg who discovered in 1908 that following fertilization of sea urchin eggs there was a sudden increase in oxygen consumption.

I am aware that many important results obtained not only using sea urchins, but using echinoderms in general have not been included here. I hope, however, to have succeeded in giving an idea of the contribution that studies on sea urchins have provided to developmental biology in the past in places like Naples Zoological Station, Woods Hole MBL, Stockholm Carolinska Institute, Sugashima Marine Station and Villefranche sur Mer Marine Station, and continue to provide especially in places like the USA (i.e. the MBL in Woods Hole, Caltech, Pennsylvania and so on), Italy (Naples and Palermo), France and Japan.

Dipartimento di Biologia Cellulare e dello
Sviluppo "Alberto Monroy", Università di Palermo
Palermo, Italy

Professor G. Giudice

Contents

Are Echinoderms of Interest to Biotechnology?

C. Petzelt

Abstract. The huge potential of echinoderms as a so far fairly untapped source of bioactive molecules is described. Examples are presented that show the usefulness of echinoderm-derived molecules for therapeutic application in selected fields of cancer research, in the control of bacterial growth as substances with new antibiotic properties, and finally in the context of technical applications such as antifouling substances. The molecules described here are but the mere beginning of a commercial exploitation of echinoderms and may incite a deeper involvement of biotechnology-oriented research in this material.

Echinoderms have been used as embryos since the dawn of cell biology as model systems to study basic phenomena such as mitosis, cell division, differentiation, and organ formation, and are linked to the great cell biologists of the 19th and 20th centuries; among many others, Boveri, Heilbrunn, Mazia, Monroy, Wilson, Hertwig, and Brachet may be cited. Today, echinoderm embryos continue to be the model of choice for many cell and molecular biologists, offering exciting overtures on the way from molecular to cell biology (e.g. Arnone et al. 1997). At the same time they serve as a sensitive test system for toxicological and environmental studies (Matranga et al. 2000).

This chapter is not intended to cover the many applications of the peculiarities of the echinoderm embryonic system in cell and molecular biology; rather it will give examples of echinoderm-derived substances that may have biotechnological value.

Without any doubt the marine environment has huge potential as a source of new compounds to be used in so far unknown strategies in the combat of many pathological situations. In the course of millions of years of evolution, many trials and errors have been made to protect the individual, either the cell

C. Petzelt (e-mail: petzelt@libmyeu.com)
Laboratoire International de Biologie Marine (LIBM), 85350 Ile d'Yeu, France. *Present address*: Experimental Anaesthesiology, University Hospital Charité, Spandauer Damm 130, FH 31, 14050 Berlin, Germany

Progress in Molecular and Subcellular Biology
Subseries Marine Molecular Biotechnology
V. Matranga (Ed.), Echinodermata

or the entire organism, and to gain advantage in the face of possible competitors. Substances have evolved that interfere with signalling or are simply poisonous to other cells. Such products, to be discussed later, could be used in the human environment to attack key problems such as cancer, bacterial infections, or, more practically, biofouling. Surprisingly, echinoderms appear to be a rather untapped source in the pursuit of the identification of new and useful products.

Successful identification depends to a large extent on the test system used and on subsequent purification techniques. Probably one of the most direct approaches is the use of whole animals or pieces of an animal to test whether there is any antibacterial or cytotoxic activity.

Sasaki et al. (1985) identified in the macromolecular fractions of aqueous extracts from the sea urchin *Strongylocentrotus nudus* antitumor activities that inhibited transplanted sarcoma 180 solid form in ICR mice. Unfortunately, this interesting work was apparently not followed up. Earlier on, it was shown in an elegant study that triterpene glycoside isolated from 19 holothurian species of the Pacific tropical zone exhibited cytotoxic activity against yeast and tumour cells, whereas bacteria were not affected. Out of these glycosides the stichoposides, theoturins, and oligosides of Holothuria of the genus *Bohadschia* were found to be the most active versus fungal and yeast microflora and tumour cells (Kuznetsova et al. 1982).

Haug et al. (2002) isolated from the sea urchin, *Strongylocentrotus droebachiensis*, the sea cucumber *Cucumaria frondosa*, and the starfish *Asterias rubens* antibacterial activities that were detected in extracts from several tissues in all species tested, but mainly in the coelomocyte and body wall extracts. High antibacterial activity was also found in gastrointestinal organs and eggs from *A. rubens* and in eggs from *C. frondosa*. If differences in hydrophobicity and sensitivity to heat and proteinase K treatment were compared between active extracts, it was observed that several different compounds were responsible for the antibacterial activities detected. Lysozyme-like activity was identified in several tissues from *A. rubens*. Haemolytic activity could be detected in all species tested, especially in the body wall extracts. These results indicate that echinoderms may serve as a useful source when searching for novel antibiotics.

Carballo et al. (2002) used two brine shrimp assays to identify potential cytotoxic substances useful in cancer therapy. They incubated whole body extracts from three echinoderms (*Holothuria impatiens*, *Pseudoconus californica*, and *Pharia pyramidata*) that showed a strong cytostatic (growth inhibition) and cytotoxic effect against two human cell lines, lung carcinoma A-549 and colon carcinoma HT-29. Palagiano et al. (1996) isolated up to 20 steroid glycosides from the starfish *Henricia downeyae* that caused growth inhibition in bacteria and fungi. In this work, it is remarkable that the biological activity originally identified in ethanolic extracts was related to single compounds whose molecular structures were even identified. Aminin et al. (1995) first identified in the Pacific brittle star *Ophiopholis aculeata* disulfated polyhydroxysteroids that turned out to be potent Ca^{2+} agonists in mammalian cell systems (Aminin et al. 1995; Agafonova et al. 2002).

Even in the main constituents of the immune systems of echinoderms, cytotoxic substances are found. Coelomocytes are intriguing entities expressing variable effector mechanisms that are elicited specifically and are repeatable after a variety of non-self challenges (Glinski and Jarosz 2000; Lin et al. 2001). Stabili et al. (1996) were able to isolate from coelomocytes of the sea urchin *Paracentrotus lividus* a bactericidal protein and purified it to a single polypeptide chain with a molecular weight of 60 kDa.

Epibiosis, the colonization of biogenic surfaces by epibiotic organisms such as bacteria, filamentous algae, and sessile invertebrates, poses a major threat to the fitness and survival of macroorganisms which could potentially be fouled. Fouling of artificially submerged structures (e.g. ship hulls) can also cause severe economic problems, establishing the need for refined bioassays to determine the efficacy of potential antifouling compounds. Palagiano et al. (1996) identified steroid glycosides from the starfish *H. downeyae* with profound antifouling activities, compounds that were also found in several species of the family Echinasteridae. Iken et al. (2003) monitored the brown algal spore swimming behaviour in the presence of echinoderm extracts in order to identify possible antifouling activities. They tested different concentrations of aqueous and organic extracts from body walls of sympatric echinoderms (starfish *Luidia* and *Astropecten* and the brittle star *Astrocyclus*). They found significant effects of those extracts on spore swimming behaviour at concentrations three orders of magnitude lower than that present naturally in the echinoderm body walls.

Sulfated fucans are among the most prominent of all the sulfated polysaccharides of non-mammalian origin that exhibit biological activities in mammalian systems. Pereira et al. (1999) investigated the anticoagulant activity of echinoderm fucans in comparison with that of several species of brown algae and found that the linear sulfated fucans from echinoderms had an anticoagulant action resembling that of mammalian dermatan sulfate, whereas the branched fucans from brown algae were direct inhibitors of thrombin. Such differences have also been described for the linear sulfated fucans derived from sea cucumbers compared to algal fucans (Mulloy et al. 2000).

Glycosphingolipids and glycopeptides are normally occurring constituents of various cell membranes in a wide variety of organisms. Surprisingly, these compounds derived from echinoderms exhibit biological function that might render them useful in medical applications. Glycosphingolipids have been isolated from sea cucumbers that had neuritogenic activity in the rat pheochromocytoma cell line PC-12, i.e. they were able to induce neurite differentiation in the same way as can be achieved by the addition of nerve growth factor (Yamada 2002). It remains to be seen whether these compounds find medical applications, but even today they are of economic interest in view of the high price of commercially available nerve growth factor.

Several unique lectins are found in echinoderms. In marine invertebrates lectins may be considered as humoral factors in the defence mechanism, as are immunoglobulins in vertebrates, resulting in activation of phagocytes. On the other hand, direct haemolytic activity has recently been found in a galactose-specific lectin from the sea cucumber *Cucumaria echinata* (Hatakeyama et al. 1999). After binding to the specific carbohydrate chains on the erythrocyte surface, these lectins damage the cell membrane, leading to cell lysis. A lectin with biological activities such as mitogenic and chemotactic characteristics was also described in the venom of the pedicellariae of the sea urchin *Toxopneustes pileolus* (Nakagawa and Kimura 1982) as had been described with other bioactive substances many years earlier by Alender (1967) occurring in the spines of *Diadema* sea urchins.

A rich source of useful venoms has been found in the crown-of-thorns starfish *Acanthaster planci*. One of its deadly venoms has been identified as a myotoxic phospholipase A (Mebs 1991), and several other candidates for such effects have been identified (Shiomi et al. 1985, 1988; Mebs 1989).

A not yet commercially exploited area is the phenomenon of bioluminescence in the brittle star *Amphipolis squamata*. The photocytes of this animal produce light dependent on cyclic nucleotide and IP3, and the system may become as useful and widespread as the luciferin–luciferase system is today (De Bremaeker et al. 2000; Deheyn et al. 2000a–e).

Finally, an example may be cited of how echinoderms may become useful even in bionics. It has been recently discovered by Aizenberg et al. (2001) that in the brittle star *Ophiocoma wendtii* single calcite crystals are arranged in such a way that they function as lenses. These lenses focus light onto nerve bundles that run behind them, which presumably receive the signal to be further processed. Thus, these thousands of lenses form a compound eye that covers the upper surface of the animal, resulting in a function similar to a digital camera that builds up the picture pixel by pixel. At present, engineers in the photonic industry try to imitate the perfect calcite lenses and their use in signal reception.

This short overview was intended to stimulate the search for bioactive compounds in echinoderms. It is surprising how little work has been done to identify promising candidates for applied research, especially in view of the widespread availability of the animals. It is to be hoped that the urgent need for new strategies in cancer research or for new candidates for antibiotics in view of the fast development in resistance of harmful bacteria will lead to a renaissance in the field. After all, the potential rewards are of considerable magnitude.

References

Agafonova IG, Aminin DL, Shubina LK, Fedorov SN (2002) Influence of polyhydroxysteroids on [Ca(2+)](i). Steroids 67:695–701

Aizenberg J, Tkachenko A, Weiner S, Addadi L, Hendler G (2001) Calcitic microlenses as part of the photoreceptor system in brittle stars. Nature 412:819–822

Alender CB (1967) A biologically active substance from spines of two diadematid sea urchins. In: Russell FE, Saunders PR (eds) Animal toxins. Pergamon Press, Oxford, p 145

Aminin DL, Agafonova IG, Fedorov SN (1995) Biological activity of disulfated polyhydroxysteroids from the Pacific brittle star *Ophiopholis aculeata*. Comp Biochem Physiol C Pharmacol Toxicol Endocrinol 112:201–204

Arnone MI, Bogarad LD, Collazo A, Kirchhamer CV, Cameron RA, Rast JP, Gregorians A, Davidson EH (1997) Green fluorescent protein in the sea urchin embryo: new experimental approaches to transcriptional regulatory analysis in embryos and larvae. Development 124:4649–4659

Carballo JL, Hernandez-Inda ZL, Perez P, Garcia-Gravalos MD (2002) A comparison between two brine shrimp assays to detect in vitro cytotoxicity in marine natural products. BMC Biotechnol 2:17–23

De Bremaeker N, Dewael Y, Baguet F, Mallefet J (2000) Involvement of cyclic nucleotides and IP3 in the regulation of luminescence of the brittlestar *Amphipholis squamata* (Echinodermata). Luminescence 15:159–163

Deheyn D, Mallefet J, Jangoux M (2000) Cytological changes during light production process in dissociated photocytes from the ophiuroid *Amphipholis squamata* (Echinodermata). Cell Tissue Res 299:115–128

Deheyn D, Mallefet J, Jangoux M (2000) Evidence of seasonal variation in luminescence performance of *Amphipholis squamata* (Ophiuroidea, Echinodermata): effect of environmental factors. J Exp Mar Biol Ecol 245:245–264

Deheyn D, Mallefet J, Jangoux M (2000) Evidence from polychromatism and bioluminescence that the cosmopolitan ophiuroid *Amphipholis squamata* (Echinodermata) might not represent a unique taxon. C R Acad Sci Paris 323:499–509

Deheyn D, Mallefet J, Jangoux M (2000) Expression of bioluminescence in *Amphipholis squamata* (Ophiuroidea: Echinodermata) in presence of various organisms: a laboratory study. J Mar Biol Assoc UK 80:179–180

Deheyn D, Jangoux M, Warnau M (2000) Alteration of bioluminescence in *Amphipholis squamata* (Ophiuroidea: Echinodermata) by heavy metal contamination: a field study. Sci Tot Environ 247:41–49

Glinski Z, Jarosz J (2000) Immune phenomena in echinoderms. Arch Immunol Ther Exp 48:189–193

Hatakeyama T, Ohuchi K, Kuroki M, Yamasaki N (1995) Amino acid sequence of a C-type lectin CEL-IV from the marine invertebrate *Cucumaria echinata*. Biosci Biotech Biochem 59:1314–1317

Haug T, Kjuul AK, Styrvold OB, Sandsdalen E, Olsen OM, Stensvag K (2002) Antibacterial activity in *Strongylocentrotus droebachiensis* (Echinoidea), *Cucumaria frondosa* (Holothuroidea), and *Asterias rubens* (Asteroidea). J Invertebr Pathol 81:94–102

Iken K, Greer SP, Amsler CD, McClintock JB (2003) A new antifouling bioassay monitoring brown algal spore swimming behaviour in the presence of echinoderm extracts. Biofouling 19:327–334

Kuznetsova TA, Anisimov MM, Popov AM, Baranova SI, Afiyatullov SS, Kapustina II, Antonov AS, Elyakov GB (1982) A comparative study in vitro of physiological activity of triterpene glycosides of marine invertebrates of echinoderm type. Comp Biochem Physiol 73:41–43

Lin W, Zhank H, Beck G (2001) Phylogeny of natural cytotoxicity: cytotoxic activity of coelomocytes of the purple sea urchin, *Arbacia punctulata*. J Exp Zool 290:741–750

Matranga V, Toia G, Bonaventura R, Müller WEG (2000) Cellular and biochemical responses to environmental and experimentally induced stress in sea urchin coelomocytes. Cell Stress Chaper 5:113–120

Mebs D (1989) Der Dornenkronenseestern im Korallenriff. Eine oekologische Katatrophe. Naturwiss Rdschau 42:480–482

Mebs D (1991) A myotoxic phospholipase A_2 from the crown-of-thorn starfish *Acanthaster planci*. Toxicon 29:289–293

Mulloy B, Mourao PA, Gray E (2000) Structure/function studies of anticoagulant sulphated polysaccharides using NMR. J Biotechnol 77:123–135

Nakagawa J, Kimura A (1982) Partial purification of a toxic substance from pedicellariae of the sea urchin *Toxopneustes pileolus*. Jpn J Pharmacol 32:966–971

Palagiano E, Zollo F, Minale L, Iorizzi M, Bryan P, McClintock J, Hopkins T (1996) Isolation of 20 glycosides from the starfish *Henricia downeyae*, collected in the Gulf of Mexico. J Nat Prod 59:348–354

Pereira MS, Mulloy B, Mourao PA (1999) Structure and anticoagulant activity of sulphated fucans. Comparison between the regular, repetitive, and linear fucans from echinoderms with the more heterogeneous and branched polymers from brown algae. J Biol Chem 274:7656–7667

Sasaki T, Uchida NA, Uchida H, Takasuka N, Kamiya H, Endo Y, Tanaka M, Hayashi T, Shimizu Y (1985) Antitumor activity of aqueous extracts of marine animals. J Pharmacobiodyn 8:969–974

Shiomi K, Itoh K, Yamanaka H, Kikuchi T (1985) Biological activity of crude venom from the crown-of-thorns starfish *Acanthaster planci*. Bull Jpn Soc Fish 51:1151–1155

Shiomi K, Yamamoto S, Yamanaja H, Kikuchi T (1988) Purification and characterization of a lethal factor in venom of the crown-of-thorns starfish (*Acanthaster planci*). Toxicon 26:1077–1081

Stabili L, Pagliara P, Roch P (1996) Antibacterial activity in the coelomocytes of the sea urchin *Paracentrotus lividus*. Comp Biochem Physiol Biochem Mol Biol 113:639–644

Yamada K (2002) Chemo-pharmaceutical studies on the glycospingolipid constituents from echinoderm, sea cucumbers, as the medicinal materials. J Pharm Soc Jpn 122:1133–1143

Cell Adhesion and Communication: A Lesson from Echinoderm Embryos for the Exploitation of New Therapeutic Tools

F. Zito, C. Costa, S. Sciarrino, C. Cavalcante, V. Poma, V. Matranga

Abstract. In this chapter, we summarise fundamental findings concerning echinoderms as well as research interests on this phylum for biomedical and evolutionary studies. We discuss how current knowledge of echinoderm biology, in particular of the sea urchin system, can shed light on the understanding of important biological phenomena and in dissecting them at the molecular level. The general principles of sea urchin embryo development are summarised, mainly focusing on cell communication and interactions, with particular attention to the cell–extracellular matrix and cell–cell adhesion molecules and related proteins. Our purpose is not to review all the work done over the years in the field of cellular interaction in echinoderms. On the contrary, we will rather focus on a few arguments in an effort to re-examine some ideas and concepts, with the aim of promoting discussion in this rapidly growing field and opening new routes for research on innovative therapeutic tools.

1 Introduction

Echinodermata are among the most familiar marine invertebrates. The phylum, characterised by the great morphological variety of its members, belongs to a branch of the animal kingdom known as deuterostomes. The Echinodermata have been extensively studied, particularly because of some aspects such as the ample fossil record extending back to the Precambrian, their ecological importance in the marine environment, the interesting mor-

F. Zito (e-mail: zito@ibim.cnr.it), C. Costa, S. Sciarrino, C. Cavalcante, V. Poma, V. Matranga
Istituto di Biomedicina e Immunologia Molecolare (IBIM) "Alberto Monroy", Consiglio Nazionale delle Ricerche, Via U. La Malfa 153, 90146 Palermo, Italy

Progress in Molecular and Subcellular Biology
Subseries Marine Molecular Biotechnology
V. Matranga (Ed.), Echinodermata

Fig. 1. A Living fossil, the pencil urchin *Eucidaris tribuloides*. **B** Presumptive crinoid fossil found in 1981, in the Black Forest, Germany

phology of adult organisms and the advantage of experimentally manipulable embryos. Approximately 13,000 echinoderm fossil species are known (Fig. 1), while living species number about 7,000 and fall into five well-defined classes:

- Crinoidea (sea lilies and feather stars), with about 600 living species. The body is oriented with the mouth facing up and they may or may not have a stalk.
- Asteroidea (starfishes or sea stars), the most familiar forms to man, with about 1,500 living species. Typically, they have their oral surface on the ventral side and usually have multiple "arms" surrounding a central disc.
- Ophiuroidea (basket stars and brittle stars), with five arms, typically flexible, radiating from a central disc, which are used for locomotion. The class includes about 2,000 living species.
- Echinoidea (sea urchins, sand dollars and sea biscuits), with a usually globular body, formed by the fusion of skeletal plates, without distinct arms. Some of them have bilateral symmetry, which occurred secondarily during evolution. About 1,700 living species are known.
- Holothuroidea (sea cucumbers), characterised by a cylindrical shape and a relatively soft body. About 1,100 living species are known.

The Concentricycloidea (sea daisies), with only two species, is an enigmatic group of echinoderms whose phylogenetic position remains elusive, although evidence suggests a relationship with asteroids (Pearse and Pearse 1994). The phylogenetic relationships among the five classes have been extensively controversial, but it seems generally accepted that the class of Crinoidea branched first and that the Echinoidea and Holothuroidea are sister clades (Fig. 2).

Although the main character of the body plan of adult echinoderms, the pentamerous symmetry, led the father of palaeontology, George Cuvier (1769–1832), to place them in the Radiata phylum, it is now widely accepted

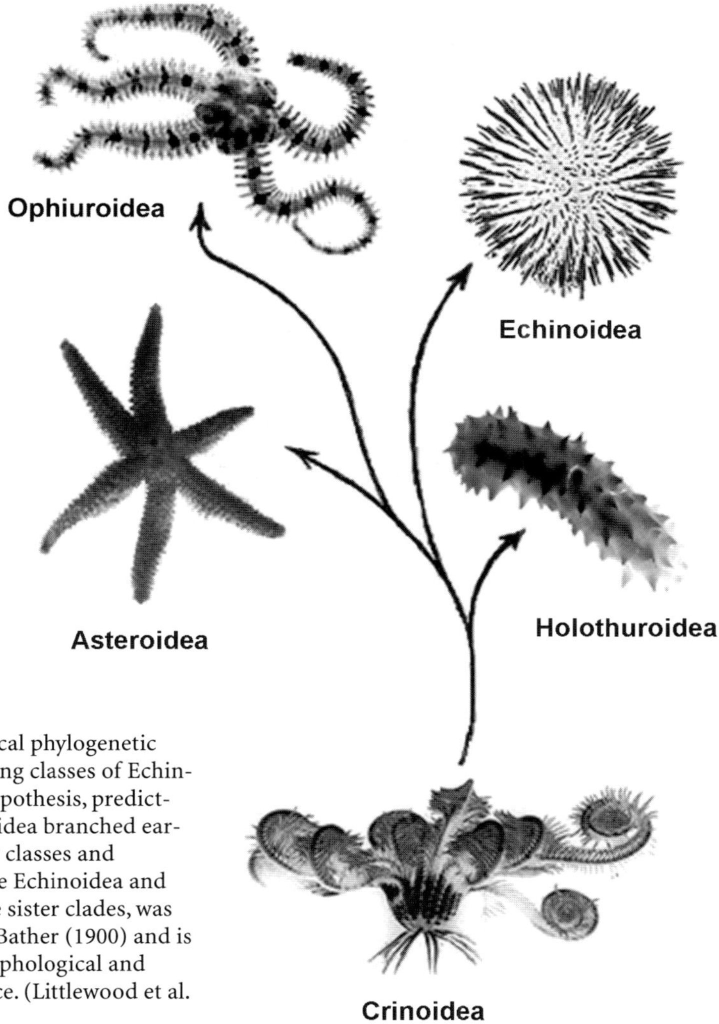

Fig. 2. Hypothetical phylogenetic tree of the five living classes of Echinodermata. This hypothesis, predicting that the Crinoidea branched earlier than the other classes and suggesting that the Echinoidea and Holothuroidea are sister clades, was first proposed by Bather (1900) and is supported by morphological and molecular evidence. (Littlewood et al. 1997)

that the phylogenetic proximity of echinoderms is to chordates. Among marine invertebrates, in fact, the echinoderm clade has a unique position in the evolutionary tree: during evolution it first appeared in the Metazoa phylum with features of deuterostomes, in direct evolutionary line with chordates and vertebrates, thus implying that there is less divergence between echinoderms and vertebrates than between echinoderms and other invertebrates. This is not surprising since, although the two phyla appear so dissimilar, molecular data and fossil records strongly support this grouping, even if the attempts to identify a common ancestor are highly speculative. As the eminent anthropologist Konrad Lorenz observed, "It is not a theory, but an irrefutable historical fact, that the living world – since its origin – has evolved from 'below' to 'above'". Actually, it is becoming clearer that the genes encoding developmental regulatory proteins already identified in other phyla, the modified interactions among them and changes in their expression patterns are at the basis of the evolution of animal morphology (Raff 1996; Wray and Lowe 2000).

The name echinoderm is derived from the Greek word meaning spiny skin and it has been the original denomination for sea urchins, since spines are their main external characteristic. Echinoderms are exclusively marine invertebrates and, with a few exceptions, are all benthic organisms (bottom-dwellers). A number of lifestyles are typical among different classes, e.g. sea star and crinoids are predators and filter-feeders respectively, while sea urchins scrape algae from rocks, and holothurians, sand dollars and ophiuroids often feed on remains.

Echinoderms usually have separate sexes with no evident sexual dimorphism. Reproduction is typically achieved by external fertilisation, with eggs and sperm freely released into the seawater. The life cycle is usually complex. In most echinoderms the embryo develops into a planktonic larva with a bilateral symmetry, that in some species goes through metamorphosis, typically radical, before reaching the final adult morphology.

A number of morphological features are unique to the Echinodermata phylum, including:

- a pentaradial symmetry in adults (higher-order radial symmetry can be observed in several cases, but it is clearly a secondary modification);
- a water vascular system, consisting of a set of water-filled canals branching from a ring canal and leading to tube feet, which has many important functions, including locomotion, cellular respiration and feeding;
- a mesodermal endoskeleton composed of calcareous plates assembled in a meshwork;
- a complex subepithelial nervous system with radial nerves running under each of the ambulacra, the row of tube feet, but without a recognisable central part;
- sensory neurons located primarily within the ectoderm of podia, without specialised sense organs;
- a circulatory system, if present, consisting of a haemal system derived from coelomic sinuses.

2 Why Study Echinoderms?

There are some features that make echinoderms so interesting to study. First, their sensitivity to environmental changes in seawater ecosystems. It is well known that the disappearance of fragile species from certain geographic areas is in direct relationship with high contamination of seawater and sediments, and echinoderms are one such fragile species. In particular, embryos, juveniles and adults of the classes Echinoidea and Asteroidea are well utilised in studies on marine pollution, since they highly resent it. Second, their ability to regenerate parts of the body (particularly Ophiuroidea and Asteroidea), based on stem cell recruitment. This phenomenon makes a fundamental contribution to the adaptive capacities of the whole species. Third, their ability to cause remarkable transformations in submarine substrates. Since the echinoderms are one of the most important marine invertebrates that do not feed on filtration, as they graze on substrate (except for holothurians and crinoids), they can induce changes in the ratios and distributions of other marine species (fishes and others), and eventually cause segregation as well as speciation. These three features of echinoderms are just some examples of the huge number of ways in which echinoderms can be of help in the understanding of important biological phenomena, and in dissecting them at the molecular level.

2.1 Echinoderms for Biomedical Research: A Simple Model to Study Biological Events Occurring in Higher Organisms

Recently, a number of studies have offered new notions that non-mammalian models may represent important future directions for studies on human diseases, since the results obtained from these models are in many cases applicable to mammals including humans. Few people realise that research on simple marine organisms has led to some of our greatest medical advances, as well as to new insights into environmental pollution. One of the advantages of using echinoderms is that they produce thousands of virtually identical embryos, and that the morphological abnormalities are readily visualised in the live organism under the light microscope. In the following text, some aspects of the medical advances achieved from such studies will be briefly outlined.

2.1.1 Screening and Testing of Toxic Substances

Sea urchin gametes, embryos and larvae can be used for fast, low-cost and reliable screening and testing of toxic substances and for detailed studies on their mechanism of action. One example is the screening for the toxicity caused by retinoids utilised in dermatological practice, since it has been shown that foetal malformation is a major form of toxicity associated with some of them (Kahn et al. 1988; Sciarrino and Matranga 1995). The sea urchin embryo has also been used as a model for screening for suspected mammalian developmental neurotoxicants and for anticancer drug testing (Nishioka et al. 2003; Qiao et al. 2003). This test system is applicable also for examination of new and known pharmacologically active substances, including their adverse effects and potential antidotes. Interestingly, a method has been developed to study the invasive properties of metastatic cells and to test the differential effects of anti-tumoural substances on their invasive capacity, which makes use of the sea urchin embryo basement membrane (Livant et al. 1995; Dyer et al. 2002). This structure is selectively permeable and can be obtained intact with the associated extracellular matrix (ECM) from sea urchin embryos (Livant et al. 1995). It has been demonstrated that all metastatic tumour cells placed in contact with these basement membranes were able to invade them and the invasion was rapid and efficient; on the other hand, as expected, non-metastatic cells failed in the invasion. These results suggest that molecules participating in basement membrane recognition and invasion have been functionally conserved during evolution and that their constitutive activity may allow metastatic cells to escape their tissues of origin (Livant et al. 1995).

2.1.2
From Echinoderm Molecules to Mammalian Diseases: How Fundamental Research Points to Clinical Trials

The analysis of genome sequences led to the finding of novel non-traditional targets involved in disease pathogenesis, whose usage has the advantage of removing infection without inducing resistance. For example, it has been shown that novel polysaccharides present on echinoderm surfaces seem able to stimulate early host defence and microbial clearance, but not the later phases of inflammatory tissue injury associated with sepsis. These are the most promising alternative or integrative treatments for pneumonia that are under development (Cazzola et al. 2004).

In the last 10 years, a number of molecules with different effects on mammalian cells have been purified from echinoderms. These include compounds with anti-coagulant activity on human blood cells, such as the peptide "plancinin" isolated from the sea star (Koyama et al. 1998), a promising drug for anti-thrombotic therapy. Other compounds display considerable cytotoxicity against a small panel of human solid tumour cell lines, such as polyhydroxysterols and saponins isolated from the sea star (Wang et al. 2004) and the glycolipid A-5 extracted from sea urchin intestine (Sahara et al. 1997, 2002). The latter compounds have been suggested as useful drugs for cancer chemotherapy.

Recently, Meijer and Raymond (2003) have reviewed the steps that lead to the identification of new drugs; that are now under evaluation for therapeutic use against cancer, neurodegenerative diseases and cardiovascular disorders. This is an example of how results obtained from basic research, i.e. studies on the cell cycle in the starfish oocyte model, can be utilised in applied medical research and treatment. The starfish cyclin-dependent kinase CDK1/cyclin B was initially identified as a universal M-phase-promoting factor and then used as a screening target to identify pharmacological inhibitors. From the first inhibitors discovered, a more selective one was optimised, which is now entering phase II clinical trials against cancers and phase I clinical tests against glomerulonephritis (Meijer and Raymond 2003).

2.1.3 Highly Conserved Proteins Associated with Important Biological Functions

The study of echinoderms also led to the identification of proteins with high levels of homology to vertebrate proteins expressed in particular syndromes or tumour cells. A sea urchin gene showing very strong sequence and structural homology with the gene coding for dystrophin, which is defective in Duchenne muscular dystrophy, has been identified. The partial characterisation of this gene helped in the construction of an evolutionary tree connecting the vertebrate dystrophin gene family with related genes in invertebrates (Wang et al. 1998). A novel protein homologue to the sea urchin fascin (an actin-bundling protein) has been found to be over-expressed in pancreatic ductal adenocarcinoma, suggesting its use as a tumour marker with potential diagnostic and therapeutic implications for pancreatic carcinoma (Maitra et al. 2002).

Sea urchin sperm homologues of polycystin-1 and polycystin-2, the proteins mutated in autosomal-dominant polycystic kidney disease, have been sequenced (Mengerink et al. 2002; Neill et al. 2004). Both proteins have been shown to co-localise exclusively to the plasma membrane over the sperm acrosomal vesicle, where they may function as a cation channel mediating the sperm acrosome reaction. These data provide the first suggestion for the role of a polycystin-1 protein in a specific cellular process (Mengerink et al. 2002).

Recently, a fasciclin-I-like protein has been purified from sea urchin ovaries and, by in vitro assays, it has been shown to be active in promoting HT1080 human fibrosarcoma cell attachment (Sato et al. 2004). Fasciclin-I is a neuronal cell adhesion molecule and up to now various proteins belonging to the family have been identified in different species, including bacteria, plants and vertebrates, and in the sea urchin embryo and eggs (Brennan and Robinson 1994; Wessel et al. 2000), although their biological function had not been characterised. However, latest findings indicate that the protein is highly conserved in evolution and suggest important biological roles (Sato et al. 2004).

There are also examples of the isolation of new human genes whose function has been hypothesised on the basis of their high homology to already known and characterised echinoderm genes. A novel human homologue of the gene coding for echinoderm microtubule-associated protein (EMAP) has been isolated from a locus of Usher syndrome type 1, an autosomal recessive genetically heterogeneous disorder. The finding of its high level of homology to the echinoderm cytoskeletal component EMAP, especially at the microtubule-binding domain, and the proposed cytoskeletal nature of Usher disease, define the human EMAP as a good candidate for the USH1a syndrome origin (Eudy et al. 1997).

3
Use of Echinoderm Embryos to Study the Basic Mechanisms of Communication Among Cells

Developmental biology is a discipline studying the mechanisms regulating embryogenesis and it is one of the most attractive fields in rapid expansion among the biological sciences. Its study is becoming essential for the comprehension of any other fields in biology, since it combines molecular and cellular biology, physiology, anatomy, immunology, research on cancer and also evolution and ecology. Development is mainly devoted to the production and organisation of all the different cellular types constituting the adult organism. The generation of different types of cells is a process known as differentiation, while the organisation of differentiated cells in tissues and organs is performed during morphogenesis.

It is well known that cells do not behave as single entities, but rather their association in multicellular structures requires precise co-ordination between release and uptake of signals. Communication and interaction among living cells are, in fact, fundamental events required for the proper development of tissues and organs. Living cells continuously receive inputs from the environment and modify their behaviour throughout a complex network of signalling pathways. This communication occurs at various levels and, to obtain co-ordinated responses, the integration of exchanged information is essential. Malfunctioning of this network of signalling pathways is often connected with pathological conditions ranging from abnormal proliferation to cell death. Understanding the molecular basis of communication among living cells is a fundamental challenge for biologists, since, besides providing better understanding of the processes controlling growth, differentiation and death, it may increase the number of discoveries of new therapies against diseases caused by inappropriate signalling.

Historically, within the phylum of Echinodermata, the sea urchin embryo has been an excellent experimental system for investigating the cellular basis of development, principally because of its relatively simple organisation and because of its optical transparency that makes the observation of morphogenesis in vivo possible. Pioneering studies on development date back to 1892, when Driesch, following the experimental approach to embryology proposed by Roux about 10 years before, utilised the sea urchin embryo in his studies with important outcomes for embryology (Driesch 1892). On the basis of his results, Driesch proposed the very modern concept of nucleus–cytoplasm interaction as an essential event for development (Driesch 1894). Later, from 1928 to 1935, Hörstadius performed some of the most remarkable experiments in the history of embryology using the sea urchin embryo (Figs. 3, 4). First, he separated each blastomere of the early embryo and followed their fates; then he was able to recombine different series of blastomeres and, from the results obtained, to propose the well-known theory on the existence of graded properties within the unfertilised egg and the early embryo (Hörstadius 1939).

The sea urchin system was also one of the first in which time-lapse microscopy was exploited extensively. For example, the classic studies of Gustafson and Wolpert (1967) led to the identification of many of the basic behaviours exhibited by cells in the embryo during morphogenetic movements. At the time, these authors remarked, "we are, however, still ignorant about the final steps in the casual chain between the genes and the shapes they control" (Gustafson and Wolpert 1967). In the past 35 years, our knowledge of the molecular basis of developmental processes and the relationship between molecules and cell behaviour has advanced considerably. The classic studies

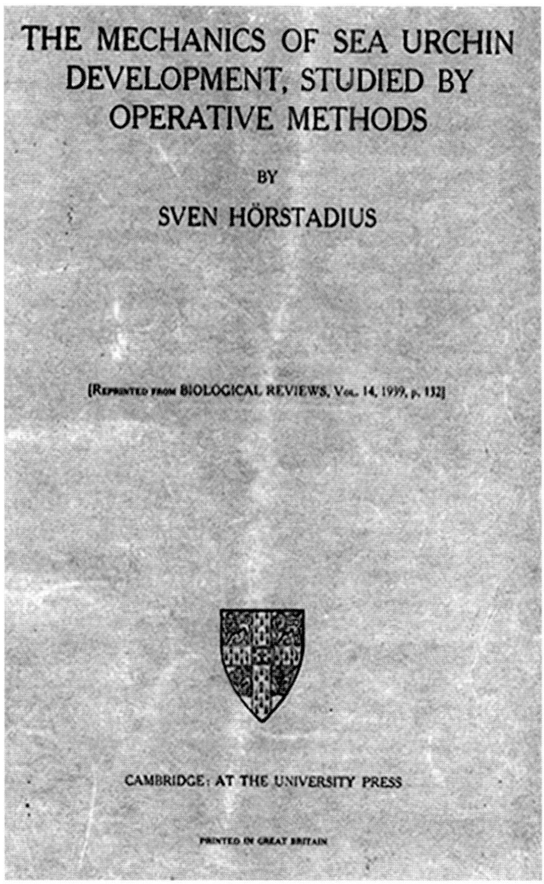
THE MECHANICS OF SEA URCHIN DEVELOPMENT, STUDIED BY OPERATIVE METHODS

BY

SVEN HÖRSTADIUS

CAMBRIDGE: AT THE UNIVERSITY PRESS

Fig. 3. Cover of the original review by Hörstadius (1939) on development of the *Paracentrotus lividus* sea urchin embryo (Kindly provided by M. Delarue, Laboratoire Biologie et Multimédia, UPMC-P6, www.snv.jussieu.fr/bmedia/sommaires/dvpt.htm)

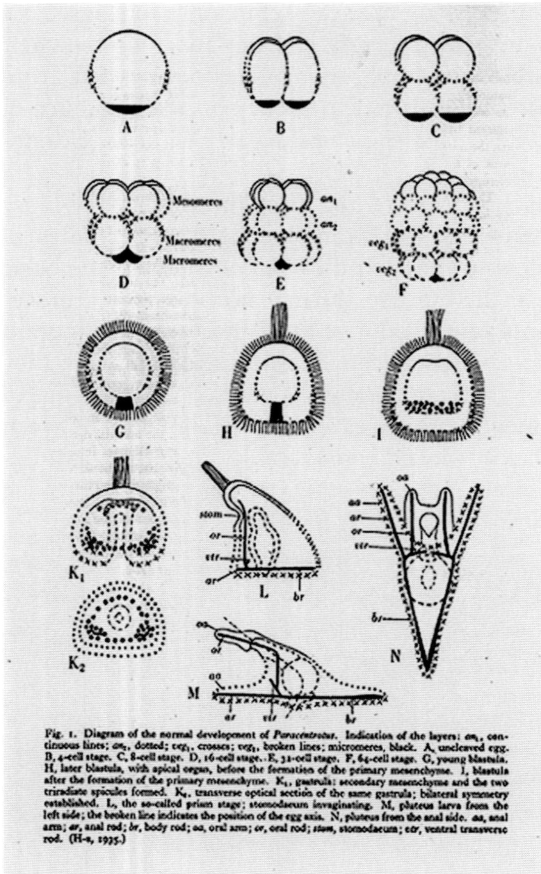

Fig. 4. Diagram of the normal development of the *P. lividus* sea urchin embryo published in 1939 by Hörstadius (Kindly provided by M. Delarue)

of Hörstadius have been extended by the lineage studies performed by Davidson and colleagues, providing a more detailed picture of the establishment of tissue territories in the early embryo (see reviews by Cameron and Davidson 1991; Cameron et al. 1991). More recently, many other laboratories in the world, thanks to new technologies developed in the field of molecular biology, have succeeded in the identification and characterisation of control genes and their key target sites and the determination of their functional significance in the early embryo, allowing the proposal of complex gene regulatory networks (see review by Davidson et al. 2002). An equivalent remarkable progress has not been seen in the cell biology field, particularly in the characterisation of the proteins involved in cell communication and cell adhesion, although some studies date back to the early 1970s.

In the following text, molecules of the sea urchin embryo involved in cell communication, such as cell–matrix and cell–cell adhesion proteins (*Pl-*

nectin, fibronectin, laminin, collagen, integrin, toposome, cadherin), growth factors (BMP2/4, univin), signal transduction molecules (kinases), and their genes (when characterised) will be described. Our intention is not to review all the work done over the years in the field of cellular interaction in echinoderms, but rather we will focus on a few arguments on the assumption of re-examining some ideas and concepts.

3.1 Sea Urchin Embryo Transparency: A Living Laboratory for Studying Development and Morphogenesis

To facilitate the reader in the following discussion on the relationship between adhesion/signalling molecules and cell differentiation occurring during embryogenesis, the development of the Mediterranean sea urchin species *Paracentrotus lividus* will be briefly described. The time scale and embryo morphology apply, with very little modifications, to many other sea urchin species.

Fertilisation is the first event that gives rise to development and triggers instantaneously several changes within the egg. A transient decrease in the membrane potential, a slight increase in the intracellular pH and free calcium ions activate several metabolic processes, causing, among other things, the exocytosis of cortical granules and the consequent elevation of the fertilisation membrane (Fig. 5A). Cleavage stages begin within 1 h after fertilisation and are characterised by a number of cell divisions occurring about every 30 min (Fig. 5B,C). While the first three divisions give rise to eight equivalent blastomeres, the fourth one produces 16 cells of different sizes: eight mesomeres at the animal pole, four macromeres and four micromeres at the vegetal pole (Fig. 5D). Cameron and colleagues were able to describe extensively the cell lineage for each of the 16 blastomeres; in addition, they showed that most of the lineage founder cells for many tissues of the later embryo are established at the 64-cell stage (Cameron and Davidson 1991; Davidson et al. 1998). It is noteworthy to underline that the lineage boundaries are detected well before the appearance of any morphological difference in the embryo and the spatially restricted patterns of gene expression already described agree with these lineage boundaries. As cell divisions proceed, the embryo develops into a blastula, which consists of one layer of epithelial cells, with a typical apical–basal polarity, surrounding a filled cavity, the blastocoel (Fig. 5E). Later, the embryo hatches from the fertilisation envelope to become a free-swimming blastula. From this stage another phase of development starts, which is characterised by a number of morphogenetic movements that will completely rearrange the embryo. The first cells to move are the primary mesenchyme cells (PMCs) that detach from the vegetal plate and ingress into the blastocoel, where they migrate using short filopodia (mesenchyme blastula stage; Fig. 5F). Some PMCs migrate towards two ventro-lateral sites and

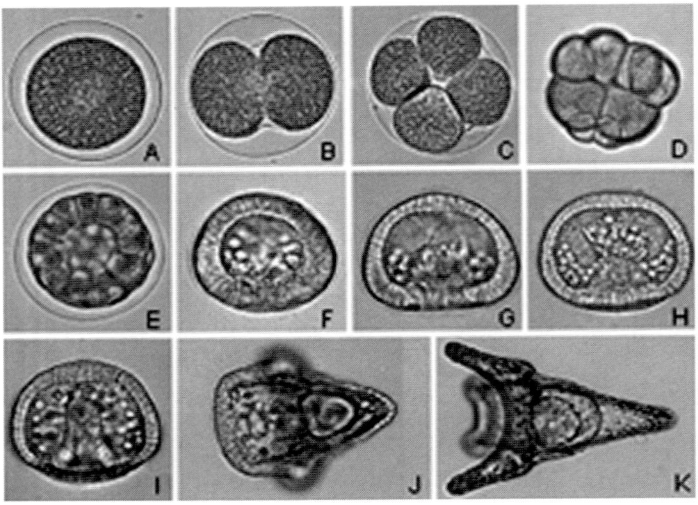

Fig. 5A–K. Development of the sea urchin embryo *Paracentrotus lividus*. **A** Fertilised egg; **B** two blastomeres; **C** four blastomeres; **D** 16 blastomeres; **E** early blastula; **F** mesenchyme blastula; **G** early gastrula; **H** middle gastrula; **I** late gastrula; **J** early pluteus; **K** pluteus

form clusters that are connected by other PMCs in a form of a subequatorial ring. The morphology of PMCs during migration was described for the first time by Gustafson and Wolpert using time-lapse microscopy (1961, 1967). Shortly after PMC ingression, the vegetal plate epithelium invaginates to form the archenteron, which is the future intestine. It is possible to distinguish at least three gastrula stages during the elongation of the archenteron towards the animal pole: early, middle and late gastrula (Fig. 5G–I). PMCs produce the skeleton, first in the form of triradiate spicules, which elongate and pattern in a complex species-specific manner. Skeleton is formed by calcium and magnesium carbonate, which deposit on an organic matrix constituted by a number of well-known spicule-specific proteins. In the laboratory, the larval stage of pluteus is obtained after 48 h post-fertilisation: it shows four arms, a complex patterned skeleton and a tripartite intestine, encircled by muscles undergoing peristaltic contractions (Fig. 5J,K). At this point, the larva, if correctly fed, is ready to continue development and form the juvenile sea urchin through metamorphosis.

From the developmental processes outlined above it is evident that cell–cell and cell–ECM contacts play a fundamental role in morphogenetic movements occurring during embryogenesis. Knowledge of the key actors involved as well as understanding of their complex interactions would eventually lead to the unravelling of the developmental machinery. Furthermore, it should be noted that cells forming the blastula stage monolayer exhibit a structural asymmetry of the cytoplasm and the plasma membrane is compartmentalised into distinct apical, basal and lateral domains, with characteristic lipid and protein compositions. Embryogenesis and morphogenesis are characterised by cell movements and complex cell rearrangements, which require appropriate interactions of cells with the underlying ECM by means of specific membrane receptors. For this reason, interest in the identification, purification and functional studies of ECM components, along with their ligands and other molecules involved in cell–matrix adhesion, has increased in recent years. In the following text, after a brief description of the appearance of the ECM upon embryo development, protein molecules found in different plasma membrane compartments of blastula cells will be described.

3.1.1
ECM Patterning in Echinoids During Embryo Development

In recent years, it has been shown that the ECM of the sea urchin embryo is a very complex structure, consisting of a number of layers organising during different developmental steps. Using specific antibodies, different storage compartments of the ECM components, like granules and vesicles, have been identified in the unfertilised egg cytoplasm. The protein contents of these compartments are exocytosed and assembled in a highly regulated fashion at different moments after fertilisation (for a review see McClay et al. 1990). The early event after sperm entry is the elevation of the fertilisation envelope above a water-filled perivitelline space (Fig. 5A; for more detailed description on ionic events occurring at this stage, see the chapter by Angelini et al., this Vol.). Then, a new ECM is secreted on the outer surface of the zygote, which gives rise to different layers, the most acknowledged of them being the apical lamina and the hyaline layer. During cleavage, other ECM molecules are also released into the newly forming basal lamina, which will underlie the blastocoelic cavity from the stage of blastula (Fig. 5E). A great number of in vitro and in vivo studies suggest that ECM has an important role during morphogenesis, serving as a mechanical support to the embryo as well as a substrate for cell movements, and providing both spatial and temporal information to adherent cells. Thus the ECM is not a fixed structure but rather is able to respond to and effect changes in its local microenvironment.

Changes in ECM composition and regulated matrix remodelling are common events associated with embryogenesis and are carried on by matrix metalloproteases (MMPs). These molecules are a growing family of metalloendopeptidases that act by cleaving the protein components of the ECM and thereby regulating its composition (for a review see Stamenkovic 2003). For many years MMPs were believed to function in facilitating cell migration simply by removing barriers such as collagen. However, recent discoveries have shed new light on the role of MMPs in embryology, physiology and disease. It is becoming increasingly clear, in fact, that MMPs are also involved in the functional regulation of non-ECM molecules, including growth factors and their receptors, cytokines and chemokines, adhesion receptors and cell-surface proteoglycans, and a variety of enzymes. MMPs therefore play an important role in the control of cellular interactions with their environment, promoting tissue turnover, both physiological, such as normal development, and pathological, such as inflammation and cancer. The sea urchin embryo was found to express a dynamic pattern of gelatinase activities associated with the external ECM, some expressed only in later stages of development, others expressed in the unfertilised egg and persisting throughout the course of embryonic development (Mayne and Robinson 1996; Robinson 1997; Flood et al. 2000). Their substrate specificity and metal ion requirements suggest that these molecules are members of the MMP class of ECM remodelling enzymes.

In recent years, a new family of molecules in the animal kingdom has been described, called ADAMs (A Disintegrin And Metalloprotease). These are multidomain transmembrane proteins containing a metalloproteinase and a disintegrin domain and all or some of the following structures: a signal peptide, a propeptide, a cysteine-rich and an epidermal growth factor (EGF)-like domain, a transmembrane region, and a cytoplasmic tail. As a consequence, ADAMs could have four potential functions: proteolysis, adhesion, fusion and intracellular signalling (Stone et al. 1999). Although the number of ADAM genes has grown rapidly, the biological functions of most members are still unclear. However, they seem to be key regulators of the cell–cell and cell–ECM interactions (see review by White 2003). Recently, a sea urchin SpADAM gene has been sequenced and the deduced protein sequence includes all the characteristic domains of vertebrate ADAMs. The structure and the types of cells in which SpADAM orthologues are expressed are then apparently conserved in deuterostomes (Rise and Burke 2002).

4 Apical Cell Surface

It has been reported that the first ECM structure developing soon after fertilisation of the egg is the external one. As viewed by electron microscopy after ruthenium staining (Lundgren 1973), the presence of at least two distinct layers is evident: an inner apical lamina more tightly associated with the apical plasma membrane, which is the thickest, and an outer layer, the hyaline layer, associated with the tips of microvilli (see Fig. 6).

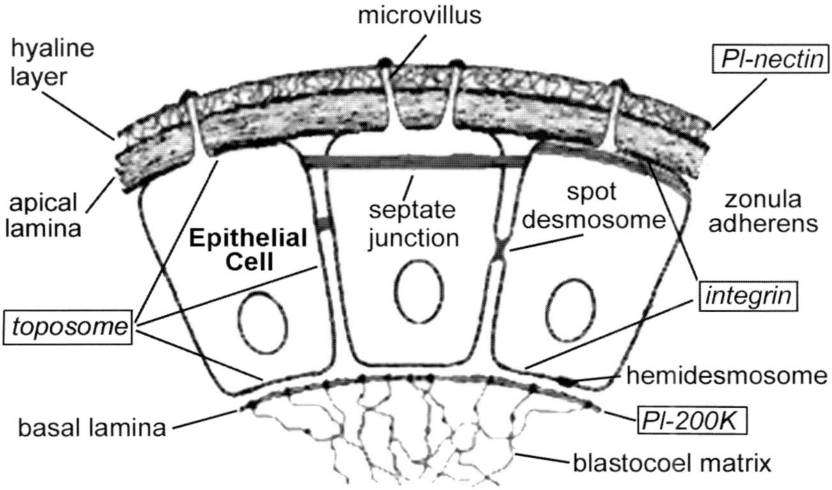

Fig. 6. Schematic diagram of the ultrastructure of the epithelium of the blastula embryo. (Adapted from Hardin 1996)

4.1 The Apical Lamina

The apical lamina is a fibrous layer surrounding the embryo, which remains on the apical surface after removal of hyalin from the hyaline layer (Hall and Vacquier 1982; Burke et al. 1998), and is composed principally of fibropellins (Bisgrove et al. 1991; Burke et al. 1991). The gene coding for one of the three known fibropellins has been sequenced and, like other extracellular molecules, it contains a series of up to 20 epidermal growth-factor-like repeats (Delgadillo-Reynoso et al. 1989; Grimwade et al. 1991; Bisgrove and Raff 1993). In vitro experiments showed that dissociated cells adhere to affinity-purified fibropellins in a temperature-, time- and dose-dependent manner, suggesting the role as a substrate for cell adhesion (Burke et al. 1998). Their in vivo functional role has been investigated using monoclonal antibodies, which were shown to interfere with the initial phase of gastrulation (Nakajima and Burke 1996). At this time no further proteins have been described as belonging to this layer.

4.2 The Hyaline Layer

Historically, the first function assigned to the hyaline layer has been to hold blastomeres together (Osanai 1960; Vacquier and Mazia 1968). In agreement with this hypothesis, Citkowicz (1971) showed that the hyaline layer forms a shell that remains structurally intact after removal of epithelial cells using hyperosmotic solutions, demonstrating that it is a structure independent from cells that lie underneath. A number of proteins constituting the hyaline layer have been identified and characterised. The major component is hyalin, a high molecular weight glycoprotein secreted by the cortical granules, which can be isolated by successive steps of solubilisation and precipitation through removal and re-addition of Ca^{2+} to seawater or other osmotically balanced media (Kane 1970). Hyalin has been extensively studied both structurally and functionally in sea urchin and sea star embryos (Vacquier 1969; Kane and Stephens 1969; Spiegel and Spiegel 1979; Adelson et al. 1992) and partial cDNAs have been isolated by screening expression libraries with monoclonal antibodies to hyalin (Wessel et al. 1998). Its sequence contains consensus calcium-binding motifs, in agreement with its capacity to interact with calcium biochemically (Robinson et al. 1992), and modular repeats similar to that known in other large ECM molecules (McClay 1991). Hyalin has been shown to support cell adhesion in vitro (McClay and Fink 1982) and to be involved in sea urchin morphogenesis in vivo (Adelson and Humphreys 1988). Interestingly, a recent study has demonstrated that two mammalian cortical granule envelope proteins share common antigenic epitope(s) with echinoderm hyalin and, like hyalin, play a role in early embryogenesis (Hoodbhoy et al. 2000).

4.2.1 Pl-Nectin: An Example of Signal(s) from Outside to Inside the Embryo

About 10 years ago, with the aim of purifying fibronectin from the sea urchin embryo, we isolated and characterised another ECM protein localised in the hyaline layer, namely *Pl*-nectin (Fig. 6). Although the protocol used for its purification was that originally developed for fibronectin, the molecular weight and other features indicated that this protein was a new molecular species. *Pl*-nectin has been isolated from the species *Paracentrotus lividus* (Matranga et al. 1992) and *Temnopleurus hardwickii* (*Th*-nectin; Yokota et al. 1994) as a collagen-binding molecule and, like other ECM components, it has been found stored in specific cytoplasmic vesicles, recently named nectosomes, in the unfertilised egg (Kato et al. 2004). The genesis and distribution of the nectosome during oogenesis, its translocation to the cortex and gradual secretion into the hyaline layer after fertilisation have recently been fully documented by using immunoelectron microscopy in the Japanese species *T. hardwickii* (Kato et al. 2004). At later stages, *Pl*-nectin has been found to be localised on the apical surface of ectoderm and endoderm cells. By in vitro assays, the protein has been shown to mediate cell-substrate adhesion in a dose-dependent fashion, suggesting a functional role during development (Matranga et al. 1992). Thus it was crucial to investigate the in vivo *Pl*-nectin function by morphogenetic assays in which embryos were cultured in the

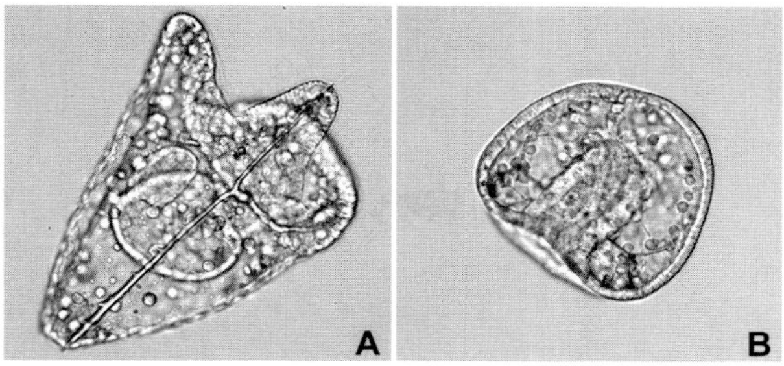

Fig. 7A, B. Skeleton defects obtained after culturing embryos in the presence of high concentrations of monoclonal antibody to *Pl*-nectin. **A** Control embryo cultured in the presence of unrelated IgGs; **B** treated embryo showing strong skeleton defects

presence of monoclonal antibodies to *Pl*-nectin. To our surprise, after this treatment, we observed the presence of a high number of embryos with dramatic skeleton defects, but with normally developed ectoderm and endoderm structures (Fig. 7; Zito et al. 1998, 2000). In addition, depending on the amount of antibody used, it was possible to obtain embryos with different degrees of skeleton defects, which were classified on an arbitrary scale (Zito et al. 2003). These data suggested that outer ectoderm–*Pl*-nectin interaction is indirectly involved in inner skeleton formation. For more detailed description on this issue, see Section 10. The partial sequence of the coding region of the *Pl*-nectin gene (accession no. AJ578435) has revealed its similarity with another component of the hyaline layer, namely echinonectin, purified from the species *Lytechinus variegatus*. This protein has a lectin-binding activity that allowed its chromatographic purification and, like *Pl*-nectin, it is an adhesive substratum for cells (Alliegro et al. 1988; 1990). Although in a previous report no cross-reaction between the two molecules was found, it is very likely that *Pl*-nectin and echinonectin belong to the same family of ECM components.

4.2.2
Other Apical ECM Components

Concerning the presence of fibronectin-like molecules in the sea urchin embryo, first evidence was obtained by immunofluorescence experiments using antibodies against human fibronectin, even if the cross-reactive material was described as being differently distributed in the embryo, i.e. on the outer cell surface of the epithelial layer (Spiegel et al. 1983), on the basement membrane (Spiegel et al. 1980; DeSimone et al. 1985), and also on the surface of migrating PMCs (Katow et al. 1982). A sea urchin fibronectin-like molecule has been purified for the first time from ovaries of *Pseudocentrotus depressus* by Iwata and Nakano (1981, 1983). The same authors have also demonstrated that this protein is involved in migration, adhesion and spicule synthesis of in vitro-cultured micromeres (Miyachi et al. 1984). Later, fibronectin-like proteins were purified from five different species living in Mediterranean and

Pacific seawaters and their biochemical and immunological relationships have been analysed for the purpose of comparative phylogenetic studies (Matranga et al. 1995). These proteins differed in their binding affinity to gelatin but shared different epitopes, suggesting that they are members of a sea urchin fibronectin superfamily. Furthermore, analysis of the different features of fibronectin-like proteins and sea urchin nectins, i.e. *Pl*-nectin and *Th*-nectin, has been carried out, since essentially the same method was utilised to purify both groups of molecules (Yokota et al. 1994). It has been shown that these two groups of proteins belong to different ECM protein families since they differ in their affinity to collagen as well as in their localisation inside the embryo.

A gene coding for a new ECM protein has been sequenced in the direct developer *Heliocidaris erythrogramma*. By the use of polyclonal antibodies raised against the protein, it has been demonstrated that it is localised on the apical surface of ectoderm, in tight association with the plasma membrane. The protein has been named apextrin and it has been proposed to be involved in apical cell adhesion (Haag et al. 1999).

Although a reasonable amount of information is currently available on the hyaline layer and apical lamina components, the spatial relationship between them is still unclear. An approach to such studies could come from experiments of co-immunoprecipitation after transient cross-linking of surface molecules in live embryos. Alternatively, electron microscopic observations of rotary shadowed purified molecules in diluted solutions, or in combination with other potential partners, would serve to solve the intricate meshwork that surrounds the embryo.

5 Basal Cell Surface

As previously reported, the basal lamina develops when new ECM molecules are released, from the blastula stage, at the basal surfaces of epithelial cells, into the constituting blastocoelic cavity (Fig. 6). A discrete number of reports have described the formation of this layer and its function in sea urchin development (Okazaki and Nijima 1964; Kawabe et al. 1981; Galileo and Morrill 1985; Amemiya 1989).

Different experimental approaches have shown the presence of few types of collagens in the sea urchin embryo, in contrast to about 20 different types of the protein found in higher metazoan phyla. By immunofluorescence, Wessel et al. (1984) showed collagen localisation on the basal lamina of the embryo, using antibodies against known vertebrate ECM components, i.e. collagen types I, III and IV. Partial biochemical characterisation of collagen-like molecules has been carried out from in vitro-cultured micromeres (Pucci-Minafra et al. 1975; Benson et al. 1990; Shimizu et al. 1990). Collagen genes have been characterised in several species of sea urchin. Using mouse type IV cDNAs, genomic clones have been isolated from *S. purpuratus* and characterised (Venkatesan et al. 1986). A putative collagen gene has also been cloned from a *P. lividus* genomic library using a *C. elegans* collagen gene as a probe (D'Alessio et al. 1989, 1990; Exposito et al. 1992). It has also been demonstrated that fibrillar collagen types I and II and non-fibrillar collagen type IV are expressed only by PMCs and secondary mesenchyme cells (SMC) (Angerer et al. 1988; D'Alessio et al. 1989; Wessel et al. 1991; Lethias et al. 1995).

The first evidence of the presence of a vertebrate laminin-like molecule as a component of the embryonic sea urchin basal lamina came from studies that utilised an monoclonal antibody specific to laminin (McCarthy and Burger 1987). Later, the isolation and characterisation of a cDNA clone encoding a region of the carboxy terminal globular domain (G domain) of the alpha-1 chain of laminin from the sea urchin *S. purpuratus* was reported (Benson et al. 1999).

Among echinoderm-specific ECM components, pamlin has been isolated from *H. pulcherrimus* embryos (Katow 1995). The author showed that PMCs isolated from mesenchyme blastulae bound exclusively to pamlin, which stimulates their migration in in vitro functional assays. Analysis of the molecular image of pamlin by transmission electron microscopy showed that the protein is composed of three subunits, which can aggregate in a complex fashion, giving rise to a large supramolecular network (Omoto and Katow 1998).

Other ECM molecules have been identified as part of the basal lamina surrounding the blastocoel (Fig. 6). A novel protein has been purified from the sea urchin species *P. lividus* and *H. pulcherrimus*, and its biochemical characterisation and functional features have been described (Tesoro et al. 1998). The protein has been named *Pl*-200K or *Hp*-200K, respectively, because of the species from which it was isolated, and, as assessed by affinity chromatography columns, it shows different binding affinities to type I collagen and heparin. *Pl*-200 K has been shown to function as an adhesive substrate, and to be involved in the regulation of sea urchin embryo skeletogenesis (Tesoro et al. 1998).

By means of specific monoclonal antibodies, a molecule with multiple calcium-binding domains, ECM3, has been shown to be arranged in fibres on the basal surface of the ectoderm. These studies established ECM3 as a strong candidate for a PMC substrate molecule and point to several possible mechanisms by which interactions between PMC filopodia and ECM3-containing fibers could provide guidance information to migrating PMCs (Hodor et al. 2000).

As in the case of external ECM, it would be interesting to know the relationship among all the ECM proteins described so far and their spatial organisation.

6 The Blastocoel Matrix

It is useful to remember that the blastocoel is not an empty cavity, but it becomes filled with a matrix whose composition is still not well known (Fig. 6). Several methods have been used to visualise and study the structure of the blastocoel matrix that forms at the blastula stage. Earlier electron microscope studies described this matrix as a "meshwork of fibrous and granular material" which is an extension of the basal lamina (Katow and Solursh 1979). More recent studies have revealed an oriented meshwork, formed by fibrillar sheets interconnected by finer filaments, that completely fill the blastocoelic cavity, in contact with the basal lamina (Cherr et al. 1992). Although detailed information about the structure of the blastocoel matrix is available, in this case little is known about the protein molecules constituting it (for review see Solursh 1986).

7 Receptors to ECM Molecules

ECM molecules are clearly important, but equally significant are the receptors by which cells interact with it, the major class being the integrin superfamily (for a review see Berman et al. 2003; Danen and Sonnenberg 2003). More than 20 of these receptors have been identified so far in different vertebrates: they are heterodimers of two transmembrane chains, α and β, and are widely expressed among the animal kingdom (Bokel and Brown 2002). The different combination of α and β subunits generates integrins with differing ligand specificity (Hynes 1992). Due to their structure, integrins can interact with ECM by their extracellular domains and transduce external signals to the inside by their cytoplasmic tails, which interact with the cytoskeleton, signalling molecules and other cellular proteins, resulting in regulation of many biological functions. These include adhesion, motility, shape, polarity, growth and differentiation (Yamada 1997). Although a reasonable number of ECM proteins have been described in the sea urchin embryo, little is known about their putative receptors. In recent years, however, it has been shown that this system expresses integrins structurally similar to those characterised in other animals (Burke 1999). Marsden and Burke (1997) reported evidence of three novel β integrin subunits that are expressed during early development of *S. purpuratus*. The full-length cDNA sequence for one of them, βG, shows 58 % similarity to vertebrate integrins, particularly to the cytoplasmic domain in which amino acids of the human β1 subunit involved in cell adhesion and signalling are conserved (Marsden and Burke 1997). Functional assays using antibodies and Fab fragments against another sea urchin β integrin, βL, demonstrated its involvement in the initial phase of gastrulation and in the organisation of actin filaments (Marsden and Burke 1998). Our preliminary results, in collaboration with R.D. Burke, indicate that a β integrin serves as a receptor for *Pl*-nectin. In fact, by immunodetection using confocal microscopy, we found that both molecules are localised on the apical cell surface of gastrula embryos. In addition, we were able to show, by immunoprecipitation and affinity chromatography, a calcium-dependent binding between the two molecules, in agreement with what is known for vertebrate integrins (unpubl. results).

An α integrin subunit has been recently cloned from *S. purpuratus* and it was found to exhibit 74–77 % sequence similarity to mammalian α_5, α_8, α_{IIb} and α_v integrins, but its function or its putative ligand(s) have not been investigated (Susan et al. 2000). In contrast, an α integrin similar to the α_5 subgroup of vertebrate integrins, namely alphaSU2, has been described from *L. variegatus*, whose ligand has been identified (Hertzler and McClay 1999). In fact, this integrin appears to be involved in the binding of epithelial cells to laminin, as it has been demonstrated by its localisation on the basal cell surface of epithelia at the mid-blastula stage and by transfection experiments using an α-integrin-deficient CHO cell line. Transfected alphaSU2-expressing CHO cells have been shown to bind to isolated sea urchin basal lamina and to purified laminin, while they bind weakly or not at all to fibronectin, type I collagen and type IV collagen (Hertzler and McClay 1999).

A hypothetical integrin β_5 subunit has been recognised in an RGDS peptide-binding receptor, FR-1R. The molecule localises to the basal side of the ectoderm and to PMCs in sand dollar embryos and it is involved in PMCs migration and gastrulation (Katow and Sofuku 2001).

8 Lateral Cell Surface

From as early as 1900, it was documented that sea urchin embryos can be dissociated into single cells after a simple treatment with Ca^{2+}-free seawater (Fig. 8A; Herbst 1900). Later, based on Herbst's observation, Giudice (1962) observed that dissociated cells were able to spontaneously reaggregate and differentiate into structures closely resembling normal larvae if Ca^{2+} and Mg^{2+} were restored in the seawater (Fig. 8B). The possibility of obtaining normal bipinnaria larvae from dissociated cells has been also demonstrated in the starfish *Asterina pectinifera* (Dan-Sohkawa et al. 1986).

8.1 Cell–Cell Adhesion and Communication: The Discovery of Toposome

The ease of dissociating and reaggregating cells of the sea urchin embryo offered an exceptional model for studies on molecules involved in cell–cell adhesion and for the development of methods for their isolation. In the late

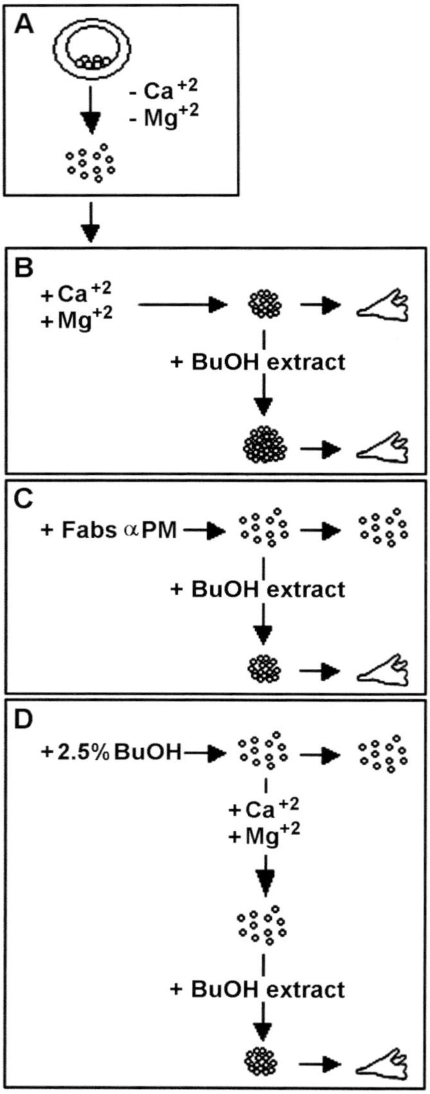

Fig. 8A–D. Schematic drawing illustrating dissociation and reaggregation experiments leading to identification of the active component mediating cell adhesion in the sea urchin embryo. *BuOH* n-Butanol; *PM* plasma membrane

1970s, it was first shown that Fab fragments (Fabs) from antibodies to plasma membranes purified from blastula embryos were able to prevent reaggregation of dissociated cells (Fig. 8C). The inhibition was reversed if soluble proteins extracted with n-butanol from purified membranes were added (Fig. 8C). In addition, these n-butanol extracts were able to strongly stimulate the rate of reaggregation of dissociated cells, suggesting the presence of some aggregating-factor(s) (Fig. 8B; Noll et al. 1979). In agreement with this hypothesis, it was observed that the exposure of dissociated cells to n-butanol completely removed the protein(s) responsible for reaggregation, since cells were not able to reaggregate even in the presence of Ca^{2+} and Mg^{2+} (Fig. 8D). However, the treatment with n-butanol did not affect the viability of the cells, since both reaggregation and embryonic development were completely restored by readdition of extracted proteins to the butanol-treated cells (Fig. 8D; Noll et al. 1979). As a consequence, an easy and low-cost procedure, i.e. the non-cytolytic treatment of live dissociated cells with diluted n-butanol, allowed the preparation of large quantities of crude extracts, from which the isolation, purification and characterisation of the active component could be attempted. In fact, the biochemical identity and biological activity of a large and oligomeric glycoprotein complex, called toposome, was later achieved (Noll et al. 1985; Matranga et al. 1986). Toposome is a 22S complex consisting of six 160-kDa subunits that are processed proteolytically as development proceeds. Cuts are revealed only after analysis by SDS-PAGE, since nicks introduced by specific enzymes, probably a cathepsin B-like protease (Yokota and Kato 1988), do not cause fragmentation of the native protein, thus securing the embryonic integrity. The need for toposome processing has been explained by postulating that a limited number of subunits could be generated by this strategy. Their differential association could then give rise to various molecular populations, each specifying a positional code guiding the cell in the embryo; from this comes the name toposome (Noll et al. 1985). Supporting this interesting hypothesis, toposome-specific monoclonal antibodies have been shown to stain cell surface structures in a pattern consistent with a positional code. The biological activity of the whole toposome complex, or parts of it, in mediating cell adhesion of dissociated cells has been tested and it was found that the oligomeric integrity of toposome is essential for its function (Matranga et al. 1986; Scaturro et al. 1998). The ultrastructural localisation of toposome has been investigated by electron microscopy of immunogold-labeled eggs and hatched blastulae (Gratwohl et al. 1991). Toposomes were seen on the surface of the egg, as well as stored in yolk granules and in the electron-dense lamellar compartment of the cortical granules. In the hatched blastula, toposomes modified by limited proteolysis in the yolk granules have been found associated with the plasma membranes, while unmodified 160-kDa toposomes, originating from the cortical granules, have been found on the outside of the hyaline layer. The latter distribution suggests that these toposomes function by attaching the apical lamina to the surface of the microvilli and thereby to the cytoskeleton of the growing embryo. Therefore, the authors proposed a different function for the two differently localised populations of toposomes (Gratwohl et al. 1991).

An important aspect of dissociation and reaggregation experiments is the effect on the transduction of signals from the exterior of the cell to the nucleus, ultimately involving DNA synthesis. An intriguing paradox is that in the sea urchin embryos, a "no-contact inhibition" was found. In fact, dissociated cells stop DNA synthesis until cell contacts are re-established, i.e. in the formation of the reaggregates (Sconzo et al. 1970; De Petrocellis and Vittorelli 1975). Later, we found that toposomes were responsible for the required signal transduction since reaggregation-inhibiting Fab restored DNA synthesis in dissociated cells (Vittorelli et al. 1980). Apparently, the binding of Fabs to the contact sites mimics cell–cell adhesion and thus stimulates DNA synthesis, in the same way that binding of Fabs to the receptors of epidermal growth factor or insulin mimics the action of those hormones (Kahn et al. 1978; Schreiber et al. 1981).

Further analysis on toposome molecules led to the characterisation of its precursor from sea urchin coelomic fluids of both male and female adults (Cervello and Matranga 1989). The authors produced, for the first time, evidence that the so-called vitellogenin (Vg), found in the coelomic fluid of both male and female sea urchin adults, and its intermediate form, the so-called major yolk protein (MYP) present in granules of unfertilised eggs, are both unprocessed precursor forms of toposome. Both proteins promote cell reaggregation of dissociated blastula cells, suggesting that processing is not required for the cell-adhesion function, but rather directs their localisation during development (Cervello and Matranga 1989). Historical work on the 22S particle assigned to the protein a nutritional role due to its accidental occurrence in granules of the egg and to the need for the non-feeding embryo to develop soon. Similarly, the so-called vitellogenin, although found in both male and female sea urchins, has been recognised as the bona fide precursor to yolk protein, asking for genetic and functional analogies with the vertebrate homologue. It was then important to find the coding sequence for the protein. Previous efforts to gain decisive evidence of a yolk-related nutritional role by cloning the gene for the 22S glycoprotein particle from *S. purpuratus* failed because they resulted in the isolation of only a short cDNA and genomic DNA fragments (Shyu et al. 1986, 1987).

Recently, full-length cDNAs from four different sea urchin species have been reported. They include the Hawaiian *Tripneustus gratilla* toposome mRNA complete coding sequence (cds), submitted back in 2001 and released in 2003 (accession no. AY026514) and the Mediterranean *P. lividus* cds submitted in 2003 and released in 2004 (accession no. AY274929). Strikingly, it has been discovered that the protein, also ironless, is a member of the transferrin family (H. Noll et al., pers. comm., work in prep.), in agreement with sequence data from reports on *Pseudocentrotus depressus* (Unuma et al. 2001), *S. purpuratus* (Brooks and Wessel 2002) and *H. pulcherrimus* (Yokota et al. 2003) cDNAs. Other ironless members of the transferrin family continue to be discovered, two of which are also membrane-associated (Morabito and Moczydlowski 1994; McNagny et al. 1996). Their functions, however, remain unknown.

8.2 Other Cell–Cell Adhesion Molecules

Among cell–cell adhesion molecules, cadherins have been fully characterised in vertebrate organisms (see review by Koch et al. 2004). These proteins are transmembrane glycoproteins that mediate homophilic calcium-dependent cell–cell adhesion in a number of cellular junctions, and their function has been shown to be critical both in normal development and in the development of the invasive and metastatic phenotype (see reviews by Wheelock and Johnsony 2003a; Hazan et al. 2004). Several pathways are activated by cadherin-mediated cell–cell interactions and numerous studies are in progress to elucidate the complex relationships among them (Wheelock and Johnsony 2003b). Although recent work on sequence analysis has shed new light on the molecular basis of cadherin adhesion, understanding the specificity of these interactions remains a major challenge (Patel et al. 2003). The first immunological evidence for the presence of a cell adhesion protein similar to the mouse E-cadherin in the sea urchin embryo has been shown in *P. lividus* (Ghersi and Vittorelli 1990; Ghersi et al. 1993). Furthermore, the use of polyclonal antibodies raised against a cloned sea urchin cadherin, which recognises at least three major polypeptides, and the cloning of a novel sea-urchin-specific cadherin molecule support the hypothesis for the presence of several cadherins in this system (Miller and McClay 1997).

8.3 Cell Junctions

The presence of cell junctions in the sea urchin embryo has been described since the 1960s. The blastular wall has the structure of a simple epithelium, similar to that of vertebrates. At their apical surfaces, the cells are joined by typical junctional complexes, including zonulae adherens or belt desmosomes, septate junctions and spot desmosomes, while hemidesmosome-like structures appear localised at their basal surface (Fig. 6; Wolpert and Mercer 1963; Spiegel and Howard 1983). Septate junctions have also been observed at the four-blastomere stage (Chang and Afzelius 1973), although they constitute a continuous layer only later in development (Gilula 1973). At least three types of desmosome, two types of septate junction and a tricellular junction have been described (Spiegel and Howard 1983). Studies on the osmotic and structural properties of the blastular wall date back to 1940, when a decrease in the permeability of the blastula wall to small molecules such as sucrose (Moore 1940) was observed, coinciding with the formation of junctional structures and with an apparent increase in the adhesion between cells.

9 Signalling Pathways

A key feature of both cell–ECM and cell–cell adhesion is the linkage of receptors or other transmembrane molecules to the cytoskeleton, which activates specific intracellular pathways. In particular, cadherins and integrins have been described to modulate and to be modulated by multiple signalling transduction pathways (Thiery 2003). These are complex networks involving a number of steps, including the production of second messengers, the activation of protein kinases and their subcellular distribution that brings them into contact with the appropriate targets. Although each adhesion protein activates a particular pathway, i.e. the tyrosine kinases and mitogen-activated protein (MAP) kinase cascade components activated by integrins and the Wnt pathway influenced by cadherins, it is becoming increasingly clear that cell adhesion proteins may share much of the same basic components of signal transduction machinery. Cell adhesion proteins, in addition to their well-established role in providing positional information, have been demonstrated to regulate also the cellular response to other extracellular stimuli, such as soluble growth factors. However, much remains to be learned concerning the mechanisms that co-ordinate positional and biochemical signals (for a review see Aplin et al. 1998). Some of the components of such signalling pathways have been described in echinoderm embryos and some pathways are going to be elucidated, but this issue will not be further extended in this chapter.

10 Growth Factors

Growth factors, like the ECM, have been shown to fulfil vital developmental roles in many embryonic systems. Among them, transforming growth factor-beta (TGF-β) superfamily members appear to have diversified greatly with the evolution of vertebrates, but, although they are believed to be widely distributed among the animal kingdom, only a few invertebrate deuterostome TGF-β molecules have been identified so far. Sequence comparisons suggest an early origin and an evolutionary conservation of the molecular conformation of the members of TGF-β superfamily (see reviews by Hogan et al. 1994; Kingsley 1994; Chin et al. 2004). Among echinoderms, the first gene encoding a member of the TGF-β superfamily to be reported was identified in the sea urchin embryo, and it was named *univin* (Stenzel et al. 1994). Sequence comparisons placed *univin* in the bone morphogenetic protein (BMP) group of the TGF-β superfamily along with the vertebrate BMPs, decapentaplegic protein from *Drosophila* and Vg-1 from *Xenopus*. Recently, we have shown the involvement of this growth factor in the ectoderm inductive signals needed for the skeleton morphogenesis in the sea urchin embryo (Zito et al. 2003). In this system, skeletogenesis is a fully documented example of ecto-mesoderm induction, since PMCs, although already determined to only synthesise the skeleton (Okazaki 1975), need signals from the surrounding ectoderm to direct skeleton growth and patterning (Armstrong et al. 1993; Ettensohn and Malinda 1993; Guss and Ettensohn 1997). As previously mentioned, we found that the inhibition of *Pl*-nectin interaction with ectoderm cells produced perturbed embryos, in which only skeleton elongation and patterning were specifically affected. Interestingly, we observed that *univin* expression was strongly inhibited in skeleton-defective embryos, while its misexpression, obtained by *univin* mRNA injection, rescued skeletogenesis. These findings are in agreement with the hypothesis of the involvement of diffusible molecules in the ecto-mesoderm induction suggested by other researchers (Kiyomoto and Tsukahara 1991; Page and Benson 1992; Ettensohn and Malinda 1993; Guss and Ettensohn 1997). Our data, in particular, support a model in which some ectodermal cells secrete processed *univin* or a related growth factor into the blastocoel, where it signals PMCs to synthesise the spicule matrix proteins required for spicule growth (Fig. 9). The ability of ectodermal cells to produce the signal depends on their association with *Pl*-nectin on the apical ECM. Since by our preliminary results the interaction between ectodermal cells and *Pl*-nectin seems mediated by an integrin receptor, it is reasonable to hypothesise the involvement of one of the signalling pathways already known for integrins in other systems. Studies aimed at the identification of such a pathway are in progress in our laboratory. Furthermore, this model predicts

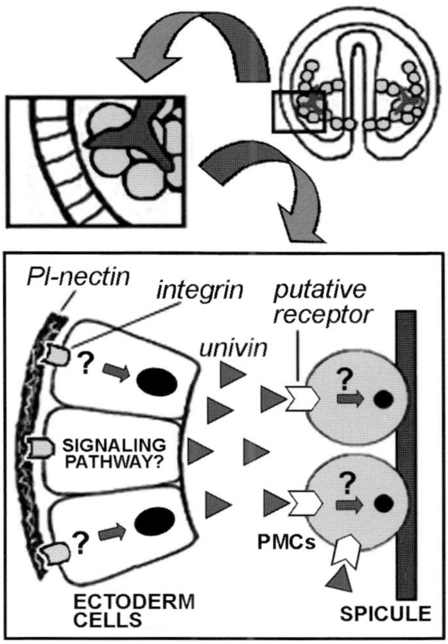

Fig. 9. Model to explain ecto-mesoderm induction in the sea urchin embryo. Ectoderm cells properly interacting with the outer *Pl*-nectin secrete into the blastocoel the growth factor *univin*, which signals PMCs to synthesise the spicule. The interaction of ectoderm cells with *Pl*-nectin, possibly mediated by an integrin receptor, activates a yet unknown signalling pathway. The model predicts also the expression of a putative TGF-β receptor on PMCs and thus a signalling pathway

that PMCs should express a TGF-β receptor. Recently, one similar to vertebrate type I receptor, Alk2, has been found to be expressed in ingressed PMCs of *S. purpuratus* embryos, which can mediate both activin and BMP signals (L. Angerer, unpubl. observ.). However, whether this receptor mediates the skeleton-promoting signal is not yet known.

Two very recent papers have shown the involvement of a MAP kinase signalling pathway in the development of the micromere lineage and mesenchyme differentiation (Fernandez-Serra et al. 2004; Rottinger et al. 2004). Whether a MAP kinase signalling is required for the ecto-mesoderm induction guiding skeletogenesis remains to be demonstrated.

Other BMPs homologues have been recently cloned in the sea urchin embryo, specifically *Sp*BMP5–7 (Ponce et al. 1999), *Tg*BMP2/4 (Hwang et al. 1999) and *Lv*BMP2/4 (Angerer et al. 2000), and *At*BMP2/4 in the starfish (Shih et al. 2002). Their developmental expression patterns have been described by Northern blotting and in situ hybridisation experiments. Functional studies altering *Lv*BMP2/4 mRNA levels showed the involvement of this growth factor in regulating cell fate allocation along the sea urchin animal–vegetal embryonic axis (Angerer et al. 2000). Evidence of the presence of growth-factor-like molecules belonging to other families than TGF-βs has also been reported. In 1987, a cDNA clone whose protein product displayed striking homology to the EGF family of proteins was identified and characterised in the sea urchin embryo (Hursh et al. 1987). The presence of other growth factors in echinoderms has been shown by indirect methods. Platelet-derived growth factor-BB (PDGF-BB) and TGF-α have been reported to be involved in development by using human recombinant proteins, which rescued sea urchin embryo abnormalities induced by ECM disrupters (Ramachandran et al. 1993), or using anti-human PDGF-B and TGF-α antibodies which affected embryo development (Govindarajan et al. 1995). Antibodies against mammalian receptors for PDGF and EGF, or a dominant/negative RNA for PDGF receptor, have been reported to be involved in early differentiation and morphogenesis of the embryo (Ramachandran et al. 1995, 1997). A new member of the fibroblast growth factor receptor (FGFR) family has been cloned in *S. purpuratus* and it has been shown to contain a series of domains characteristic of the family, as three immunoglobulin-like motifs, an acid box, a transmembrane domain, a relatively long juxtamembrane sequence, a split tyrosine kinase domain and two conserved intracellular tyrosine residues (McCoon et al. 1996). These authors detected *Sp*FGFR protein only in muscle cells of the embryo, suggesting that its function may be required to support the proliferation, migration and/or differentiation of myoblasts, rather than being involved in commitment to a muscle fate (McCoon et al. 1998).

11 Concluding Remarks

In the last 30 years, a number of adhesion molecules have been identified in the echinoderm model system, some homologous to already known vertebrate adhesion molecules, while others are specific to the phylum. However, it is noteworthy to remark here that some of the echinoderm-specific cell-adhesion proteins share epitope(s) with mammalian proteins. These include, for example, hyalin, whose homology has not been found in vertebrates, but some of its antigenic epitopes are shared with two mammalian cortical granule envelope proteins (Hoodbhoy et al. 2000). Taken all together, the findings described up to now reveal all the possible scenarios, such as a great conservation of some basic processes as well as the development of new genes and processes during evolution. We can find cell-adhesion proteins developed early in metazoan evolution that have been conserved and have conserved their developmental roles. There are other cases in which only some of the functional epitopes have been conserved and new proteins have been assembled from new arrangements of old domains. Eventually, entirely new proteins evolved, which do not show any close equivalent in invertebrates, and examples can be found in some uniquely vertebrate functions such as vascular biology, neurobiology and neural crest migration (Hynes and Zhao 2000). Nevertheless, one message still remains clear: that is, the phylogenetic closeness of echinoderms to chordates. Support for this relationship, other than by molecular data, is provided by fossil records. A new group of small fossils have just been discovered in China (Shu et al. 2004). These fossils, dating back to the Lower Cambrian (about 520 million years ago), have been interpreted to be the most primitive echinoderms yet known, and have been supposed to be the bridge linking echinoderms and other deuterostomes.

All the topics discussed up to now imply as a consequence that the knowledge of the functional genome of echinoderms might lead to the definition of basic pathways involved in development as well as in diseases. Further, important outcomes for the understanding of echinoderm and chordate evolution will be obtained from the comparative analysis of the expression patterns of regulatory genes in Echinodermata (Wray and Lowe 2000). Currently, a systematic analysis of the echinoderm genome, namely that of the sea urchin *S. purpuratus*, is in progress (http://sugp.caltech.edu/), which is expected to provide new understanding on the organisation of the genome, as well as new insights into the possible application of echinoderm genes in biotechnology and human therapy. An analogous analysis of the genome of the Mediterranean species *P. lividus* has recently started within the NoE Marine Genomics Europe, joining together the European scientific laboratories (www.marine-genomics-europe.org).

It is hoped that studies on the Echinodermata will result not only in important cell biological insights, but also in development of further therapeutic tools.

Acknowledgements. C.C. and V.P. were supported by the Italian Ministry for Education, University and Research, Master MAMBO, prot. MIUR no. 1473/663. We wish to thank Drs. A. Bonomolo and R. Bonaventura for help in preparation of the figures.

References

Adelson DL, Humphreys T (1988) Sea urchin morphogenesis and cell-hyalin adhesion are perturbed by a monoclonal antibody specific for hyalin. Development 104:391–402

Adelson DL, Alliegro MC, McClay DR (1992) On the ultrastructure of hyalin, a cell adhesion protein of the sea urchin embryo extracellular matrix. J Cell Biol 116:1283–1289

Alliegro MC, Ettensohn CA, Burdsal CA, Erickson HP, McClay DR (1988) Echinonectin: a new embryonic substrate adhesion protein. J Cell Biol 107:2319–2327

Alliegro MC, Burdsal CA, McClay DR (1990) In vitro biological activities of echinonectin. Biochemistry 29:2135–2141

Amemiya S (1989) Development of the basal lamina and its role in migration and pattern formation of primary mesenchyme cells in sea urchin embryos. Dev Growth Differ 31:131–145

Angerer LM, Chambers SA, Yang Q, Venkatesan M, Angerer RC, Simpson RT (1988) Expression of a collagen gene in mesenchyme lineages of the *Strongylocentrotus purpuratus* embryo. Genes Dev 2:239–246

Angerer LM, Oleksyn DW, Logan CY, McClay DR, Dale L, Angerer RC (2000) A BMP pathway regulates cell fate allocation along the sea urchin animal–vegetal embryonic axis. Development 127:1105–1114

Aplin AE, Howe A, Alahari SK, Juliano RL (1998) Signal transduction and signal modulation by cell adhesion receptors: the role of integrins, cadherins, immunoglobulin-cell adhesion molecules, and selectins. Pharmacol Rev 50(2):197–263

Armstrong N, Hardin J, McClay DR (1993) Cell–cell interactions regulate skeleton formation in the sea urchin embryo. Development 119:833–840

Bather FA (1900) The Echinodermata. In: Lankester RR (ed) A treatise on zoology, part III, A and C. Black, London

Benson S, Smith L, Wilt F, Shaw R (1990) The synthesis and secretion of collagen by cultured sea urchin micromeres. Exp Cell Res 188:141–146

Benson S, Page L, Ingersoll E, Rosenthal E, Dungca K, Signor D (1999) Developmental characterization of the gene for laminin alpha-chain in sea urchin embryos. Mech Dev 81:37–49

Berman AE, Kozlova NI, Morozevich GE (2003) Integrins: structure and signaling. Biochemistry (Mosc) 68:1284–1299

Bisgrove BW, Raff RA (1993) The SpEGF III gene encodes a member of the fibropellins: EGF repeat-containing proteins that form the apical lamina of the sea urchin embryo. Dev Biol 157:526–538

Bisgrove BW, Andrews ME, Raff RA (1991) Fibropellins, products of an EGF repeat-containing gene, form a unique extracellular matrix structure that surrounds the sea urchin embryo. Dev Biol 146:89–99

Bokel C, Brown NH (2002) Integrins in development: moving on, responding to, and sticking to the extracellular matrix. Dev Cell 3:311–321

Brennan C, Robinson JJ (1994) Cloning and characterization of HLC-32, a 32-kDa protein component of the sea urchin extraembryonic matrix, the hyaline layer. Dev Biol 165:556–565

Brooks JM, Wessel GM (2002) The major yolk protein in sea urchins is a transferrin-like, iron binding protein. Dev Biol 245:1–12

Burke RD (1999) Invertebrate integrins: structure, function and evolution. Int Rev Cytol 191:257–284

Burke RD, Myers RL, Sexton TL, Jackson C (1991) Cell movements during the initial phase of gastrulation in the sea urchin embryo. Dev Biol 146:542–557

Burke RD, Lail M, Nakajima Y (1998) The apical lamina and its role in cell adhesion in sea urchin embryos. Cell Adhes Commun 5:97–108

Cameron RA, Davidson EH (1991) Cell type specification during sea urchin development. Trends Gen 7:212–218

Cameron RA, Fraser SE, Britten RJ, Davidson EH (1991) Macromere cell fates during sea urchin development. Development 113:1085–1091

Cazzola M, Page CP, Matera MG (2004) Alternative and/or integrative therapies for pneumonia under development. Curr Opin Pulm Med 10:204–210

Cervello M, Matranga V (1989) Evidence of a precursor–product relationship between vitellogenin and toposome, a glycoprotein complex mediating cell adhesion. Cell Differ Dev 26:67–76

Chang DC, Afzelius BA (1973) Electron microscopic study of membrane junctions of *Arbacia punctulata* blastomeres. Biol Bull 145:428

Cherr GN, Summers RG, Baldwin JD, Morrill JB (1992) Preservation and visualization of the sea urchin embryo blastocoelic extracellular matrix. Microsc Res Tech 22:11–22

Chin D, Boyle GM, Parsons PG, Coman WB (2004) What is transforming growth factor-beta (TGF-beta)? Br J Plast Surg 57:215–221

Citkowicz E (1971) The hyaline layer: its isolation and role in echinoderm development. Dev Biol 24:348–362

D'Alessio M, Ramirez F, Suzuki HR, Solursh M, Gambino R (1989) Structure and developmental expression of a sea urchin fibrillar collagen gene. Proc Natl Acad Sci USA 86:9303–9307

D'Alessio M, Ramirez F, Suzuki HR, Solursh M, Gambino R (1990) Cloning of a fibrillar collagen gene expressed in the mesenchymal cells of the developing sea urchin embryo. J Biol Chem 265:7050–7054

Danen EH, Sonnenberg A (2003) Integrins in regulation of tissue development and function. J Pathol 201:632–641

Dan-Sohkawa M, Yamanaka H, Watanabe K (1986) Reconstruction of bipinnaria larvae from dissociated embryonic cells of the starfish, *Asterina pectinifera*. J Embryol Exp Morphol 94:47–60

Davidson EH, Cameron RA, Ransick A (1998) Specification of cell fate in the sea urchin embryos: summary and some proposed mechanisms. Development 125:3269–3290

Davidson EH, Rast JP, Oliveri P, Ransick A, Calestani C, Yuh CH, Minokawa T, Amore G, Hinman V, Arenas-Mena C, Otim O, Brown CT, Livi CB, Lee PY, Revilla R, Rust AG, Pan Z, Schilstra MJ, Clarke PJ, Arnone MI, Rowen L, Cameron RA, McClay DR, Hood L, Bolouri H (2002) A genomic regulatory network for development. Science 295:1669–1678

Delgadillo-Reynoso MG, Rollo DR, Hursh DA, Raff RA (1989) Structural analysis of the uEGF gene in the sea urchin *Strongylocentrotus purpuratus* reveals more similarity to vertebrate than to invertebrate genes with EGF-like repeats. J Mol Evol 29:314–327

De Petrocellis B, Vittorelli ML (1975) Role of cell interactions in development and differentiation of the sea urchin *Paracentrotus lividus*. Changes in the activity of some enzymes of DNA biosynthesis after cell dissociation. Exp Cell Res 94:392–400

DeSimone DW, Spiegel E, Spiegel M (1985) The biochemical identification of fibronectin in the sea urchin embryo. Biochem Biophys Res Commun 133:183–188

Driesch H (1892) The potency of the first two cleavage cells in echinoderm development. Experimental production of partial and double formations. In: Willier BH, Oppenheimer JM (eds) Foundations of experimental embryology . Hafner, New York

Driesch H (1894) Analytische Theorie de organischen Entwicklung. W Engelmann, Leipzig

Dyer ES, Paulsen MT, Markwart SM, Goh M, Livant DL, Ljungman M (2002) Phenylbutyrate inhibits the invasive properties of prostate and breast cancer cell lines in the sea urchin embryo basement membrane invasion assay. Int J Cancer 101:496–499

Ettensohn CA, Malinda KM (1993) Size regulation and morphogenesis: a cellular analysis of skeletogenesis in the sea urchin embryo. Development 119:155–167

Eudy JD, Ma-Edmonds M, Yao SF, Talmadge CB, Kelley PM, Weston MD, Kimberling WJ, Sumegi J (1997) Isolation of a novel human homologue of the gene coding for echinoderm microtubule-associated protein (EMAP) from the Usher syndrome type 1a locus at 14q32. Genomics 43:104–106

Exposito JY, D'Alessio M, Solursh M, Ramirez F (1992) Sea urchin collagen evolutionarily homologous to vertebrate pro-alpha 2(I) collagen. J Biol Chem 267:15559–15562

Fernandez-Serra M, Consales C, Livigni A, Arnone MI (2004) Role of the ERK-mediated signaling pathway in mesenchyme formation and differentiation in the sea urchin embryo. Dev Biol 268:384–402

Flood J, Mayne J, Robinson JJ (2000) Identification and characterization of gelatin-cleavage activities in the apically located extracellular matrix of the sea urchin embryo. Biochem Cell Biol 78:455–462

Galileo DS, Morrill JB (1985) Patterns of cells and extracellular material of the sea urchin *Lytechinus variegatus* (Echinodermata; Echinoidea) embryo, from hatched blastula to late gastrula. J Morphol 185:387–402

Ghersi G, Vittorelli ML (1990) Immunological evidence for the presence in sea urchin embryos of a cell adhesion protein similar to mouse uvomorulin (E-cadherin). Cell Differ Dev 31:67–75

Ghersi G, Salamone M, Dolo V, Levi G, Vittorelli ML (1993) Differential expression and function of cadherin-like proteins in the sea urchin embryo. Mech Dev 41:47–55

Gilula NB (1973) Septate junction development in sea urchin embryos. J Cell Biol 55:172

Giudice G (1962) Reconstitution of whole larvae from disaggregated cells of sea urchin embryos. Dev Biol 5:402–411

Govindarajan V, Ramachandran RK, George JM, Shakes DC, Tomlinson CR (1995) An ECM-bound, PDGF-like growth factor and a TGF-alpha-like growth factor are required for gastrulation and spiculogenesis in the *Lytechinus* embryo. Dev Biol 172:541–551

Gratwohl EK, Kellenberger E, Lorand L, Noll H (1991) Storage, ultrastructural targeting and function of toposomes and hyalin in sea urchin embryogenesis. Mech Dev 33:127–138

Grimwade JE, Gagnon ML, Yang Q, Angerer RC, Angerer LM (1991) Expression of two mRNAs encoding EGF-related proteins identifies subregions of sea urchin embryonic ectoderm. Dev Biol 143:44–57

Guss KA, Ettensohn CA (1997) Skeletal morphogenesis in the sea urchin embryo: regulation of primary mesenchyme gene expression and skeletal rod growth by ectoderm-derived cues. Development 124:1899–1908

Gustafson T, Wolpert L (1961) Studies on the cellular basis of morphogenesis in sea urchin embryos: directed movements of primary mesenchyme cells in normal and vegetalized larvae. Exp Cell Res 24:64–79

Gustafson T, Wolpert L (1967) Cellular movement and contact in sea urchin morphogenesis. Biol Rev 42:442–498

Haag ES, Sly BJ, Andrews ME, Raff RA (1999) Apextrin, a novel extracellular protein associated with larval ectoderm evolution in *Heliocidaris erythrogramma*. Dev Biol 211:77–87

Hall HG, Vacquier VD (1982) The apical lamina of the sea urchin embryo: major glycoproteins associated with the hyaline layer. Dev Biol 89:168–178

Hardin J (1996) The cellular basis of sea urchin gastrulation. Curr Topics Dev Biol 33:159–262

Hazan RB, Qiao R, Keren R, Badano I, Suyama K (2004) Cadherin switch in tumour progression. Ann N Y Acad Sci 1014:155–163

Herbst C (1900) Über das auseinandergehen wilhelm in furchungs- und gewebezellen in kalkfreiem medium. Arch Entwicklungsmech Org 9:424–463

Hertzler PL, McClay DR (1999) alphaSU2, an epithelial integrin that binds laminin in the sea urchin embryo. Dev Biol 207:1–13

Hodor PG, Illies MR, Broadley S, Ettensohn CA (2000) Cell–substrate interactions during sea urchin gastrulation: migrating primary mesenchyme cells interact with and align extracellular matrix fibers that contain ECM3, a molecule with NG2-like and multiple calcium-binding domains. Dev Biol 222:181–194

Hogan BL, Blessing M, Winnier GE, Suzuki N, Jones CM (1994) Growth factors in development: the role of TGF-beta related polypeptide signalling molecules in embryogenesis. Dev (Suppl):53–60

Hoodbhoy T, Carroll EJ Jr, Talbot P (2000) Relationship between p62 and p56, two proteins of the mammalian cortical granule envelope, and hyalin, the major component of the echinoderm hyaline layer, in hamsters. Biol Reprod 62:979–987

Hörstadius S (1939). The mechanics of sea urchin development, studied by operative methods. Biol Rev 14:132–179

Hursh DA, Andrews ME, Raff RA (1987) A sea urchin gene encodes a polypeptide homologous to epidermal growth factor. Science 237:1487–1490

Hwang SL, Chen CA, Chen C (1999) Sea urchin TgBMP2/4 gene encoding a bone morphogenetic protein closely related to vertebrate BMP2 and BMP4 with maximal expression at the later stages of embryonic development. Biochem Biophys Res Commun 258:457–463

Hynes RO (1992) Integrins: versatility, modulation, and signaling in cell adhesion. Cell 69:11–25

Hynes RO, Zhao Q (2000) The evolution of cell adhesion. J Cell Biol 150:F89–F95

Iwata M, Nakano E (1981) Fibronectin from the ovary of the sea urchin, *Pseudocentrotus depressus*. Wilhelm Roux Arch Dev Biol 194:377–384

Iwata M, Nakano E (1983).Characterization of sea-urchin fibronectin. Biochem J 215:205–208

Kahn CR, Baird KL, Jarrett DB, Flier JS (1978) Direct demonstration that receptor crosslinking or aggregation is important in insulin action. Proc Natl Acad Sci USA 75:4209–4213

Kahn TA, Blumer J, Silverman RA, Bickers DR (1988) Screening for the developmental toxicity of retinoids: use of the sea urchin model. Fund Appl Toxicol 11:511–518

Kane RE (1970) Direct isolation of the hyaline layer protein released from the cortical granules of the sea urchin egg at fertilization. J Cell Biol 45:615–622

Kane RE, Stephens RE (1969) A comparative study of the isolation of the cortex and the role of the calcium-insoluble protein in several species of sea urchin egg. J Cell Biol 41:133–144

Kato KH, Abe T, Nakashima S, Matranga V, Zito F, Yokota Y (2004) 'Nectosome': a novel cytoplasmic vesicle containing nectin in the egg of the sea urchin, *Temnopleurus hardwickii*. Dev Growth Differ 46:239–447

Katow H (1995) Pamlin, a primary mesenchyme cell adhesion protein, in the basal lamina of the sea urchin embryo. Exp Cell Res 218:469–278

Katow H, Sofuku S (2001) An RGDS peptide-binding receptor, FR-1R, localizes to the basal side of the ectoderm and to primary mesenchyme cells in sand dollar embryos. Dev Growth Differ 43:601–610

Katow H, Solursh M (1979) Ultrastructure of blastocoel material in blastulae and gastrulae of the sea urchin *Lytechinus pictus*. J Exp Zool 210:561–567

Katow H, Yamada KM, Solursh M (1982) Occurrence of fibronectin on the primary mesenchyme cell surface during migration in the sea urchin embryo. Differentiation 22:120–124

Kawabe R, Armstrong P, Pollock E (1981) An extracellular fibrillar matrix in gastrulating sea urchin embryos. Dev Biol 85:509–515

Kingsley DM (1994) The TGF-beta superfamily: new members, new receptors, and new genetic tests of function in different organisms. Genes Dev 8:133–146

Kiyomoto M, Tsukahara J (1991) Spicule formation-inducing substance in sea urchin embryo. Dev Growth Differ 33:443–450

Koch AW, Manzur KL, Shan W (2004) Structure-based models of cadherin-mediated cell adhesion: the evolution continues. Cell Mol Life Sci 61:1884–1895

Koyama T, Noguchi K, Aniya Y, Sakanashi M (1998) Analysis for sites of anticoagulant action of plancinin, a new anticoagulant peptide isolated from the starfish *Acanthaster planci*, in the blood coagulation cascade. Gen Pharmacol 31:277–282

Lethias C, Exposito JY, Descollonges Y, Penin F, Garrone R (1995) Expression of a new fibrillar collagen class in the sea urchin embryo during spiculogenesis. Biol Cell 84:98

Littlewood DTJ, Smith AB, Clough KA, Emson RH (1997) The interrelationships of the echinoderm classes: morphological and molecular evidence. Biol J Linn Soc 61:409–438

Livant DL, Linn S, Markwart S, Shuster J (1995) Invasion of selectively permeable sea urchin embryo basement membranes by metastatic tumor cells, but not by their normal counterparts. Cancer Res 55:5085–5093

Lundgren B (1973) Surface coating of the sea urchin larva as revealed by ruthenium red. J Submicrosc Cytol 5:61–70

Maitra A, Iacobuzio-Donahue C, Rahman A, Sohn TA, Argani P, Meyer R, Yeo CJ, Cameron JL, Goggins M, Kern SE, Ashfaq R, Hruban RH, Wilentz RE (2002) Immunohistochemical validation of a novel epithelial and a novel stromal marker of pancreatic ductal adenocarcinoma identified by global expression microarrays: sea urchin fascin homolog and heat shock protein 47. Am J Clin Pathol 118:52–59

Marsden M, Burke RD (1997) Cloning and characterization of novel beta integrin subunits from a sea urchin. Dev Biol 181:234–245

Marsden M, Burke RD (1998) The beta L integrin subunit is necessary for gastrulation in sea urchin embryos. Dev Biol 203:134–148

Matranga V, Kuwasaki B, Noll H (1986) Functional characterization of toposomes from sea urchin blastula embryos by a morphogenetic cell aggregation assay. EMBO J 5:3125–3132

Matranga V, Di Ferro D, Zito F, Cervello M, Nakano E (1992) A new extracellular matrix protein of the sea urchin embryo with properties of a substrate adhesion molecule. Wilhelm Roux Arch Dev Biol 201:173–178

Matranga V, Yokota Y, Zito F, Tesoro V, Nakano E (1995) Biochemical and immunological relationships among fibronectin-like proteins from different sea urchin species. Wilhelm Roux Arch Dev Biol 204:413–417

Mayne J, Robinson JJ (1996) Purification and metal ion requirements of a candidate matrix metalloproteinase: a 41 kDa gelatinase activity in the sea urchin embryo. Biochem Cell Biol 74:211–218

McCarthy RA, Burger MM (1987) In vivo embryonic expression of laminin and its involvement in cell shape change in the sea urchin *Sphaerechinus granularis*. Development 101:659–671

McClay DR (1991) The role of cell adhesion during gastrulation in the sea urchin. In: Keller R, Clark WH Jr, Griffin F (eds) Gastrulation: movements, patterns, and molecules. Plenum Press, New York, pp 313–327

McClay DR, Fink RD (1982) Sea urchin hyalin: appearance and function in development. Dev Biol 92:285–293

McClay DR, Alliegro MC, Black SD (1990) The ontogenetic appearance of extracellular matrix during sea urchin development. In: Adair WS, Mecham R (eds) Organization and assembly of plant and animal extracellular matrix. Academic Press, New York, pp 1–15

McCoon PE, Angerer RC, Angerer LM (1996) SpFGFR, a new member of the fibroblast growth factor receptor family, is developmentally regulated during early sea urchin development. J Biol Chem 271:20119–20125

McCoon PE, Blackstone E, Angerer RC, Angerer LM (1998) Sea urchin FGFR muscle-specific expression: posttranscriptional regulation in embryos and adults. Dev Biol 200:171–181

McNagny KM, Rossi F, Smith G, Graf T (1996) The eosinophil-specific cell surface antigen, EOS47, is a chicken homologue of the oncofetal antigen melanotransferrin. Blood 87:1343–1352

Meijer L, Raymond E (2003) Roscovitine and other purines as kinase inhibitors. From starfish oocytes to clinical trials. Acc Chem Res 36:417–425

Mengerink KJ, Moy GW, Vacquier VD (2002) suREJ3, a polycystin-1 protein, is cleaved at the GPS domain and localizes to the acrosomal region of sea urchin sperm. J Biol Chem. 277:943–948

Miller JR, McClay DR (1997) Characterization of the role of cadherin in regulating cell adhesion during sea urchin development. Dev Biol 192:323–339

Miyachi Y, Iwata M, Sato H, Nakano E (1984) Effect of fibronectin on cultured cells derived from isolated micromeres of the sea urchin, *Hemicentrotus pulcherrimus*. Zool Sci 1:265–271

Moore AR (1940) Osmotic and structural properties of the blastular wall in *Dendraster excentricus*. J Exp Zool 84:73–79

Morabito MA, Moczydlowski E (1994) Molecular cloning of bullfrog saxiphilin: a unique relative of the transferrin family that binds saxitoxin. Proc Natl Acad Sci USA 91:2478–2482

Nakajima Y, Burke RD (1996) The initial phase of gastrulation in sea urchins is accompanied by the formation of bottle cells. Dev Biol 179:436–446

Neill AT, Moy GW, Vacquier VD (2004) Polycystin-2 associates with the polycystin-1 homolog, suREJ3, and localizes to the acrosomal region of sea urchin spermatozoa. Mol Reprod Dev 67:472–477

Nishioka D, Marcell V, Cunningham M, Khan M, Von Hoff DD, Izbicka E (2003) The use of early sea urchin embryos in anticancer drug testing. Method Mol Med 85:265–276

Noll H, Matranga V, Cascino D, Vittorelli L (1979) Reconstitution of membranes and embryonic development in dissociated blastula cells of the sea urchin by reinsertion of aggregation-promoting membrane proteins extracted with butanol. Proc Natl Acad Sci USA 76:288–292

Noll H, Matranga V, Cervello M, Humphreys T, Kuwasaki B, Adelson D (1985) Characterization of toposomes from sea urchin blastula cells: a cell organelle mediating cell adhesion and expressing positional information. Proc Natl Acad Sci USA 82:8062–8066

Okazaki K (1975) Spicule formation by isolated micromeres of the sea urchin embryo. Am Zool 15:567–581

Okazaki K, Nijima L (1964) Basement membrane in sea urchin larvae. Embryologia 8:89–100

Omoto T, Katow H (1998) Initial analysis of the molecular image of pamlin, a sea urchin cell adhesion protein, by transmission electron microscopy. Dev Growth Differ 40:287–295

Osanai K (1960) Development of the sea urchin egg with the inhibited breakdown of the cortical granules. Sci Rep Tohoku Univ 36:77–87

Page L, Benson S (1992) Analysis of competence in cultured sea urchin micromeres. Exp Cell Res 203:305–311

Patel SD, Chen CP, Bahna F, Honig B, Shapiro L (2003) Cadherin-mediated cell–cell adhesion: sticking together as a family. Curr Opin Struct Biol 13:690–698

Pearse VB, Pearse JS (1994) Echinoderm phylogeny and the place of the concentricycloids. In: David B, Guille A, Feral JP, Roux M (eds) Echinoderms through time. Balkema, Rotterdam, pp 121–126

Ponce MR, Micol JL, Peterson KJ, Davidson EH (1999) Molecular characterization and phylogenetic analysis of SpBMP5-7, a new member of the TGF-beta superfamily expressed in sea urchin embryos. Mol Biol Evol 16:634–645

Pucci-Minafra I, Minafra S, Gianguzza F, Casano C (1975) Amino acid composition of collagen extracted from the spicules of sea urchin embryos (*Paracentrotus lividus*). Boll Zool 42:201–204

Qiao D, Nikitina LA, Buznikov GA, Lauder JM, Seidler FJ, Slotkin TA (2003) The sea urchin embryo as a model for mammalian developmental neurotoxicity: ontogenesis of the high-affinity choline transporter and its role in cholinergic trophic activity. Environ Health Perspect 111:1703–1705

Raff RA (1996) The shape of life. Genes, development, and the evolution of animal form. University of Chicago Press, Chicago

Ramachandran RK, Seid CA, Lee H, Tomlinson CR (1993) PDGF-BB and TGF-alpha rescue gastrulation, spiculogenesis, and LpS1 expression in collagen-disrupted embryos of the sea urchin genus *Lytechinus*. Mech Dev 44:33–40

Ramachandran RK, Govindarajan V, Seid CA, Patil S, Tomlinson CR (1995) Role for platelet-derived growth factor-like and epidermal growth factor-like signaling pathways in gastrulation and spiculogenesis in the *Lytechinus* sea urchin embryo. Dev Dyn 204:77–88

Ramachandran RK, Wikramanayake AH, Uzman JA, Govindarajan V, Tomlinson CR (1997) Disruption of gastrulation and oral–aboral ectoderm differentiation in the *Lytechinus pictus* embryo by a dominant/negative PDGF receptor. Development 124:2355–6364

Rise M, Burke RD (2002) SpADAM, a sea urchin ADAM, has conserved structure and expression. Mech Dev 117:275–281

Robinson JJ (1997) Characterization of a metalloproteinase: a late stage specific gelatinase activity in the sea urchin embryo. J Cell Biochem 66:337–345

Robinson JJ, Hall D, Brennan C, Kean P (1992) Hyalin, a sea urchin extraembryonic matrix protein: relationship between calcium binding and hyalin gelation. Arch Biochem Biophys 298:129–134

Rottinger E, Besnardeau L, Lepage T (2004) A Raf/MEK/ERK signaling pathway is required for development of the sea urchin embryo micromere lineage through phosphorylation of the transcription factor Ets. Development 131:1075–1087. Erratum in: Development 131:2233

Sahara H, Ishikawa M, Takahashi N, Ohtani S, Sato N, Gasa S, Akino T, Kikuchi K (1997) In vivo anti-tumor effect of 3′-sulphonoquinovosyl 1′-monoacylglyceride isolated from sea urchin (*Strongylocentrotus intermedius*) intestine. Br J Cancer 75:324–332

Sahara H, Hanashima S, Yamazaki T, Takahashi S, Sugawara F, Ohtani S, Ishikawa M, Mizushina Y, Ohta K, Shimozawa K, Gasa S, Jimbow K, Sakaguchi K, Sato N, Takahashi N (2002) Anti-tumor effect of chemically synthesized sulfolipids based on sea urchin's natural sulphonoquinovosylmonoacylglycerols. Jpn J Cancer Res 93:85–92

Sato K, Nishi N, Nomizu M (2004) Characterization of a fasciclin I-like protein with cell attachment activity from sea urchin (*Strongylocentrotus intermedius*) ovaries. Arch Biochem Biophys 424:1–10

Scaturro G, Zito F, Matranga V (1998) The oligomeric integrity of toposome is essential for its morphogenetic function. Cell Biol Int 22:321–326

Schreiber AB, Lax I, Yarden Y, Eshhar Z, Schlessinger J (1981) Monoclonal antibodies against receptor for epidermal growth factor induce early and delayed effects of epidermal growth factor. Proc Natl Acad Sci USA 78:7535–7539

Sciarrino S, Matranga V (1995) Effects of retinoic acid and dimethylsulfoxide on the morphogenesis of the sea urchin embryo. Cell Biol Int 19:675–680

Sconzo G, Pirrone AM, Mutolo V, Giudice G (1970) Synthesis of ribosomal RNA in disaggregated cells of sea urchin embryos. Biochim Biophys Acta 199:441–446

Shih LJ, Chen CA, Chen CP, Hwang SP (2002) Identification and characterization of bone morphogenetic protein 2/4 gene from the starfish *Archaster typicus*. Comp Biochem Physiol B Biochem Mol Biol 131:143–151

Shimizu K, Amemiya S, Yoshizato K (1990) Biochemical and immunological characterization of collagen molecules from echinothurioid sea urchin *Asthenosoma ijimai*. Biochim Biophys Acta 1038:1039–1046

Shu DG, Conway Morris S, Han J, Zhang ZF, Liu JN (2004) Ancestral echinoderms from the Chengjiang deposits of China. Nature 430:422–428

Shyu AB, Raff RA, Blumenthal T (1986) Expression of the vitellogenin gene in female and male sea urchin. Proc Natl Acad Sci USA 83:3865–3869

Shyu AB, Blumenthal T, Raff RA (1987) A single gene encoding vitellogenin in the sea urchin *Strongylocentrotus purpuratus*: sequence at the 5′ end. Nucleic Acids Res 15:10405–10417

Solursh M (1986) Migration of sea urchin primary mesenchyme cells. In: Browder L (eds) The cellular basis of morphogenesis. Plenum, New York, pp 391–431

Spiegel E, Howard L (1983) Development of cell junctions in sea-urchin embryos. J Cell Sci 62:27–48

Spiegel E, Spiegel M (1979) The hyaline layer is a collagen-containing extracellular matrix in sea urchin embryos and reaggregating cells. Exp Cell Res 123:434–441

Spiegel E, Burger MM, Spiegel M (1980). Fibronectin in the developing sea urchin. J Cell Biol 87:309–313

Spiegel E, Burger MM, Spiegel M (1983) Fibronectin and laminin in the extracellular matrix and basement membrane of sea urchin embryos. Exp Cell Res 144:47–55

Stamenkovic I (2003) Extracellular matrix remodelling: the role of matrix metalloproteinases. J Pathol 200:448–464

Stenzel P, Angerer LM, Smith BJ, Angerer RC, Vale WW (1994) The univin gene encodes a member of the transforming growth factor-beta superfamily with restricted expression in the sea urchin embryo. Dev Biol 166:149–158

Stone AL, Kroeger M, Sang QX (1999) Structure–function analysis of the ADAM family of disintegrin-like and metalloproteinase-containing proteins. J Protein Chem 18:447–465

Susan JM, Just ML, Lennarz WJ (2000) Cloning and characterization of alphaP integrin in embryos of the sea urchin *Strongylocentrotus purpuratus*. Biochem Biophys Res Commun 272:929–935

Tesoro V, Zito F, Yokota Y, Nakano E, Sciarrino S, Matranga V (1998) A protein of the basal lamina of the sea urchin embryo. Dev Growth Differ 40:527–535

Thiery JP (2003) Cell adhesion in development: a complex signaling network. Curr Opin Genet Dev 13:365–371

Unuma T, Okamoto H, Konishi K, Ohta H, Mori K (2001) Cloning of cDNA encoding vitellogenin and its expression in red sea urchin, *Pseudocentrotus depressus*. Zool Sci 18:559–565

Vacquier VD (1969) The isolation and preliminary analysis of the hyaline layer of sea urchin eggs. Exp Cell Res 54:140–142

Vacquier VD, Mazia D (1968) Twinning of sea urchin embryos by treatment with dithiothreitol. Roles of cell surface interactions and of the hyaline layer. Exp Cell Res 52:459–468

Venkatesan M, de Pablo F, Vogeli G, Simpson RT (1986) Structure and developmentally regulated expression of a *Strongylocentrotus purpuratus* collagen gene. Proc Natl Acad Sci USA 83:3351–3355

Vittorelli ML, Matranga V, Feo S, Giudice G, Noll H (1980) Inverse effects of thymidine incorporation in dissociated blastula cells of the sea urchin *Paracentrotus lividus* induced by butanol treatment and Fab addition. Cell Differ 9:63–70

Wang J, Pansky A, Venuti JM, Yaffe D, Nudel U (1998) A sea urchin gene encoding dystrophin-related proteins. Hum Mol Genet 7:581–588

Wang W, Hong J, Lee CO, Im KS, Choi JS, Jung JH (2004) Cytotoxic sterols and saponins from the starfish *Certonardoa semiregularis*. J Nat Prod 67:584–591

Wessel GM, Marchase RB, McClay DR (1984) Ontogeny of the basal lamina in the sea urchin embryo. Dev Biol 103:235–245

Wessel G, Etkin M, Benson S (1991) Primary mesenchyme cells of the sea urchin embryo require autonomously produced nonfibrillar collagen for spiculogenesis. Dev Biol 148:261–272

Wessel GM, Berg L, Adelson DL, Cannon G, McClay DR (1998) A molecular analysis of hyalin-a substrate for cell adhesion in the hyaline layer of the sea urchin embryo. Dev Biol 193:115–126

Wessel GM, Zaydfudim V, Hsu YJ, Laidlaw M, Brooks JM (2000) Direct molecular interaction of a conserved yolk granule protein in sea urchins. Dev Growth Differ 42:507–517

Wheelock MJ, Johnsony KR (2003a) Cadherins as modulators of cellular phenotype. Annu Rev Cell Dev Biol 19:207–235

Wheelock MJ, Johnsony KR (2003b) Cadherin-mediated cellular signaling. Curr Opin Cell Biol 15:509–514

White JM (2003) ADAMs: modulators of cell–cell and cell–matrix interactions. Curr Opin Cell Biol 15:598–606

Wolpert L, Mercer EH (1963) An electron microscope study of the development of the blastula of the sea urchin embryo and its radial polarity. Exp Cell Res 30:280–300

Wray GA, Lowe CJ (2000) Developmental regulatory genes and echinoderm evolution. Syst Biol 49:28–51

Yamada KM (1997) Integrin signaling. Matrix Biol 16:137–141

Yokota Y, Kato KH (1988) Degradation of yolk granules in sea urchin eggs and embryos. Cell Differ 23:191–199

Yokota Y, Matranga V, Zito F, Cervello M, Nakano E (1994) Nectins in sea urchin eggs and embryos. J Mar Biol Assoc UK 74:27–34

Yokota Y, Unuma T, Moriyama A, Yamano K (2003) Cleavage site of a major yolk protein (MYP) determined by cDNA isolation and amino acid sequencing in sea urchin, *Hemicentrotus pulcherrimus*. Comp Biochem Physiol B Biochem Mol Biol 135:71–81

Zito F, Tesoro V, McClay DR, Nakano E, Matranga V (1998) Ectoderm cell–ECM interaction is essential for sea urchin embryo skeletogenesis. Dev Biol 196:184–192

Zito F, Nakano E, Sciarrino S, Matranga V (2000) Regulative specification of ectoderm in skeleton disrupted sea urchin embryos treated with monoclonal antibody to *Pl*-nectin. Dev Growth Differ 42:499–506

Zito F, Costa C, Sciarrino S, Poma V, Russo R, Angerer LM, Matranga V (2003) Expression of univin, a TGF-beta growth factor, requires ectoderm–ECM interaction and promotes skeletal growth in the sea urchin embryo. Dev Biol 264:217–227

Cell Signalling During Sea Urchin Development: A Model for Assessing Toxicity of Environmental Contaminants

C. Angelini, M.G. Aluigi, M. Sgro, S. Trombino, H. Thielecke, C. Falugi

Abstract. The early development of sea urchins has been thoroughly studied since the beginning of the 20th century thanks to the particular features of the model involving cell signalling, making it easy to follow the complex cell-to-cell interactions that lead to development. In this chapter, the prominent role of cell-to-cell communication in developmental events is discussed, as well as the role of intracellular ion changes that are in turn regulated by signal molecules belonging to the cholinergic system. The results seem to indicate that the zygote stage is the most suitable to study the role of the cholinergic system, as at this stage, a calcium spike can be evoked by exposure to acetylcholine (ACh) or to muscarinic drugs, at any time before the nuclear breakdown. The described outcomes also open a path to a new way of considering biomarkers. In fact, most environmental factors have the capacity to interfere with the cholinergic system: stress, wounds, inflammation and pollution in general. In particular, this offers a way to investigate the presence in the environment and the degree of aggressiveness of neurotoxic contaminants, such as organophosphate and carbamate pesticides, largely used in European countries for many purposes, including agricultural pest control and medical treatment. These drugs exert their function by interfering with the regulation of the cholinergic system and the consequent electrical events. Thus, the sea urchin zygote could represent a reliable model to be used in biosensors with the capacity to translate the effect of neurotoxic pesticides, and generally of stress-inducing contaminants, in living cell responses, such as electrical responses.

C. Angelini, M.G. Aluigi, M. Sgro, S. Trombino, C. Falugi (e-mail: falugi@unige.it)
Dipartimento di Biologia Sperimentale, Ambientale ed Applicata (DIBISAA), University of Genoa, Viale Benedetto XV, 16132 Genova, Italy
H. Thielecke
Department of Biohybrid Systems, Fraunhofer Institute for Biomedical Engineering, St. Ingbert, Germany

Progress in Molecular and Subcellular Biology
Subseries Marine Molecular Biotechnology
V. Matranga (Ed.), Echinodermata

1 Basic Research

1.1 Cell-to-Cell Signalling During Sea Urchin Development

The early development of sea urchins has been thoroughly studied since the beginning of the 20th century. The great challenge of understanding development has been achieved thanks to the particular characteristics of the model involving cell signalling: echinoderm eggs are about 80–100 μm in diameter, and may be observed with the aid of a simple stereomicroscope. Each stage of the development, including fertilisation, occurs outside the maternal body, the eggs and embryos are transparent and all the cell dynamics can be followed in vivo. During fertilisation and egg activation and the first developmental events, the cortical reaction responsible for the block to polyspermy is revealed by a dramatic elevation of the envelope, called the fertilisation layer (Fig. 1; see Giudice 1973, 1986 for reviews). Moreover, a large number of eggs are released for each spawning, and development proceeds synchronously up to metamorphosis.

For this reason, sea urchin development represents one of the few models selected by developmental biologists for both basic and applied research. The advantage of this model is its simple organisation, making it easy to follow the complex cell-to-cell interactions that lead to development with non-invasive methods which in other animal models are possible only in cultured cell lines.

The development of echinoderms, and in particular of sea urchins, has been studied at each stage, including fertilisation. The most significant dis-

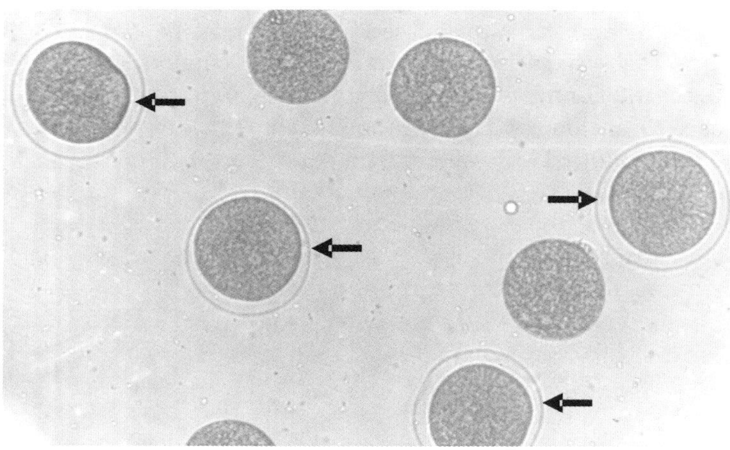

Fig. 1. *Paracentrotus lividus* unfertilised and fertilised eggs. Fertilised eggs show the elevated fertilisation layer (*arrows*)

coveries on this subject have been possible thanks to this model: for instance, most of the electric events responsible for fertilisation were first studied during early sea urchin development (see Epel 1975, 1980 for reviews). In a recent paper, Quiao et al. (2003) have shown that sea urchin developmental events are exportable to mammalian development.

In recent years, there has been increasing awareness concerning the prominent role of cell-to-cell communication in developmental events: in particular, neurotransmitter system molecules were found to have a relevant role in modulating cell-to-cell interaction mediated by ion fluxes or ion intracellular changes. Buznikov and colleagues (since the 1970s: Buznikov et al. 1972, 1996, 2001a; Buznikov 1980, 1990; Quiao et al. 2003) have detailed the history of classic neurotransmitter involvement (biogenic amines, acetylcholine and GABA) in early developmental events from segmentation to the formation of highly specialised structures, and have adapted the discoveries obtained from sea urchin embryos to about every other animal model (Buznikov 1990; Buznikov et al. 1996; Quiao et al. 2003). Ryberg and colleagues (Gustafson et al. 1972; Ryberg 1973, 1974, 1979) used the localisation of acetylcholinesterase activity (AChE, E.C. 3.1.1.7., the lytic enzyme of the neurotransmitter acetylcholine) as a marker to investigate cholinergic neurodevelopment, and found a diffuse net of neuro-epithelial cells, not organised in a real nervous system, while Nakajima (1986, 1988; Nakajima et al. 1993) identified a system of organised catecholaminergic ganglia. The presence and role of a serotonergic system were studied by Thorndyke and colleagues (Thorndyke et al. 1992; Beer et al. 2001). Professor Drews' group (Drews 1975) localised cholinesterase (ChE) in sea urchin mesenchyme cells undergoing morphogenetic movements, and compared this function with the one found in high-vertebrate developing structures. Following his discoveries, some authors interpreted the presence of AChE in cells and tissues as a "symptom" of cell migration (Weinberger et al. 1984).

Professor Minganti's group focused attention on the role of cholinergic neurotransmitter system molecules in cell-to-cell communication mediated by ion fluxes, and confirmed the exportability of the results to other animal models as well, including chordates and high vertebrates (Minganti et al. 1981). We have followed Minganti's line of research with studies on the development of the Mediterranean sea urchin, *Paracentrotus lividus*, from fertilisation up to metamorphosis. Cholinergic signalling system molecules have been found at each developmental stage, playing different roles according to the temporal windows and the degree of differentiation, including the neuromuscular function in the rudiment and in the juvenile (Vidal et al. 1993; Falugi et al. 1999; 2002), localised in radial muscles, in the tube feet and at the basis of the spines (Fig. 2).

Here, we present a summary of studies on fertilisation and early cell cycles, some conducted by our group, along with information from some of the outstanding papers available in the literature.

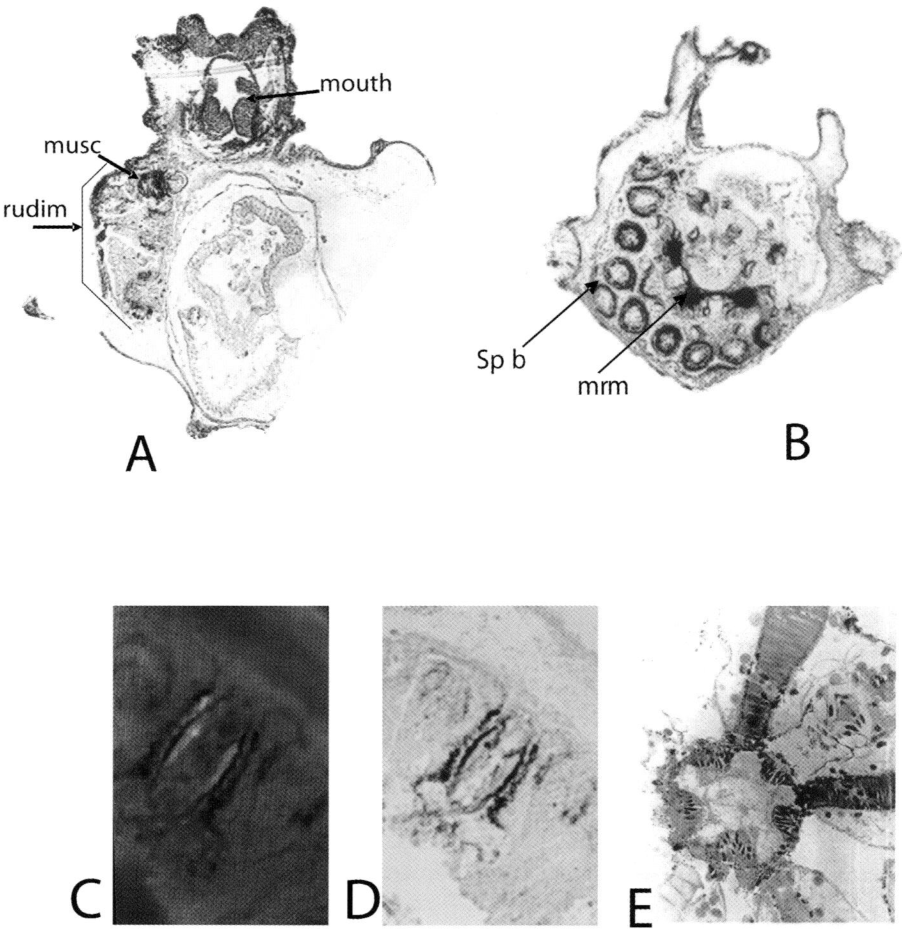

Fig. 2A–E. *Paracentrotus lividus* larva 18 days old, competent for metamorphosis. **A** Section of the larva containing a sagittal section of the rudiment. **B** Section of the larva containing an oblique section of the rudiment. Histochemical reaction to acetylcholinesterase, revealed by *dark precipitation*. **C, D** Co-localisation of formaldehyde-induced fluorescence suggesting presence of biogenic amines and AChE (*dark staining*) at basis of forming spines. **E** AChE reaction in the radial mouth muscles. *musc* Muscles under the forming spines of the rudiment (*rudim*); *mrm* mouth radial muscle; *Sp b* spine bases. Resin section, 5 μm thick; *bar* 100 μm. (Karnovsky and Roots 1964)

1.2 Fertilisation

Fertilisation has been the most investigated point in the history of research, and the sea urchin has been the most investigated model in this respect: authors such as Epel (Epel 1975; Kuo et al. 2000) and Jaffe, both Laurinda and

Lionel (Jaffe 1976, 1980, 1983; Jaffe and Robinson 1978; Turner et al. 1984), investigated the electrophysiological events; Vacquier (1969, 1981; Vacquier and Moy 1977) and Lennarz (Schmell et al. 1977; Ohlendieck et al. 1994; Ohlendieck and Lennarz 1996; Just and Lennarz 1997) investigated sperm–egg reception and adhesion; Shen (Rakow and Shen 1990; Shen 1995) and Chambers (Chambers and De Armendi 1979; Longo et al. 1986; McCulloh et al. 1987; Crossley et al. 1988; Ivonnet and Chambers 1997) investigated calcium dynamics in egg responses; Schuel (e.g. Schuel et al. 1991) investigated sperm dynamics; and Whitaker (Whitaker and Irvine 1984; Crossley et al. 1988; Wilding et al. 1996; Harrison et al. 2002) investigated the mechanisms of calcium release from inner stores. Each author's approach included morphological, biochemical and electrophysiological methods, sometimes with different hypotheses and interpretations.

Essentially, fertilisation occurs in two phases: the first phase is characterised by sperm activation events; the second, by the egg activation events.

1.2.1
First Phase: Sperm Activation

Sperms are immobilised in male gonads and genital ducts, as the membrane in these sites is stabilised by cholesterol and glycoproteins (Gilbert 1994) and by the sperm fluid, which has a low pH and contains calmodulins (Gilbert 1994).

During the first phase of fertilisation, the sperm motility is activated, the sperm are able to swim towards the egg, and are capacitated to the acrosome reaction, during which the acrosome vesicle releases the enzymes it contains into the water surrounding the egg envelope. During these phases, electrical events caused by inward–outward ion fluxes take place.

Actually, in each of these events, calcium is made available for the sperm since it is abundant in seawater as well as in the peri-ovular jelly coat; the latter probably also plays a role in modifying the sperm membrane, and also in activating the acrosome reaction (Darszon et al. 1987; Garcia-Soto et al. 1987; Liévano et al. 1990). Moreover, K^+ and Ca^{2+} channels are activated by speract, the attractant molecule released by the egg, thus leading the sperm to find the egg of the same species (Babcock et al. 1992).

At the same time, Na^+, abundant in the marine environment, causes an increase in the intracellular pH level, which in turn activates flagellum dynein, maintained low by the low pH of the sperm fluid. This event seems to be guided by the activation of nicotinic receptors, present in the membrane of the flagellum, as shown by experiments with α-bungarotoxin, a type of snake venom which is specific for these receptors (Nelson 1976; Falugi et al. 1993b), and by histochemical staining of agonist and antagonist drug binding (Baccetti et al. 1995). In each case, ionic changes play a role in the phase in which swimming is activated as well as in the phase in which the egg membrane is "attacked"; these events take place thanks to: (1) an enhanced swimming activity and (2) exocytosis of the acrosome granule which, releasing lytic enzymes, exposes the egg membrane, free from accessory envelopes (jelly coat), and perforates the primary envelope (vitelline membrane). Simple neurotransmitter signal molecules play a crucial role in these processes as well.

Cholinergic system molecules were discovered in sperm cells of different animal species (Bishop et al. 1977; Sastry et al. 1979; Baccetti et al. 1995). These findings led researchers to investigate the possible role played by neurotransmitter system molecules in these cells. In particular, the presence in sperm of cholinergic molecules has been reported and studied by Nelson (1973, 1978, 1990) and Nelson et al. (1970), who supplied evidence for the implication of the cholinergic system in sperm propulsion. The activity of the enzyme acetylcholinesterase (AChE, E.C. 3.1.1.7.) was evaluated by biochemical methods in the sperm flagellum of the sea urchin *P. lividus* (Cariello et al. 1986). AChE is the lytic enzyme of the cholinergic system, and its function is to cleave ACh into choline and acetate, thus restoring ACh receptor excitability. In sperm of the same sea urchin species, AChE was also localised by histochemical methods on the head membrane (Falugi et al. 1991). ACh receptors were also found and localised in the sperm cell structures: muscarinic receptors mainly in the acrosome, nicotinic receptors both in the acrosome and in the flagellum membrane (Baccetti et al. 1995). The two different families of receptors seem to play different roles in the different moments of fertilisation: the nicotinic ones (nAChRs, see Stroud et al. 1990), gating Na^+ channels on activation by their specific ligands (ACh and cholinomimetic agonist drugs), may result in a change of the Na^+ influx and consequently H^+ efflux, increasing the internal pH and activating the dynein of the flagellum. The muscarinic ones, known in five different molecular forms, associated with G-protein in the intracellular domain, trigger a second messenger transduction cascade (Birdsall et al. 1978; Watson and Arkinstall 1994), activating intracellular dynamics related to fertilisation (Falugi et al. 1993a).

In addition, the effects of cholinomimetic drugs on *P. lividus* sperm motility and fertilisation ability suggested an involvement of the nAChRs and mAChRs in these phenomena (Falugi et al. 1993b). This was also clear from experiments with potentiometric dyes, which helped to evidence the pattern of membrane depolarisation and fluidity during swimming and acrosome reaction caused by the presence of either eggs, calcium ionophore A 23187 or nicotine (Falugi et al. 1993a).

The localisation of the studied molecules in the different cell districts is useful in order to understand their function. The presence of molecules immunologically related to AChRs was found to be constant in gametes of different organisms, and their localisation seems to be related to the different mode of interaction between sperm and egg in the different species. In particular, *P. lividus* sperm presented molecules immunologically correlated to muscarinic receptors mainly at the acrosome and nicotinic receptors at the acrosome and along the flagellum (Fig. 3), thus confirming previous

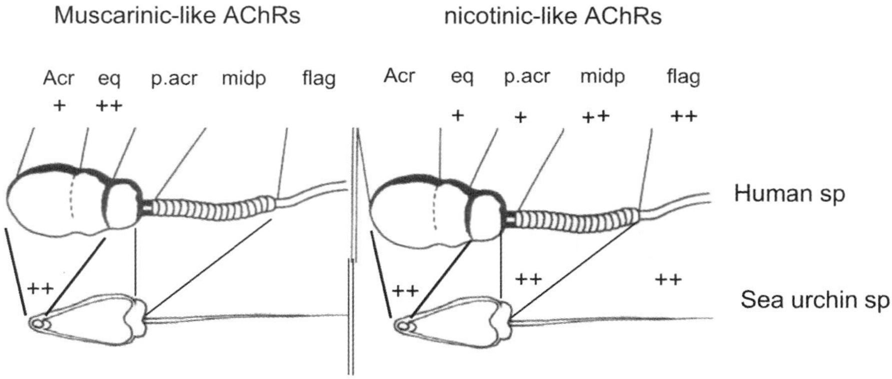

Fig. 3. Comparison of the localisation of muscarinic and nicotinic receptors in human and sea urchin sperm (*sp*). *Acr* Acrosome; *eq* equatorial ring; *p.acr* post-acrosomal region; *midp* midpiece; *flag* flagellum

results (Dwivedi and Long 1989; Baccetti et al. 1995). This localisation suggests that the nicotinic receptors play a role in sperm propulsion, as previously suggested by in vivo pharmacological experiments, with agonist and antagonist drugs (Nelson 1976; Dwivedi and Long 1989; Falugi et al. 1993b). As to the receptors present in the acrosome, their function may be related to sperm–egg interaction, including acrosome reaction (AR) and membrane fusion. Similarly, Bray et al. (2002), recently investigating whether the nAChR may have a role in the acrosome reaction (AR) of human sperm, have obtained strong evidence that ACh is capable of initiating the AR by activating an α7 subunit-containing nAChR and that this receptor is essential for the AR initiated by purified recombinant human *zona pellucida* (ZP) protein (rhZP3), thus playing a potential role in the initiation of the AR by intact ZP in vivo. In the sea urchin, this reaction could also be enhanced by molecules immunologically related to choline acetyltransferase (ChAT, E.C. 2.3.1.6), the biosynthetic enzyme of ACh (Mautner 1986) present on the surface of mature eggs (Angelini et al. 2004), as this enzyme is capable of autonomously synthesising ACh, active on the ACh receptors located in the acrosome (Fig. 4).

The function of muscarinic AChRs localised in the acrosome of *P. lividus* sperm might be related to the fusion of the membrane between the gametes. This assumption arises from the comparison between their localisation in sea urchin sperm and that in mammalian sperm, respectively: to be precise, in sea urchin sperm, the first fusion between the sperm and the egg membrane takes place at the tip of the acrosome, where these ACh receptors are also localised, while mammalian sperm present the receptors at the acrosome equatorial ring (Baccetti et al. 1995), where the first contact and fusion with the egg takes place (see Longo 1987 for comparison among the different models).

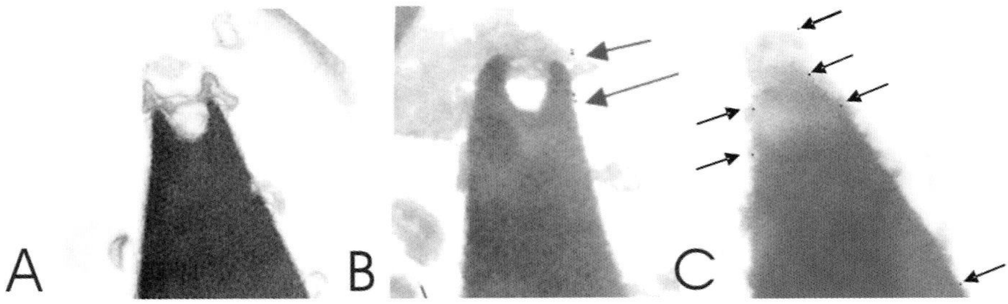

Fig. 4A–C. ChAT ultrastructural immunogold reaction in *Paracentrotus lividus* sperm. **A** Control; **B, C** reacted sperm (×12,000)

A matter of debate over recent years has been the fact that the binding sites for pharmacological agents may differ between in vitro and in vivo experiments, as suggested from experiments on mouse sperm (Florman and Storey 1982). However, experiments on sea urchin fertilisation are normally carried out in seawater, the natural medium in which this event takes place, so as to eliminate any possible difference between in vivo and in vitro experiments.

1.2.2
Second Phase: Egg Activation

It is well known that sea urchin egg activation and fertilisation are led and followed by electrical phenomena, especially as far as ion dynamics are concerned. Following the contact of the egg's membrane with that of the sperm, an early depolarisation event of the egg membrane takes place, caused by an influx of Na^+ ions (see Epel 1975, 1980), which makes the membrane permissive to the fusion (McCulloh et al. 1987).

This spike is immediately followed by an explosive increase in intracellular $[Ca^{2+}]$, released from intracellular stores (Giudice 1973; Longo 1987). This event evokes a strong membrane depolarisation (about 90 mV) and activates the egg, causing the so-called fast block to polyspermy as well as the cortical reaction. The egg, thus activated, begins to "breathe" and starts its metabolic activities, including DNA synthesis and the trigger for the first cell cycles (Giudice 1973; Longo 1987).

As far as the mode of regulation and unfolding of these ion dynamics is concerned, a great deal of research has been carried out since the beginning of the last century and, according to Ohlendieck and Lennarz (1996), the number of investigations on the subject has greatly exceeded that of the works concerning the philosopher's stone over the past centuries! However, recently some progress has been made towards the comprehension of these phenomena.

1.2.2.1
First Step: Na^+ Influx

This event was described by Epel (1980): as soon as the sperm touches the egg membrane, a minor influx of Na^+ takes place throughout the egg membrane; at the same time, sperm–egg membrane fusion takes place, immediately followed by calcium release from intracellular stores. The meaning of this first Na^+ influx was apparently forgotten by researchers in the following years, who rather chose to focus on the calcium wave responsible for the block to polyspermy.

Just recently, this event has been re-examined, by considering old papers on the effects of nicotine exposure during sea urchin egg fertilisation and by observing that the ionic dynamics (first Na^+ influx, immediately followed by explosive Ca^{2+} release) are very similar to what happens in the neuromuscular synapses. Nicotine had been reported by Jaffe (1980) to cause polyspermy; thus it is very likely that nicotinic receptors may be present just as in the neuromuscular post-synaptic membrane. To be precise, nicotinic receptors were histochemically revealed by curare-prevented α-bungarotoxin binding (Falugi and Prestipino 1988; Falugi et al. 1989) at the egg surface. More recently, by exposing sea urchin unfertilised eggs to acetylcholine together with eserine (Physostigmine, Sigma), we have observed, after the addition of sperm, a membrane depolarisation of about 40 mV (starting from a rest potential of –70 mV, to the spike of –30 mV), while the normal final depolarisation in control eggs is +0 mV, with a depolarisation ranging between 70 and 90 mV. Moreover, the exposure to acetylcholine prior to fertilisation significantly increased the percentage of polyspermic eggs as compared to controls (Harrison et al. 2002; Angelini et al. 2004). The pharmacological properties of these receptors were verified with electrophysiological methods by Chambers and co-workers, who confirmed the presence of nicotinic receptors in the unfertilised sea urchin egg, and characterised them as being identical to the ones present in the neuromuscular synapses (Ivonnet and Chambers 1997). According to these authors, the presence of nAChR channels at the surface of unfertilised eggs can account for the capacity of nicotine to impair the block to polyspermy, by lowering the positive shift of the egg's membrane potential caused by sperm.

This could also explain our results described above (Angelini et al. 2004) Actually, the same effect of maintaining the fertilisation potential of the egg at low levels could be exerted by an increased quantity of ACh, which in this case could maintain the membrane fusibility for a longer period of time than under physiological conditions. This relationship between the first step of depolarisation due to Na^+ influx and membrane fusibility seems to be reliable. In fact, as previously demonstrated (Longo 1987; McCulloh et al. 1987; McCulloh and Chambers 1992), a low membrane depolarisation allows sperm entry and membrane fusion in the sea urchin *Lytechinus pictus* and *Strongylocentrotus purpuratus*.

What is the source of ACh able to activate the nicotinic receptors present on the egg surface? The answer to this question could be the discovery of molecules related to choline-acetyltransferase (ChAT, the biosynthetic enzyme of acetylcholinesterase) in the acrosome and head membrane of *P. lividus* sperm, reported by immunogold methods and electron microscopy (Piomboni et al. 2001). To be precise, we found that ChAT immunoreactive molecules, revealed by immunogold particles, were mainly localised in the sperm head of *P. lividus*. In immotile sperm ("fixed-dry sperm"), gold particles appeared to be particularly concentrated in the acrosome vesicle. However, when sperm began to swim actively as a result of their suspension in seawater containing eggs, the gold particles were observed in the membrane all along the head region. This may suggest that the sperm autonomously synthesises ACh, and that this function is active as far as the first sperm–egg interaction takes place (Angelini et al. 2004). The cholinergic molecules present in the sperm acrosome might play a role in sperm–egg interaction, as hypothesised by Ibanez et al. (1991) for mammalian sperm, where ChAT molecules and their mRNAs were found in the equatorial ring, which is where the first contact with the egg takes place.

Therefore, our hypothesis is that ACh released by the sperm surface (where ChAT is exposed) may, at the first contact, excite the nicotinic AChRs present in the egg membrane, thus evoking the first event of Na^+-induced depolarisation, responsible for membrane fusion (Angelini et al. 2004; Fig. 5).

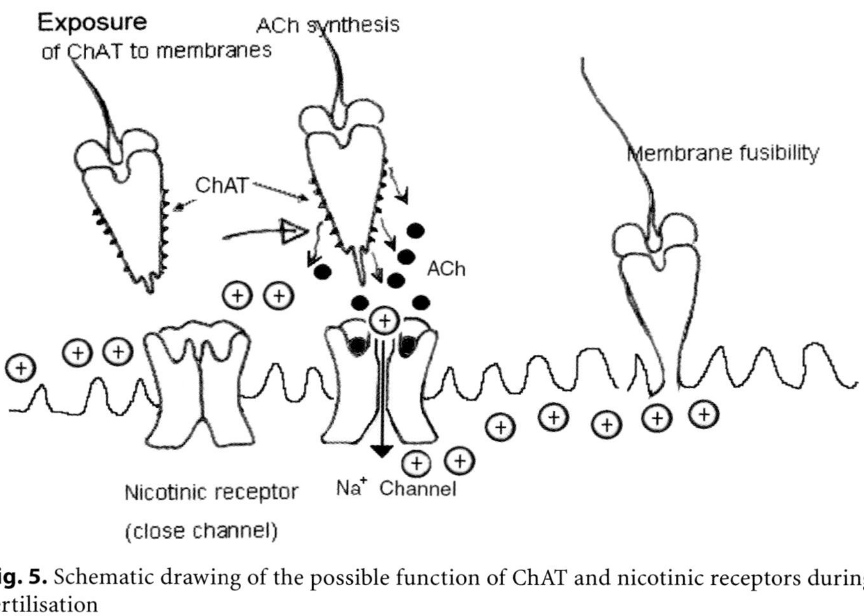

Fig. 5. Schematic drawing of the possible function of ChAT and nicotinic receptors during fertilisation

1.2.2.2
Calcium Dynamics

Although the egg membrane is effectively an excitable membrane, its similarity with the synapses is only limited to the nicotinic receptors, because the influx of Na^+ ions caused by acetylcholine in the sea urchin egg is not sufficient to evoke the release of calcium ions and the rapid block to polyspermy, as we demonstrated in cooperation with the laboratory of Prof. Whitaker (Harrison et al. 2002), thanks to the use of cholinomimetic agonists and antagonists on *P. lividus* and *Lytechinus pictus* eggs and zygotes. We did not find any evidence of an involvement of cholinergic molecules in this process, as nicotinic AChR agonists were not capable of evoking any $[Ca^{2+}]$ variation in unfertilised eggs by themselves, nor were the antagonists capable of preventing the calcium spike provoked by sperm. Also, muscarinic agonists and antagonists were completely inactive in the course of these events.

Thus, in the fast block to polyspermy a different kind of signal and receptor molecule must be active for the Ca^{2+} ion release. The latter was elicited by a signal transduction cascade, since Whitaker and Irvine (1984) had shown that an inositol-triphosphate (IP3) injection in the cytoplasm was sufficient to activate the sea urchin unfertilised egg and to cause the cortical reaction.

Actually, in August 2000, a paper by Kuo et al. was published, reporting that IP3 release was due to an atypical neurotransmitter, nitric oxide (NO). These authors showed that nitric oxide synthase (NOs) is present in the post-acrosomal area of the sperm, and that it is activated by contact with the jelly coat, to synthesise NO, which is subsequently carried inside the egg together with the sperm material. This would be the signal that triggers the transduction cascade, including the release of IP3; these events result in an explosive release of calcium from the inner stores, identified as cortical vesicles of the endoplasmic reticulum (see Giudice 1986; Shen 1995 for extensive reviews) and, recently, also vacuolar-like structures (Churchill et al. 2002).

1.3
First Cell Cycle

Therefore, cholinergic molecules do not seem to be involved in the block to polyspermy and cortical reaction. Nevertheless, several authors have found a complete set of cholinergic system molecules in the eggs both before and after fertilisation. Among these, our group has localised nicotinic AChR molecules in the sea urchin *P. lividus* by use of curare-prevented FITC-α-bungarotoxin binding at the surface of unfertilised and fertilised eggs of *P. lividus* (Falugi et al. 1989), alongside muscarinic AChR molecules (Piomboni et al. 2001), as well as other cholinergic molecules, such as acetylcholinesterase (Piomboni et al. 2001) and maternal choline acetyltransferase (ChAT) molecules, assembled on the oolemma during egg maturation and fertilisation processes (Angelini et al. 2004). Actually, muscarinic receptor molecules are present in the egg starting from the first maturation events, but are assembled on the egg membrane only after fertilisation. This was demonstrated by both ultrastructural localisation of anti-mAChR immunoreactivity and Western blot of cytoplasm homogenates compared with membrane extracts of unfertilised and fertilised eggs (Piomboni et al. 2001). A similar distribution pattern was found for ChAT (Angelini et al. 2004). Moreover, the cortical granules display a content of active molecules of AChE, identified by the enzyme reaction product as suggested by Karnovsky and Roots (1964). These molecules are synthesised within the endoplasmic reticulum, processed in the Golgi vesicles and from there exposed at the membrane surface (Piomboni et al. 2001). During the cortical reaction, these molecules are extruded into the perivitelline space (Fig. 6).

Such a coordinated distribution of signal molecules suggests a function for the cholinergic system, displayed after fertilisation. The key to the understanding of this possible function lies in the role played by muscarinic AChRs, which consists of triggering intracellular signal transduction (Watson and Arkinstall 1994). This appears to be the case in all the systems in which they have been identified. It has been recently reported that mAChRs are functional only when assembled on the cell surface, while inactive subunits are contained in ER-derived vesicles (Mei and Xiong 2003). This is true for ChAT as well, as it is known that ChAT activity is regulated by its binding to the cholinergic membrane (Dobransky and Rylett 2003) and by interaction with other cellular proteins related to this membrane (Gabrielle et al. 2003). In other words, the enzyme is not active when it is inside the cytoplasm, while it is active when it is exposed on the cholinergic membrane.

Furthermore, by FURA-2 experiments, it was shown that ACh and carbamylcholine, the natural signal and the cholinomimetic agonist of the mAChR, respectively, are capable of evoking a $[Ca^{2+}]_i$ spike. This occurs when the exposure is performed at a time preceding the nuclear envelope breakdown (NEB) occurring at the first cell cleavage, i.e. approximately 15 min for *P. lividus* and 40 min for *L. pictus* after fertilisation. In contrast, atropine (antagonist of mAChRs, non-selective among the different molecular forms of the receptors) prevented further Ca^{2+} spikes caused by agonist molecules. The ability to respond to them was acquired again after rinsing the atropine present in the egg-containing water (Harrison et al. 2002); for this reason, we hypothesised that mAChRs could be involved in the regulation of intracellular ionic dynamics related to the cleavage of the first cell cycles (Harrison et al. 2002). All our findings seem to suggest a model of autocrine regulation of these dynamics. To be precise, those ChAT molecules that after fertilisation are exposed at the egg surface may autonomously synthesise ACh. The egg cannot receive ACh from elsewhere, because after fertilisation the zygote is enclosed in the fertilisation membrane, which is in turn completely surrounded by seawater, and transported by waves and currents. The ACh pro-

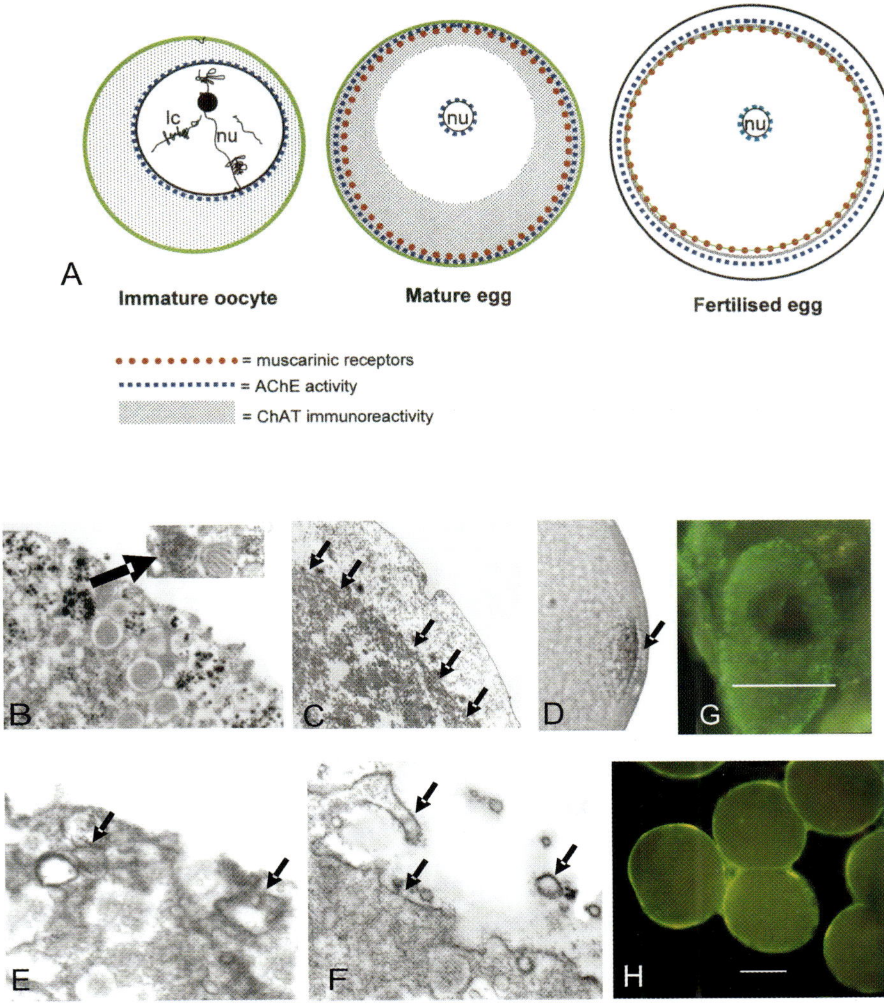

Fig. 6. A Schematic drawings of the relative localisation of AChE activity, ChAT immunoreactivity and muscarinic receptors during maturation phases of sea urchin eggs. *lc* Lampbrush chromosomes; *nu* nucleus. **B** AChE histochemical localisation in the cortical granules of *Paracentrotus lividus* mature egg. **C** Cortical granules extrusion (*arrows*) during the cortical reaction. **D** AChE activity in the elongated microvilli of the fertilisation cone. **E** Muscarinic receptors included in cortical vesicles in unfertilised eggs. **F** Muscarinic receptor immunoreactivity at the membrane surface of fertilised egg. **G** ChAT immunoreactivity in immature ovarian egg. **H** ChAT immunoreactivity in fertilised eggs. *Bars* 50 μm. The electron microscopy figures were taken at instrument magnification ×20,000

duced by the ChAT assembled on the egg membrane is released into the perivitelline space and reaches the mAChRs that at this stage are also exposed at the surface (Piomboni et al. 2001). As a result, muscarinic activation evokes a $[Ca^{2+}]$ response, which further enhances development. The hypothesis that calcium may be involved in the regulation of the first cell cycle may be also supported by the findings of Steinhardt (1990a, 1990b), who proposed that intracellular $[Ca^{2+}]_i$ changes might be possible signals for the regulation of different mitosis stages, in both animal and plant cells. In particular, together with coworkers, he forwarded the hypothesis that calcium might be responsible for the regulation of the NEB at the moment of cleavage. The above-cited authors showed that a calcium transient immediately precedes the NEB, which may also be prevented by subtracting calcium from the cell (Baitinger et al. 1990).

According to these discoveries, we proposed a model that could help to explain the function of inter- and intracellular signals mediated by cholinergic molecules during fertilisation and the first cell cycles – both gametes possess cholinergic molecules: ChAT, which is necessary in order to produce ACh,

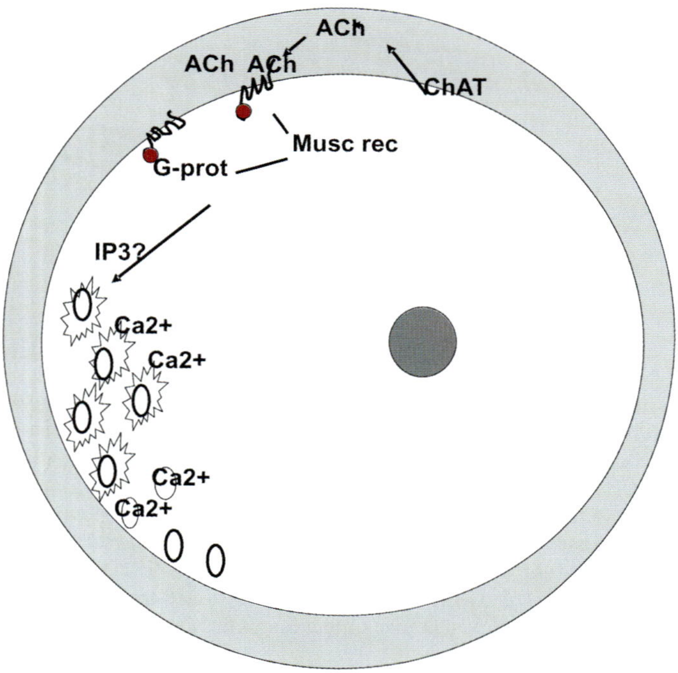

Fig. 7. Schematic drawing of the hypothesis about "autocrine" behaviour of acetylcholine in fertilised eggs: ChAT synthesises ACh, which activates the muscarinic receptors exposed at the surface after fertilisation, triggering the transduction cascade causing Ca^{2+} release from inner stores. *Musc rec* Muscarinic receptors; *G-prot* G-protein

AChE, which cleaves ACh bound to the receptors, and ACh nicotinic and muscarinic receptors (Figs. 6, 7). Moreover, their activities are displayed at different times: during the first phase, the nicotinic receptors in the sperm tail are involved in promoting swimming [α-bungarotoxin blocks their movement (see Nelson 1976; Falugi et al. 1993b); then the receptors present in the acrosome (both nAChRs and mAChRs) may be excited by the ACh synthesised by the ChAT present on the egg surface.

During the second phase, when sperm activate the eggs, ACh, autonomously synthesised by ChAT on the sperm head, activates the nicotinic receptors present in the egg membrane (characterised by Ivonnet and Chambers 1997) and the first step of depolarisation caused by Na^+ influx mediated by the nicotine receptors takes place. This makes the egg membrane permissive to fusion with the sperm membrane.

Successively, the cholinergic molecules are not involved in the calcium wave, which is released by intracellular stores at fertilisation through IP3 release by membrane phospholipases; this, in turn, is caused by the NO, inserted into the egg intracellular domain by the sperm ingression (Kuo et al. 2000). However, the cortical reaction following the calcium wave causes the exposure of muscarinic receptors at the egg surface, and AChE is released into the perivitelline space together with the other cortical granule components. After this exposure, mAChRs become active and excitable, evoking further calcium spikes (as demonstrated by Harrison et al. 2002) when bound by the ACh, autonomously synthesised by ChAT exposed on the egg membrane. This event might be related to the calcium spike that was reported to take place during the nuclear envelope breakdown (Steinhardt 1990a, b; Wilding et al. 1996).

1.4 Cleavages

From cleavage and until the blastula stage, AChE activity, which had disappeared during fertilisation, starts to be present and active once more (Fig. 8A), starting from the perinuclear envelope; this, at least in vertebrate muscle cells (Miki and Mizoguti 1981), is the first site of synthesis. Also, at these stages, the final destination of the molecules is at the blastomere surface where presumably they play a role in cell-to-cell interaction (Fig. 8B). Extensive studies conducted by Buznikov and coworkers demonstrated that, during these events, the typical neurotransmitters mainly involved in regulating cell-to-cell interaction are represented by biogenic amines (Buznikov et al. 2001); these are responsible for inducing messages among blastomeres (Shmukler and Buznikov 1998).

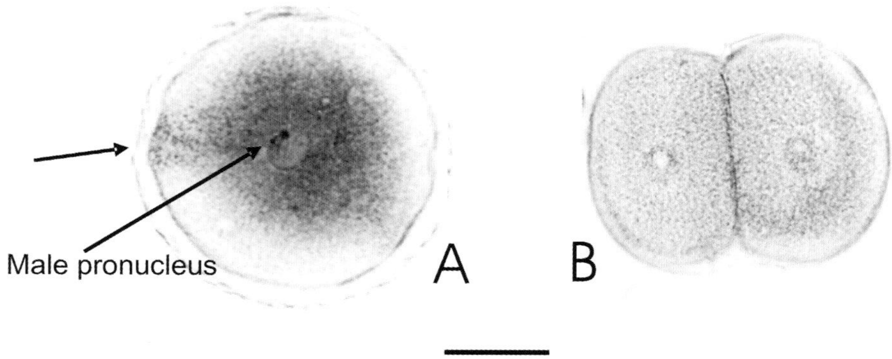

Fig. 8A, B. AChE histochemical reaction in *Paracentrotus lividus*. **A** Zygote, showing *dark labelling* around the nucleus, in the cytoplasm and in the sperm pathway from the fertilisation cone (*arrow*) to the point of contact with the female pronucleus. **B** Two-blastomere stage, showing the enzyme activity product around the nuclei and in the furrow between the blastomeres. In toto mount, *bar* 80 µm

1.5 Ongoing Research

At this point, our interest is towards the molecular forms of muscarinic receptors involved in the different stages of egg activation and cell cycling. We are now investigating the changes that take place when passing from the "embryonic" to the "adult" molecular forms of the m2 receptor. These changes are possibly due to alternative splicing during developmental events (Angelini et al. 2004). Another aim is to specify the involvement of the ryanodine receptor possibly by identifying its mRNA transcription and pharmacological characterisation during the events supported by calcium activation. From our preliminary results, the activation of egg and calcium spiking during the first cell cycle seems to be due to different release mechanisms: the first, brought about by sperm activation, seems to be mainly due to IP3 release, as demonstrated by Whitaker and Irvine (1984), while the second, susceptible to muscarinic receptor activation, seems to be mainly due to calcium release mediated by ryanodine receptor. Actually, from preliminary results, ryanodine receptors seem to be mainly expressed during the first cell cycle, while the amount of these molecules found and characterised by both incorporation in artificial membranes and Northern blot analysis indicates a very scarce presence during the other stages: unfertilised egg and up to the four-cell stage (Angelini et al. 2000; Prestipino, pers. comm.).

2 Applied Research

The described results open a path to a new way of considering biomarkers. In fact, most environmental factors have the capacity to interfere with the cholinergic system: stress, wounds, inflammation and pollution generally (Kaufer et al. 1998), in a number of organisms from invertebrates (Dauberschmidt et al. 1996) and fish (Fulton and Key 2000), to high vertebrates, such as birds (Misawa et al. 1981), up to mammalians, including man (Wessler 2003). In particular, some broadly employed pollutants and drugs specifically affect the cholinergic system, and consequently the other typical neurotransmitter systems, such as those depending on biogenic amines (Buznikov et al. 2001b). These are the so-called neurotoxic compounds.

2.1 Neurotoxic Contaminants

A class of contaminants diffused all over the world is represented by neurotoxic substances, in particular organophosphate (OP) and carbamate (CB) pesticides. These are largely used in European countries for many purposes, for example in agriculture, gardening and even domestic pest control. Among these, OP compounds have been used for a great number of other purposes, e.g. as chemical weapons and medical compounds (anxiolytics, antispasmoics, regulators of eye pressure, and drugs for Alzheimer's disease, etc.).

Neurotoxic pesticides inhibit cholinesterase activities (acetylcholinesterase, AChE, E.C. 3.1.1.7 and pseudocholinesterase, BuChE, E.C. 3.1.1.8.) and, consequently, the status of the cholinergic neurotransmitter system.

2.1.1 Health Risks

These insecticides are strongly suspected to be harmful, and clearly they are in the case of acute intoxication, when organisms come into contact with high doses, generally as a result of accidents or occupational exposure. However, there is also the possibility of subtle chronic (low-dose, long-term) damage due to aerosol diffusion, or to residuals in crops and vegetables, possibly reinforced by the presence of traces of other pollutants, such as other neurotoxic substances, heavy metals and hydrocarbons, among others. On the other hand, the no-effective concentration for humans (NOEC), indicated by the pharmaceutical companies and databases, is not surely ascertained, because it is obtained by experimental exposure of animals, generally rats or mice, and then by estimating it as several-fold lower. Moreover, the doses that do not affect adults may heavily strike embryonic differentiation, a very sensitive stage in the life of an organism (Romero et al. 1989)

2.1.2 *Mode of Function of Organophosphates and Carbamates*

These neurotoxic compounds target the cholinergic neuromuscular system, where they inhibit the activity of acetylcholinesterase (AChE, E.C. 3.1.1.7.) and pseudocholinesterases (E.C. 3.1.1.8.). These are the lytic enzymes of signal choline esters, including acetylcholine (ACh), and consequently their inhibition causes an overflow of ACh at receptor sites; this, in turn, affects intracellular responses. In this way, neurotoxic compounds may cause alteration of all functions of the cholinergic neurotransmission system, and of other neurotransmitters, the release of which is influenced by the cholinergic system (Buznikov et al. 2001). The pharmacology of OPs has been extensively studied, since they were initially conceived as chemical weapons during World War II (Karczmar et al. 1970). Some differences between the two classes of neurotoxic pesticides are known: (1) OPs irreversibly link the AChE molecule by the phosphate group (Sultatos 1994), while carbamates compete for the substrate acetylcholine (ACh) (Mineau 1991); and (2) OPs can leave residuals in the environment, while carbamates only leave small inorganic molecules (e.g. carbon dioxide) (Hayes and Laws 1991).

2.1.3 *Persistence in the Environment and Crops*

The persistence of these compounds in the environment is generally considered fairly short, but there is evidence that they may persist in sediments over long periods of time, such as in the River Rhine (Dauberschmidt et al. 1996), where lethality of fish, molluscs and aquatic birds lasted for months and kilometres downstream after the actual contamination had taken place. According to Ragnarsdottir (2000), an OP pesticide presenting a short half-life in the laboratory increases to 1 year under conditions of low pH and temperature. The same author reported that OPs are detected in soils years after application, probably due to sorption of the OPs onto soil particles, making them unavailable for microbial metabolism (Ragnarsdottir 2000). Insofar as living organisms are concerned, the effects of such compounds may last much longer, because the AChE of blood may be affected for up to several months (see the case report by Romero et al. 1989).

2.2 Biosensors

For the reasons explained above, at present, a great deal of effort is concentrated on creating "biosensors", capable of perceiving neurotoxic compounds in the environment, as well as in food and water.

Most of the biosensors are represented by devices that have the capacity to measure, with high sensitivity, the activity of acetylcholinesterase in the presence of suspected inhibitors and in particular OP or carbamate compounds. These high-technology instruments can measure the presence and amount of neurotoxic compounds in environmental matrices, or in rough material and elaborated foods. The biosensors used for this purpose are generally based on highly sensitive molecular forms of AChE, immobilised in devices capable of recording changes in activity in real time, and by transferring them to screens or other recording devices (Crew et al. 2004; Pritchard et al. 2004), or by use of mutated bacteria or yeast (Wu et al. 2002).

In order to evaluate the effects on living organisms, and their health risks, another kind of biosensor is now being studied. Actually, the effects of neurotoxic compounds, and particularly of OPs and carbamates, are not only directed towards AChE activity, but may also affect ACh receptors, directly (Abdallah et al. 1992; Sultatos 1994). Thus, the best biosensor in this case should be a biosensor with the capacity to translate the effect of neurotoxic pesticides in living cells to cell responses, such as the electrical ones. The advantages of this biosensor are represented by the fact that complex cell response is taken into account, and that the AChE molecules and the ACh receptors are in their natural environment, and follow their natural transduction cascades up to the cell response. For this purpose, a model of this biosensor is being devised in the laboratory of the Fraunhofer Institute for Biomedical Engineering (IBMT), based on microtechnology.

2.3 Effects of Neurotoxic Compounds on *P. lividus* Ion Dynamics

From this point of view, the early phases of sea urchin development offer very good material for the preparation of biosensors based on living cells, because the effects on the cholinergic system are per se translated into changes in membrane potential. Actually, by exposing *P. lividus* zygotes to diazinon, an OP compound, we found that the increase in $[Ca^{2+}]_i$, either caused by muscarinic drug exposure or the physiological one preceding the NEB, was prevented. The K^+ outward channel probably linked to excitation of the muscarinic receptor m2 was also blocked by highly diluted phentoate, another OP compound commonly used as an insecticide (Percivale 2003).

This offers the possibility of adapting this model to a biosensor capable of recording the impedance changes by using *P. lividus* zygotes (the moment when the cholinergic system is mostly dependent on electrical variations) (Fig. 9).

The employment of these cells is at present innovative, as well as bioethically compatible. In addition, it may solve some controversial points, such as:

1. The problem of experiments on animals, which are more expensive, besides causing pain, which is particularly evident in higher organisms. Gametes from sea urchins may be obtained easily by intra-oral 1/1,000 ACh injection: this treatment allows gamete spawning without sacrificing the adults, which survive in good health (as demonstrated by further natural spawning in a few weeks: M. Franzoni and M. Sgro, preliminary results, pers. comm.).

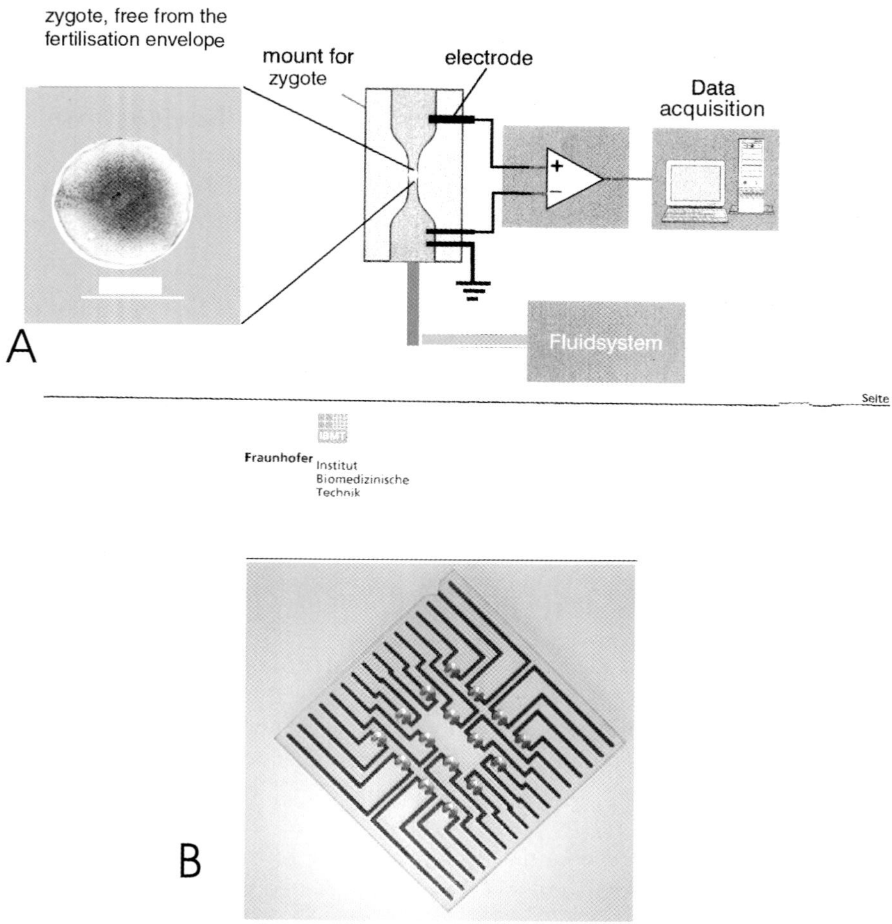

Fig. 9. A Schematic drawing of the living cell-based biosensor (IBMT). **B** The micro-device for impedance recording

2. The improved knowledge of developmental biology, within the emerging knowledge that neurotransmitter molecules are not limited to neuromuscular structures, but are generally involved in cell-to-cell communication, leading to interaction between developing cells and tissues.
3. Exporting the results to other organisms, including man, by comparing the effects on *P. lividus* zygotes with the effects on human stem cells (addressed to development).
4. Allowing us to establish conversion parameters among the different cell sources, in order to use the most suitable and available for each situation of risk assessment.

Acknowledgements. This work was supported by the EC within the project SENS-PESTI, QLK4-CT-2002-02264. We thank Dr. Mario Franzoni and Dr. Marco Sgro for allowing us to use their preliminary data on rearing sea urchin specimens; and Dr. Mario Mori for continuous supply of living sea urchins in optimal reproductive conditions.

References

Abdallah EAM, Jett DA, Eldefrawi ME, Eldefrawi AT (1992) Differential effects of paraoxon on the M3 muscarinic receptor and its effector system in rat submaxillary gland cells. J Biochem Toxicol 7:125–132

Angelini C, DelGaudio R, Falugi C, Pernigotti R, Geraci G, Prestipino G, Pesando L (2000) Calcium signaling in sea urchin zygotes. In: Proc 15th Meeting Sea Urchin Development, 29 Sept–3 Oct, Woods Hole, Massachusetts

Angelini C, Baccetti B, Piomboni P, Trombino S, Aluigi MG, Stringara S, Gallus L, Falugi C (2004) Acetylcholine synthesis and possible functions during sea urchin development. Eur J Histochem 49(3):235–244

Babcock DF, Bosma MM, Battaglia DE, Darszon A (1992) Early persistent activation of sperm K^+ channels by the egg peptide speract. Proc Natl Acad Sci USA 89:6001–6005

Baccetti B, Burrini AG, Collodel G, Falugi C, Moretti E, Piomboni P (1995) Localisation of two classes of acetylcholine receptor-like molecules in sperms of different animal species. Zygote 3:207–217

Baitinger C, Alderton J, Poenie M, Schulman H, Steinhardt RA (1990) Multifunctional Ca^{2+}/calmodulin-dependent protein kinase is necessary for nuclear envelope breakdown. J Cell Biol 111(5):1763–1773

Beer AJ, Moss C, Thorndyke M (2001) Development of serotonin-like and SALMFamide-like immunoreactivity in the nervous system of the sea urchin *Psammechinus miliaris*. Biol Bull 200(3):268–280

Birdsall NJM, Burgen ASV, Hulme EC (1978) The binding of agonists to brain muscarinic receptors. Mol Pharmacol 14:723–736

Bishop MR, Rama Sastry BW, Stavinoha B (1977) Identification of acetylcholine and propionylcholine in bull spermatozoa by integrated pyrolysis, gas chromatography and mass spectrometry. Biochim Biophys Acta 500:440–444

Bray C, Son J-H, Meizel S (2002) A nicotinic acetylcholine receptor is involved in the acrosome reaction of human sperm initiated by recombinant human sperm initiated by recombinant human ZP3. Biol Reprod 67:782–788

Buznikov GA (1980) Biogenic monoamines and acetylcholine in protozoa and metazoan embryos. In: Salanki J, Turpaev M (eds) Neurotransmitters: comparative aspects. Akademial Kiado, Budapest, pp 7–28

Buznikov GA (1990) Neurotransmitters in embryogenesis. In: Turpaev TM (series ed) Physiology and general biology, section F of Soviet Scientific Reviews. Harwood, London

Buznikov GA, Sakharova AV, Manukhin BN, Markova LN (1972) The role of neurohumors in early embryogenesis. IV. Fluorimetric and histochemical study of serotonin in cleaving eggs and larvae of sea urchins. J Embryol Exp Morphol 27(2):339–351

Buznikov GA, Shmukler YB, Lauder JM (1996) From oocyte to neuron: do neurotransmitters function in the same way throughout development? Cell Mol Neurobiol 16(5):533–559

Buznikov GA, Lambert HW, Lauder JM (2001a) Serotonin and serotonin-like substances as regulators of early embryogenesis and morphogenesis. Cell Tissue Res 305(2):177–186

Buznikov GA, Nikitina LA, Bezuglov VV, Lauder JM, Padilla S, Slotkin TA (2001b) An invertebrate model of the developmental neurotoxicity of insecticides: effects of chlorpyrifos and dieldrin in sea urchin embryos and larvae. Environ Health Perspect 109(7):651–661

Cariello L, Romano G, Nelson L (1986) Acetylcholinesterase in sea urchin spermatozoa. Gamete Res 14:323–332

Chambers EL, De Armendi J (1979) Membrane potential, action potential and activation potential of eggs of the sea urchin, *Lytechinus variegatus*. Exp Cell Res 122:203–218

Churchill GC, Okada Y, Thomas JM, Genazzani AA, Patel S, Galione A (2002) NAADP mobilizes Ca($^{2+}$) from reserve granules, lysosome-related organelles, in sea urchin eggs. Cell 111(5):703–708

Crew A, Wedge R, Hart JP, Marty JL, Fournier D (2004) A samperometric biosensor array to measure organophosphate concentration in raw products. In: Proc 8th World Congr on Biosensors, 24–26 May, Granada, Spain

Crossley I, Swann K, Chambers E, Whitaker M (1988) Activation of sea urchin eggs by inositol phosphates is independent of external calcium. Biochem J 252(1):257–262

Darszon A, Garcìa-Soto J, González M, Guerrero A, Sánchez JA, Liévano A (1987) Ion transport mechanism in sea urchin sperm. In: Mohri H (ed) New horizon in sperm cell research. Japanese Scientific Society Press/Gordon and Breach Scientific, Tokyo/New York, pp 169–183

Dauberschmidt C, Dietrich DR, Schattler C (1996) Toxicity of organophosphorous insecticides in the zebra mussel, *Dreissena polymorpha*. P Arch Environ Contam Toxicol 30:373–378

Dobransky T, Rylett RJ (2003) Functional regulation of choline acetyltransferase by phosphorylation. Neurochem Res 28(3–4):537–542

Drews U (1975) Cholinesterase in embryonic development. Prog Histochem Cytochem 7:1–52

Dwivedi C, Long NJ (1989) Effect of cholinergic agents on human spermatozoan motility. Biochem Med Metabol Biol 42(1):66–70

Epel D (1975) The program and mechanism of fertilization in the echinoderm egg. Am Zool 15:507–522

Epel D (1980) Ionic triggers in the fertilization of sea urchin eggs. Ann N Y Acad Sci 339:74–85

Falugi C, Prestipino G (1988) Localization of putative nicotinic cholinoreceptors in the early development of *Paracentrotus lividus*. Cell Mol Biol 35(2):147–161

Falugi C, Pera C, Prestipino G (1989) Localization of α-bungarotoxin receptors and AChE activity in eggs and embryos of *Paracentrotus lividus* Lamk. Acta Embryol Morphol Exp 10(2):113–119

Falugi C, Borgiani L, Faraldi G, Tagliafierro G, Toso F, Drews U (1991) Localization and possible functions of neurotransmitter systems in sperms of different animal species. In:

Baccetti B (ed) Comparative spermatology, 20 years later. vol 75. Serono Symposia Publications, Raven Press, New York, pp 475–479

Falugi C, Moretti E, Lammerding-Köppel M, Drews U (1993a) A cholinergic system is involved in gamete interaction. In: Facchinetti F, Henderson HW, Pierantoni R, Polzonetti Magni A (eds) Cellular communication in reproduction. Society for Endocrinology, Bristol, pp 101–104

Falugi C, Pieroni M, Moretti E (1993b) Cholinergic molecules and sperm functions. J Submicrosc Cytol Pathol 25(1):63–69

Falugi C, Facondini L, Angelini C, Diaspro A, Robello M, Pedrotti ML (1999) Onset of a cholinergic-like system during neurogenesis of the sea urchin, *Paracentrotus lividus*. In: Candia Carnevali MD, Bonasoro G (eds) Echinoderm Research 1998. AA Balkema, Rotterdam, pp 197–202

Falugi C, Diaspro A, Angelini C, Pedrotti L, Raimondo M, Robello M (2002) Three dimensional mapping of cholinergic molecules by confocal laser scanning microscopy in sea urchin larvae. Micron 33(3):233–239

Florman HM, Storey BT (1982) Characterization of cholinomimetic agents that inhibit in vitro fertilization of the mouse. Evidence for a sperm-specific binding site. J Androl 3:157–164

Fulton MH, Key PB (2000) Acetylcholinesterase inhibition in estuarine fish and invertebrates as an indicator of organophosphorus insecticide exposure and effects. Environ Toxicol Chem 20(1):37–45

Gabrielle P, Jeana M, Lorenza EC (2003) Cytosolic choline acetyltransferase binds specifically to cholinergic plasma membrane of rat brain synaptosomes to generate membrane-bound enzyme. Neurochem Res 28(3–4):543–549

Garcia-Soto J, Gonzalez-Martinez M, De la Torre L, Darszon A (1987) Internal pH can regulate Ca^{2+} uptake and the acrosome reaction in sea urchin sperm. Dev Biol 120:112–120

Gilbert SF (ed) (1994) Developmental biology. Sinauer Associates, Sunderland

Giudice G (ed) (1973) Developmental biology of the sea urchin embryo. Academic Press, New York

Giudice G (ed) (1986) The sea urchin embryo. Springer, Berlin Heidelberg New York

Gustafson T, Ryberg E, Treufeldt R (1972) Acetylcholine and contractile activity in the echinopluteus. A study of the cellular basis of larval behaviour. Acta Embryol Exp 1972: 199–223

Harrison PK, Falugi C, Angelini C, Whitaker MJ (2002) Muscarinic signalling affects intracellular calcium concentration during the first cell cycle of sea urchin embryos. Cell Calcium 31(6):289–297

Hayes WJ Jr, Laws ER Jr (1991) Organic phosphorous pesticides. In: Hayes WJ Jr, Laws ER Jr (eds) Handbook of pesticide toxicology. Academic Press, San Diego

Ibanez CF, Pelto-Huikko M, Soder O, Ritzen EM, Hersh LB, Hokfelt T, Persson H (1991) Expression of choline acetyltransferase mRNA in spermatogenic cells results in an accumulation of the enzyme in the postacrosomal region of mature spermatozoa. Proc Natl Acad Sci USA 88:3676–3680

Ivonnet PI, Chambers EL (1997) Nicotinic acetylcholine receptors of the neuronal type occur in the plasma membrane of sea urchin eggs. Zygote 5(3):277–287

Jaffe LA (1976) Fast block to polyspermy in sea urchins is electrically mediated. Nature 261:68–71

Jaffe LA (1980) Electrical polyspermy block in sea urchins: nicotine and low-sodium experiments. Dev Grow Diff 22:503–507

Jaffe LA, Robinson KR (1978) Membrane potential of the unfertilized sea urchin egg. Dev Biol 62:215–228

Jaffe LR (1983) Sources of calcium in egg activation: a review and hypothesis. Dev Biol 99:265–276

Just ML, Lennarz WJ (1997) Reexamination of the sequence of the sea urchin egg receptor for sperm: implications with respect to its properties. Dev Biol 184(1):25–31

Karczmar AG, Usdin E, Willis JH (1970) Anticholinesterase agents. In: Karczmar AG, Usdin E, Willis JH (eds) International encyclopedia of pharmacology and therapeutics, vol 13. Pergamon Press, Oxford

Karnovsky MJ, Roots L (1964) A "direct colouring" thiocholine method for cholinesterase. J Histochem Cytochem 12:219–221

Kaufer D, Friedman A, Seidman S, Soreq H (1998) Acute stress facilitates long-lasting changes in cholinergic gene expression. Nature 393(6683):373–377

Kuo RK, Baxter GI, Thompson SH, Stricker SA, Patton C, Bonaventura J, Epel D (2000) NO is necessary and sufficient for egg activation at fertilization. Nature 406:633–636

Liévano A, Vega-SaendzeMiera EC, Darszon A (1990) Ca^{++} channels from the sea urchin sperm plasma membrane. J Gen Physiol 95:273–296

Longo FJ (1987) Fertilization. Chapman and Hall, London

Longo FJ, Lynn JW, McCulloh DH, Chambers EL (1986) Correlative ultrastructural and electrophysiological studies of sperm–egg interactions of the sea urchin, *Lytechynus variegatus*. Dev Biol 118:155–166

Mautner HG (1986) 8-Choline acetyltransferase. In: Boulton AA, Baker GB, Yu PH (eds) Neuromethods. Vol 5. Neurotransmitter enzymes. Humana Press, Clifton, New Jersey, pp 273–304

McCulloh DH, Lynn JW, Chambers EL (1987) Membrane depolarization facilitates sperm entry, large fertilization cone formation, and prolonged current responses in sea urchin oocytes. Dev Biol 124:171–190

Mei L, Xiong WC (2003) Two birds with one stone: a novel motif for ACh receptor assembly quality control. Trends Neurosci 26(4):178–181

Miki A, Mizoguti H (1981) Acetylcholinesterase activity in the myotome of the early chick embryo. Cell Tissue Res 227:23–40

Mineau P (1991) Cholinesterase inhibiting insecticides. Their impact on wildlife and environment, vol II. Chemicals in agriculture . Elsevier, Amsterdam

Minganti A, Falugi C, Raineri M, Pestarino M (1981) Acetylcholinesterase in the embryonic development: an invitation to a hypothesis. Acta Embryol Morphol Exp 2:30–31

Misawa M, Doull J, Kitos PA, Uyeki EM (1981) Teratogenic effects of cholinergic insecticides in chick embryos. I. Diazinon treatment on acetylcholinesterase and choline acetyltransferase activities. Toxicol Appl Pharmacol 57(1):20–29

Nakajima Y (1986) Development of the nervous system of sea urchin embryos: formation of the ciliary bands and the appearance of two types of ectoneural cells in the pluteus. Dev Growth Differ 28:531–542

Nakajima Y (1988) Serotonergic nerve cells of starfish larvae. In: Burke RD, Mladenov PV, Lambert P, Parsley RL (eds) Echinoderm biology. AA Balkema, Rotterdam

Nakajima Y, Burke RD, Noda Y (1993) The structure and development of the apical ganglion in the sea urchin pluteus larvae of *Strongylocentrotus droebachiensis* and *Mespilia globulus*. Dev Growth Differ 35:531–538

Nelson L (1973) Preliminary evidence for cholinoceptive sites in the excitability of spermatozoa. Nature 242:401–402

Nelson L (1976) α-Bungarotoxin binding by cell membranes. Blockage of sperm cell motility. Exp Cell Res 101:221–224

Nelson L (1978) Chemistry and neurochemistry of sperm motility control. Fed Proc 37(11):2543–2547

Nelson L (1990) Pesticide perturbation of sperm cell function. Bull Environ Contam Toxicol 45(6):876–882

Nelson L, McGrady V, Fangboner ME (1970) Control of flagellar movement. In: Baccetti B (ed) Comparative spermatology. Academic Press, New York, pp 465–478

Ohlendieck K, Lennarz W (1996) Molecular mechanisms of gamete recognition in sea urchin fertilization. Curr Top Dev Biol 32:39–110

Ohlendieck K, Partin JS, Lennarz WJ (1994) The biologically active form of the sea urchin egg receptor for sperm is a disulfide-binded homo-multimer. J Cell Biol 125(4):817–824

Percivale G (2003) Caratterizzazione funzionale in membrane modello del recettore della rianodina nele fasi precoci dello sviluppo embrionale del ricco di mare *Paracentrotus lividus*. Thesis (Prestipino G, Tutor), University of Genova, Italy, 148 pp

Piomboni P, Baccetti B, Moretti E, Gambera L, Angelini C, Falugi C (2001) Localization of molecules related to cholinergic signaling in eggs and zygotes of the sea urchin, *Paracentrotus lividus*. J Submicrosc Cytol Pathol 1–2:187–193

Pritchard JA, Piano M, Pittson RW, Embury L, Drago GA (2004) A disposable on-site detection system for organophosphates using screen-printed biosensors. In: Proc 8th World Congr on Biosensors, 24–26 May, Granada, Spain

Quiao D, Nikitina LA, Buznikov GA, Lauder JM, Sadler FJ, Slotkin TA (2003) The sea urchin embryo as a model for mammalian developmental neurotoxicology: ontogenesis of the high-affinity choline transporter and its role in cholinergic trophic activity. Environ Health Perspect 111(14):1730–1735

Ragnarsdottir KV (2000) Environmental fate and toxicology of organophosphate pesticides. J Geol Soc 157(4):859–876

Rakow TL, Shen S (1990) Multiple stores of calcium are released in the sea urchin eggs during fertilization. Proc Natl Acad Sci USA 87:9285–9289

Romero P, Barnett PG, Midtling JE (1989) Congenital anomalies associated with maternal exposure to oxydemeton-methyl. Environ Res 50:256–261

Ryberg E (1973) The localization of cholinesterases and non-specific esterases in the echinopluteus. Zool Scripta 2:163–170

Ryberg E (1974) The localization of biogenic amines in the echinopluteus. Acta Zool 325:171–189

Ryberg E (ed) (1979) Studies on the structural and chemical basis for regulation of muscular and ciliary responses to environmental stimuli in echinoplutei. Östen Grankvist, Stockholm

Sultatos LG (1994) Mammalian toxicology of organophosphorous insecticides. J Toxicol Environ Health 43(3):271–289

Sastry BVR, Bishop MR, Sen TK (1979) Distribution of α-bungarotoxin-binding proteins in fractions from bull spermatozoa. Biochem Pharmacol 28:1271–1274

Schmell E, Earles BJ, Breaux C, Lennarz WJ (1977) Identification of a sperm receptor on the surface of the eggs of the sea urchin *Arbacia punctulata*. J Cell Biol 72:35–46

Schuel H, Berkerey D, Schuel R, Chang MC, Zimmerman AM, Zimmerman S (1991) Reduction of fertilizing capacity of sea urchin sperm by cannabinoids derived from marihuana. I. Inhibition of the acrosome reaction induced by egg jelly. Mol Reprod Dev 29:51–59

Shen SS (1995) Mechanisms of calcium regulation in sea urchin eggs and their activities during fertilisation. In: Moscona AA, Monroy A (eds) Current topics in developmental biology, vol 30. (Series eds: Pedersen RA, Schatten GP). Academic Press, San Diego, pp 65–102

Shmukler YB, Buznikov GA (1998) Functional coupling of neurotransmitters with second messengers during cleavage divisions: facts and hypotheses. Perspec Dev Neurobiol 5:469–480

Steinhardt A (1990a) Intracellular free calcium and the first cell cycle of the sea urchin embryo (*Lytechinus pictus*). J Reprod Fert Suppl 42:191–197

Steinhardt RA (1990b) Calcium regulation of the first cell cycle of the sea urchin embryo. Ann N Y Acad Sci 582:199–206

Stroud RM, McCarty MP, Shuster M (1990) Nicotinic acetylcholine receptors superfamily of ligand-gated ion channels. Biochemistry 29(50):11009–11023

Thorndyke MC, Crawford BD, Burke RD (1992) Localisation of a SALMFamide neuropeptide in the larval nervous system of the sand dollar, *Dendraster excentricus*. Acta Zool 73:207–212

Turner PR, Sheetz MP, Jaffe LA (1984) Fertilization increases the phosphoinositide content of sea urchin eggs. Nature 310:414–415

Vacquier V (1981) Dynamic changes in egg cortex. Dev Biol 84:1–26

Vacquier VD (1969) The isolation and preliminary analysis of the hyaline layer of sea urchin eggs. Exp Cell Res 54:140–142

Vacquier VD, Moy GW (1977) Isolation of bindin, the protein responsible for adhesion of sperm to sea urchin eggs. Proc Natl Acad Sci USA 74:2456–2460

Vidal AM, del Castillo J, Smith DS (1993) Contractile properties of the articular capsule or ligament, in the primary spines of the sea urchin *Eucidaris tribuloides*. Comp Biochem Physiol 106(3):643–647

Watson S, Arkinstall S (eds) (1994) The G-protein linked receptor. (Facts Book 6.) Academic Press, London

Weinberger C, Penner PL, Brick IY (1984) Polyingression, an important morphogenetic movement in chick gastrulation. Am Zool 24:545–554

Wessler I, Kilbinger H, Bittinger F, Unger R, Kirkpatrick CJ (2003) The non-neuronal cholinergic system in humans: expression, function and pathophysiology. Life Sci 72:2055–2061

Whitaker MJ, Irvine RF (1984) Inositol 1,4,5-triphosphate microinjection activates sea urchin eggs. Nature 312:636–639

Wilding M, Wright EM, Patel R, Ellis Davies G, Whitaker M (1996) Local perinuclear calcium signals associated with mitosis-entry in sea urchin embryos. J Cell Biol 135(1):191–199

Wu CF, Valdes JJ, Rao G, Bentley WE (2002) Enhancement of organophosphorus hydrolase yield in *Escherichia coli* using multiple gene fusions. Biotechnol Bioeng 75(1):100–103

Echinoderm Reactive Oxygen Species (ROS) Production Measured by Peroxidase, Luminol-Enhanced Chemiluminescence (PLCL) as an Immunotoxicological Tool

G. Coteur, B. Danis, P. Dubois

Abstract. The importance of reactive oxygen species (ROS) production in invertebrate immunity prompted the use of this response in immunotoxicological studies in several taxa including marine organisms. In this chapter, we review the effects of environmental factors and contaminants such as heavy metals and polychlorinated biphenyls (PCBs) on the production of ROS by the main immune effector cells of echinoderms, the so-called amoebocytes. ROS production was measured by the peroxidase, luminol-enhanced chemiluminescence (PLCL) method. This method was found to predominantly reflect the production of superoxide anions and peroxides, among which hydrogen peroxide and peroxynitrite are the main species detected. Exogenous factors such as water temperature and salinity can influence this immune response in echinoderms. However, gender, handling stress and parasitism by a castrating ciliate apparently did not affect it. The impact of metals on ROS production differed greatly according to the duration and routes of exposure; in vitro and short-term in vivo exposures to metals caused an inhibition of this immune response, while the opposite effect was observed in a long-term in vivo exposure study. On the other hand, PCBs systematically had a stimulatory effect on ROS production independent of the echinoderm species or exposure routes. From the study of complex field contaminations, it appeared that contaminants released in the environment, such as metals, modulate starfish amoebocyte ROS production. This impact potentially represents a threat to the sustainability of natural populations of echinoderms and thereby to the stability of benthic ecosystems.

G. Coteur (e-mail: gcoteur@ulb.ac.be), B. Danis, P. Dubois
Laboratoire de Biologie Marine (CP 160/15), Université Libre de Bruxelles, 50, Av. F.D. Roosevelt, 1050 Bruxelles, Belgium

Progress in Molecular and Subcellular Biology
Subseries Marine Molecular Biotechnology
V. Matranga (Ed.), Echinodermata

1 Introduction

In contrast to vertebrates, invertebrates defend themselves against offending microorganisms only by means of non-specific, innate mechanisms. The efficiency of this primitive immune system is witnessed by the fact that 95 % of existing animals are invertebrates. This system relies on both cellular and humoral components. The immune cells are involved in phagocytosis, encapsulation, hydrolytic enzyme secretion and the respiratory burst (Chia and Xing 1996). The latter is a mechanism by which a phagocytic cell drastically increases its oxygen consumption upon encounter with foreign material. The oxygen is converted into the very reactive superoxide anion O_2- by a membrane-bound enzyme, the NAD(P)H-oxidase (Babior 1984). This free radical in turn yields several radical or non-radical oxidants such as the hydroxyl radical (·OH–), hydrogen peroxide (H_2O_2) or singlet oxygen ($^1O_2$). The reduction of hydrogen peroxide by the myeloperoxidase (MPO) enzyme may also lead to the formation of hypochlorous acid (HClO). All these oxidants are collectively called reactive oxygen species (ROS). The transduction of the stimulatory signal from the membrane to the NAD(P)H-oxidase involves phospholipase C (PLC) and protein kinase C (PKC) and requires the presence of calcium ions (Lennartz 1999; Torreilles et al. 1999). ROS are thought to destroy foreign particles by oxidising macromolecules such as lipids, proteins or nucleic acid, resulting in lipid peroxidation, enzymatic activity disruption and DNA damage (Buechter 1988). They might also act by inducing an osmotic imbalance in the lysosomes, ending in the activation of hydrolytic enzymes (Reeves et al. 2002).

The importance of ROS production in invertebrate immunity prompted the use of this response in immunotoxicological studies in several taxa including marine organisms. Most of these studies are dedicated to bivalve molluscs or arthropods (Baier-Anderson and Anderson 2000). In these groups, metals generally inhibit ROS production, although, in some instances, an increase in this immune response was observed after both in vitro (Anderson et al. 1997) and in vivo (Dyrynda et al. 1998) contaminations. Decreased ROS production was also observed in bivalves after experimental polycyclic aromatic hydrocarbons (PAH) exposure and in mussels collected in the field after an oil spill (Larson et al. 1989; Dyrynda et al. 1997). Similarly, Pipe et al. (1995) determined the production of ROS by immune cells of mussels collected in contaminated sites and found a negative correlation between the immune response and the level of organic contamination. The observed effects appear to vary greatly according to the contaminant and the invertebrate species studied and the exposure mode. The mechanism by which the production of ROS is affected remains unclear and probably depends on the contaminant studied. Metals are known to take part in the formation of certain reactive species (Kehrer 2000) but can also act by inactivating divalent cation-requiring enzymes involved in ROS production. Other contaminants such as pentachlorophenol are uncouplers of oxidative phosphorylation and may act by interfering with the activity of the NAD(P)H-oxidase (Baier-Anderson and Anderson 1997).

Despite their ecological and economical importance, echinoderms have been largely ignored in the field of immunotoxicology. Numerous representatives of this phylum are key species of marine benthic ecosystems. By their abundance or their trophic role, they often act as structuring species of their ecosystem. Moreover, since the marine sediments represent a reservoir for many contaminants, echinoderms that live within or upon the sediments are particularly at risk concerning the impact of marine pollution. ROS production in this phylum has been reported by a few authors (Ito et al. 1992, Wheatley et al. 1998). Therefore, we investigated the effects of environmental factors and contaminants such as heavy metals and polychlorinated biphenyls (PCBs) on the production of ROS by the main immune effector cells of echinoderms, called amoebocytes. These results are reviewed and discussed in the present work along with a description of the method used to measure ROS production.

2 Measurement of Reactive Oxygen Species (ROS) Production

Luminol-enhanced chemiluminescence (LCL) has been used for a long time for the measurement of the respiratory burst of vertebrate phagocytic cells. This method is considered to be sensitive to both the formation of oxygen radicals and the activity of the endogenous peroxidase, myeloperoxidase (MPO). The signal obtained using invertebrate immune cells is often much lower compared to their vertebrate counterparts (Torreilles et al. 1996). This can be improved by adding an exogenous peroxidase (horseradish peroxidase, HRP) to the measuring medium (Coteur et al. 2002). Adding HRP increases the signal intensity and makes it more specific to the presence of peroxides. We used this system to measure ROS production by echinoderm amoebocytes and determined the specificity of this method. It was found that peroxidase, luminol-enhanced chemiluminescence (PLCL) is predominantly affected by superoxide anions and peroxides, among which hydrogen peroxide and peroxynitrite appear to be the main species detected (Coteur et al. 2002).

The detailed protocol of this method is as follows. Experimental animals are wrapped in a towel to absorb external seawater and then bled by cutting the tip of one arm and draining 3 ml of coelomic fluid in an equal volume of sterile anticoagulant buffer [Ca^{2+}, Mg^{2+}-free artificial seawater (CMFASW) (Noble 1970) with 1.2×10^{-2} M EDTA] at 4 °C. The amoebocyte concentration of this suspension is determined as follows: 200 µl of the cell suspension from each animal is distributed in wells of a UV-transparent microplate (96-wells, UVstar, Greiner), cells are mildly agitated for 5 s and the absorbance at 280 nm is determined using a microplate reader (Spectrafluor Plus, TECAN; three flashes per well). This is related to the cell concentration by using the following predetermined relation (Coteur et al. 2002): A_{280}=0.157 × cell concentration (×106 cells ml^{-1})+0.067. The suspension is then centrifuged (400 *g*, 10 min, 4 °C) and the supernatant replaced by CMFASW (without EDTA), the volume of which is adjusted to obtain a concentration of $1 \pm 0.25 \times 10^6$ cells ml^{-1}. This concentration is used for normalising chemiluminescence measurements. For PLCL measurements, 100 µl of a luminol/HRP solution [freshly prepared by 100-fold dilution in artificial seawater (ASW) of a stock solution of luminol 10 mg ml^{-1}, HRP 5 mg ml^{-1}, dimethylsulfoxide (DMSO) 1 M in ASW] is added to 20 µl of a bacteria suspension (*Micrococcus luteus*, 5×10^9 bacteria ml^{-1} in ASW) or an equivalent volume of ASW (unstimulated controls) in replicate wells of an opaque white microplate (96 wells, Lumitrac, Greiner). The chemical background (i.e. the chemiluminescence of the bacteria and the solutions alone) is then measured using a microplate reader (Spectrafluor Plus, TECAN) with the following settings: luminescence mode, integration time = 0.5 s, photomultiplier gain = 180. The reading time for 80 wells is about 75 s. Subsequently, 80 µl of amoebocyte suspension in CMFASW is added (resulting in a final amoebocyte concentration of $4 \pm 1 \times 10^5$ cells ml^{-1}). (Thus, the resulting final solution contains 60 % of the nominal Ca^{2+} and Mg^{2+} concentrations of seawater.) The microplate is stored at 13±1 °C (i.e. the average field seawater temperature) and PLCL is measured every 10 min over a 2-h period. Measurements are normalised with the actual amoebocyte concentration in each sample and expressed as the sum of all 10-min interval measurement [in relative light units (RLU), an internal scale of the instrument] for 10^6 cells ml^{-1} (total chemiluminescence) for resting or stimulated amoebocytes.

3 Modulation of ROS Production by Environmental Factors

In order to distinguish the influence of environmental pollution from natural variability of echinoderm immune responses, it is crucial to delineate the endogenous or exogenous factors that affect these responses. For instance, injury, acute temperature shock, acidic pH or heavy metal shock were shown to affect the sea urchin immune system (Matranga et al. 2000, 2002). Moreover, a year-round study of a particular immune response is necessary in order to identify periods that might be characterised by increased variability compared to others. Therefore, we investigated the effects of environmental factors and the annual variability of ROS production by amoebocytes of the starfish *Asterias rubens* (Coteur et al. 2004).

The ROS production by amoebocytes is efficiently stimulated by the presence of bacteria or bacterial wall components such as lipopolysaccharide (LPS) but not by the soluble stimulant phorbol-myristate-acetate (PMA, a

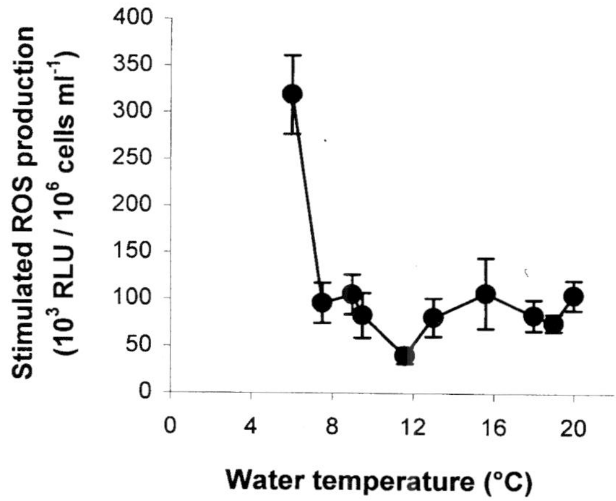

Fig. 1. Bacteria-stimulated reactive oxygen species (*ROS*) production (mean ± SE, n=10) in starfishes collected monthly in Ambleteuse (France) in function of the water temperature measured at each sampling occasion. *RLU* Relative light units

protein kinase C activator) (Coteur et al. 2002). Other factors such as water temperature affected the production of ROS both in experimental conditions and in the field. The ROS production in starfish maintained in very cold waters (≤ 6 °C) was dramatically increased compared to starfish held at (even slightly) higher temperatures and this effect was also observed in the field during the coldest month of the year (Fig. 1).

It is thus a threshold-type effect that precludes the use of ROS production in the field during this period since very high natural variations can occur in limited time. Another factor that partially determines the immune response is the water salinity as ROS production decreases with increasing salinity. However, handling stress, gender and parasitism by a castrating ciliate did not influence amoebocyte ROS production in our experimental setup (Coteur et al. 2004).

4 Impact of Metal Contaminations

In order to have a comprehensive picture of the effects of metals on ROS production, this immune response was measured after in vitro exposure of *A. rubens* amoebocytes or after in vivo experimental contamination of starfishes. Furthermore, natural populations exposed to these contaminants were monitored in the field.

A first experiment was designed to screen the potential of several metals to affect ROS production in vitro. Amoebocytes were exposed to different concentrations of selected metals prior to measuring the unstimulated ROS production. Among all metals tested, only mercury and silver were shown to modulate the resting amoebocyte ROS production (linear regressions:

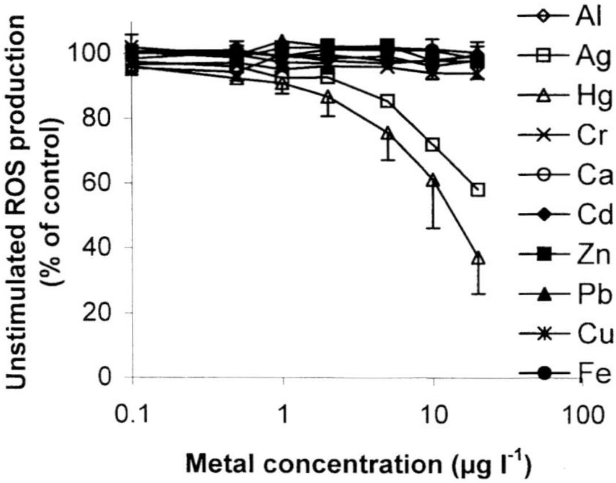

Fig. 2. Reactive oxygen species (*ROS*) production by unstimulated starfish amoebocytes exposed to different metal solution (percentages of control, means and SD are indicated for Hg and Ca as an example; n=6)

p<0.001, R^2=0.277 and p=0.004, R^2=0.165, for Hg and Ag respectively) (Fig. 2). The lack of ROS production inhibition by other metals (Al, Cr, Cd, Zn, Pb, Fe and Cu) is somewhat surprising since Larson et al. (1989) found that, among several types of xenobiotics, Cu was the most immunosuppressive in vitro. However, these authors used a contaminating copper concentration range of 200–2,500 µg l^{-1}. In our hands, the maximal copper concentration tested was 2 µg l^{-1} which is closer to environmentally relevant concentrations. In order to compare the sensitivity of *A. rubens* amoebocytes with that of bivalve hemocytes, we calculated the IC50 values (the metal concentration inducing 50 % of response suppression) for mercury and silver. These values (14.8 and 23.0 µg l^{-1} for Hg and Ag respectively) are 60–600 times lower than those obtained using the phagocytic activity of hemocytes of the clams *Spisula polynyma* and *Mya arenaria* (Brousseau et al. 2000; Fournier et al. 2000). It thus seems that starfish amoebocyte ROS production is highly sensitive to in vitro contamination by silver and mercury but not to other metals (at least at or close to environmentally relevant concentrations).

The impact of in vivo metal contaminations was studied along a metal contamination gradient occurring in a Norwegian fjord, the Sørfjord, where wastes of three smelters, built in the innermost part of the fjord, were discharged in this area for more than 60 years. Concentrations of mercury in fish and of cadmium and lead in mussels from this fjord resulted in public advisories to regulate human consumption (North Sea Task Force 1993).

ROS production was measured both in starfishes sampled along the gradient (long-term – life-long – contaminations) and in starfishes transferred up the gradient (short-term experimental contamination) (Coteur et al. 2003a). The production of ROS in starfishes from field populations increased along the pollution gradient in direct relation with the contamination of the starfishes by cadmium, lead and zinc (e.g. zinc, Fig. 3).

In contrast, when starfishes were transferred from the control site to the contaminated head of the fjord, the temporary accumulation of some metals such as zinc or cadmium in starfishes was accompanied by an inhibition of ROS production (Fig. 4).

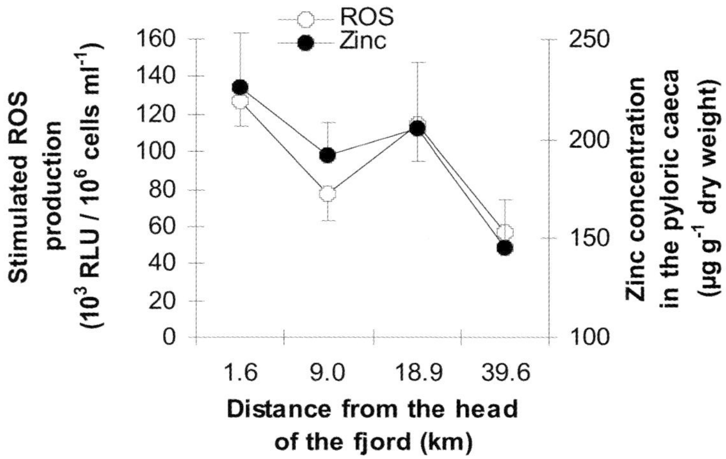

Fig. 3. Bacteria-stimulated reactive oxygen species (*ROS*) production (mean ± SE, n=10) and zinc concentration in the pyloric caeca (mean ± SE, n=5) of starfishes collected along a metal pollution gradient in a Norwegian fjord. *RLU* Relative light units

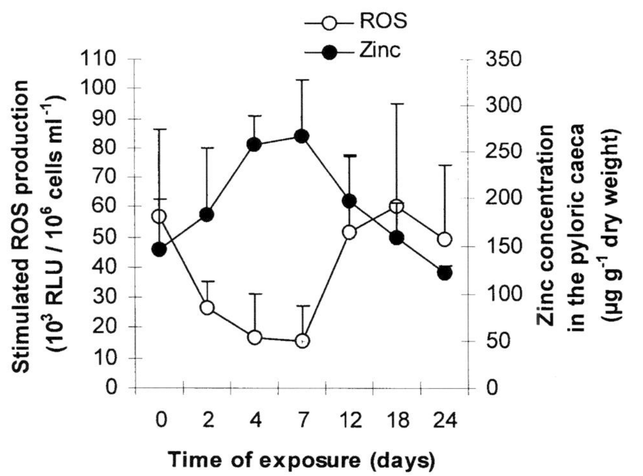

Fig. 4. Effects of short-term transfer experiments up the gradient of metal contamination in a Norwegian fjord. Bacteria-stimulated reactive oxygen species (*ROS*) production (mean ± SE, n=10) and zinc concentration in the pyloric caeca (mean ± SE, n=5) of starfishes transferred from the uncontaminated site (opening of the fjord) to the most contaminated site (head of the fjord) over time. *RLU* Relative light units

Thus, it seems that, depending on the duration of exposure (several days or life-long), the effects of the same contaminants can range from a severe inhibition to a marked increase in ROS production. We hypothesised that the impact of metals in field conditions would occur in three phases: short-term inhibitory effects exerted by direct action of metals on the immune cells are followed by a recovery due to the induction of protective mechanisms and, eventually, when these mechanisms are overwhelmed, indirect stimulatory effects on the immune responses appear due to a global disruption of the animal physiology.

For studies on the impact of metal contamination on stress marker production, see Matranga et al., (this Vol.).

5 Impact of PCB Contaminations

Polychlorinated biphenyls are persistent organic contaminants of human origin which accumulate in the environment. These contaminants are highly toxic, particularly the non-ortho-substituted and mono-ortho-substituted congeners, the so-called coplanar PCB congeners (cPCBs) that can display a configuration very close to that of the highly toxic 2,3,7,8-tetrachlorodibenzodioxin (2,3,7,8-TCDD).

The effects of PCBs were tested by injecting coplanar or non-coplanar congeners in the coelomic cavity of the sea urchin *Paracentrotus lividus* (Coteur

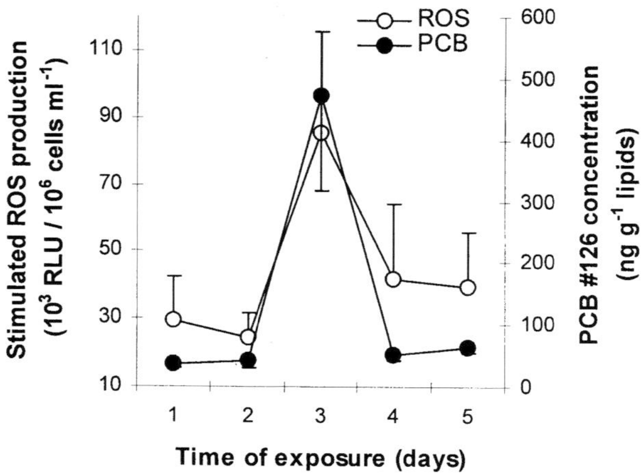

Fig. 5. Bacteria-stimulated reactive oxygen species (*ROS*) production (mean ± SE, n=4) and PCB congener #126 (mean ± SD, n=4) concentrations in the body wall of starfishes maintained on sediments spiked with a mixture of PCB congeners, over exposure time. *RLU* Relative light units

et al. 2001) and of the starfish *A. rubens* (Danis et al., pers. comm.). In both studies, PCBs were found to increase the production of ROS; coplanar congeners being the most effective in this respect. At very high concentrations, however, a steep drop in ROS production was observed probably due to direct cellular toxicity (Danis et al., pers. comm.). Similarly, when starfishes were maintained on a sediment spiked with a mixture of PCB congeners, the amoebocyte ROS production followed closely the transient accumulation of coplanar congeners, such as congener #126, in the starfish tissues (Fig. 5).

Likewise, the contamination by coplanar congener #77 was found to increase the ROS production of starfishes exposed through water, sediments or food (Danis et al. 2003).

6 Impact of Complex Contaminations

We investigated amoebocyte ROS production in natural populations of starfishes exposed to complex, multi-elemental contaminations in order to assess the general impact of contaminants on the immune function of

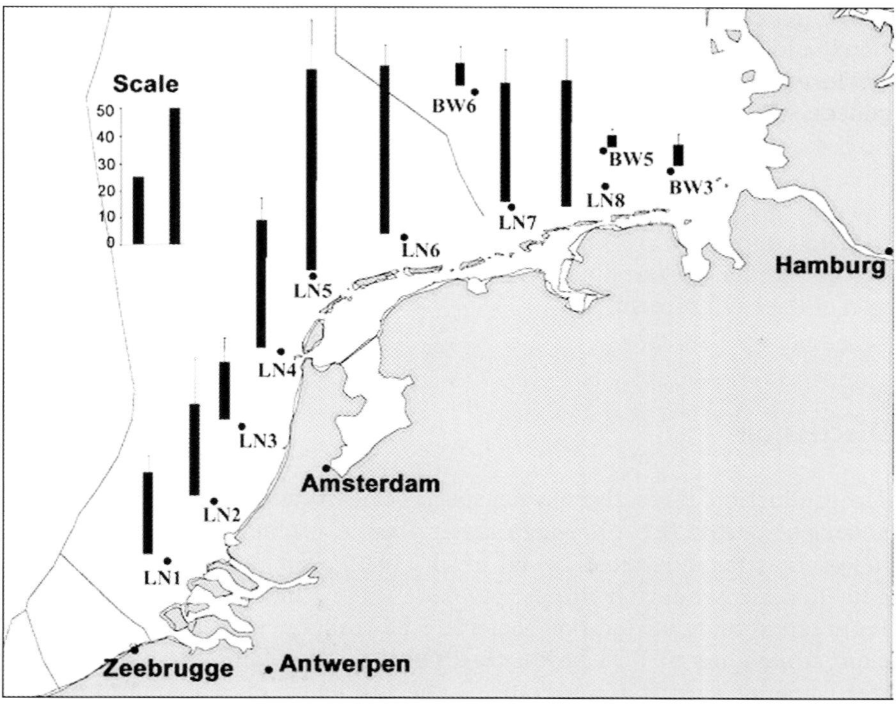

Fig. 6. Bacteria-stimulated reactive oxygen species (ROS) production (10^3 RLU/10^6 cells ml^{-1}; mean ± SE, n=10) in starfishes collected from different stations in the North Sea. *RLU* Relative light units

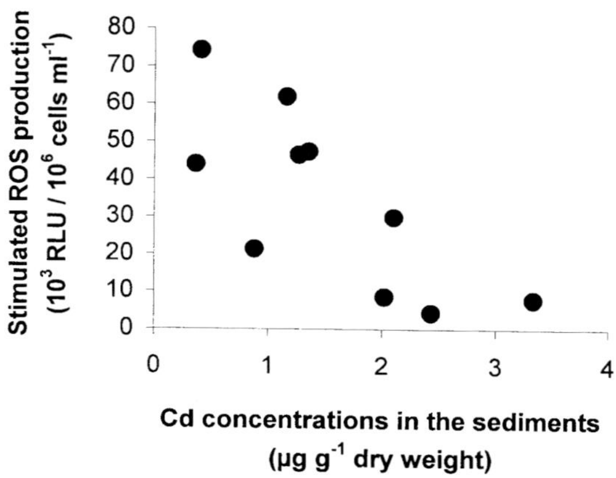

Fig. 7. Bacteria-stimulated reactive oxygen species (*ROS*) production (mean ± SE, n=10) in starfishes collected from different stations in the North Sea in function of the cadmium concentration in the sediments of these stations. *RLU* Relative light units

starfishes in the field. Starfishes were collected at different North Sea stations presenting a large range of metal and PCB contamination levels (Coteur et al. 2003b). Significant differences were found in ROS production of starfishes collected at the different stations. In particular, ROS production was significantly inhibited at the stations located off the Elbe/Weser estuaries (BW stations) and off the North Sea Canal (LN3) (Fig. 6).

Moreover, ROS production was negatively linked with the levels of metal contamination in the starfish tissues and in the sediments (Fig. 7). However, no link was found between the immune response and the non-coplanar PCB concentrations in the samples.

From this study it can be concluded that the contamination levels observed in field conditions are sufficient to induce immunomodulation in echinoderms and that this modulation is particularly linked with metal contamination of the environment.

7 Discussion

The production of reactive oxygen species constitutes one of the main mechanisms of destruction of foreign material in invertebrates. In echinoderms, it was first demonstrated by Ito et al. (1992) who found that sea urchin amoebocytes produce hydrogen peroxide when stimulated by erythrocytes. From our studies, bacteria or bacterial wall components also appear as efficient stimulators of ROS production. On the other hand, the parasitism of starfishes by a castrating ciliate or the challenge of sea urchins by the amoeba *Paramoeba invadens* did not modulate this oxidative activity (Wheatley et al. 1998; Coteur et al. 2004). Exogenous factors such as water temperature and salinity can influence this immune response in echinoderms. It has been hypothesised that this increased ROS production is linked to the elevated concentration of dissolved oxygen, a major source of ROS, at low water temperature (Regoli et al. 2000). Sea urchin coelomocytes react by overexpressing a heat shock protein (Hsp70-like protein) when subjected either to a rise or to a drop in temperature, the latter inducing a greater stress response than the former (Matranga et al. 2000). The sharp rise in Hsp70 expression in coelomocytes at low temperatures might represent a protective response towards the ROS produced by these cells, since Hsp70 is an efficient antioxidant (Chong et al. 1998). From our studies, it is concluded that if a particular period is to be selected for comparing ROS production in the field, we would suggest either October to December, which is characterised by constant (but low) resting and stimulated ROS production (the latter being around 50×10^3 RLU/10^6 cells ml^{-1}), or the period between April and August, during which stimulated ROS production is high and constant (around 100×10^3 RLU/10^6 cells ml^{-1}) but somewhat offset by an increased variation in resting ROS production.

When these precautions are taken, ROS production appears as a very efficient tool for immunotoxicological studies since it is modulated by the most abundant and toxic contaminants. However, concerning metals, the effect differs greatly according to the exposure type: the duration and routes of exposure appear crucial for determining the impact on ROS production. This variability in the impact of metals makes it necessary to always include background-exposed starfishes whether in experimental conditions or in the field in order to work on relative instead of absolute grounds.

In contrast to metals, PCBs have systematically the same effect on ROS production independent of the echinoderm species or exposure routes in experimental conditions. Either the stimulatory effect of PCB could be explained by a direct action of PCBs on the intrinsic mechanism of ROS production by amoebocytes as suggested by Coteur et al. (2001), or PCB detoxification mechanisms could take place in amoebocytes; these processes lead to the production of a certain amount of oxygen free radicals, adding to the ROS produced spontaneously by the immune cells.

A contaminant-induced increase or decrease in ROS production both represent a hazard to the health of individuals. Decreased production will lead to recurrent infections since invading microorganisms will not be efficiently destroyed. This was illustrated in molluscs in which contaminants that reduced ROS production by oyster hemocytes exacerbated the severity of infections (see Baier-Anderson and Anderson 2000). It was also shown that some mollusc pathogens have the ability to inhibit ROS production and consequently to survive in their host (LaPeyre et al. 1995). On the other hand, increased ROS production can lead to damages to self tissues since ROS are, by their very nature, totally non-specific and are able to alter a wide range of self-macromolecules such as DNA and enzymes (Buechter 1988).

From the study of complex field contaminations, it appears that contaminants released in the environment, such as metals, modulate starfish amoebocyte ROS production. This impact potentially represents a threat to the sustainability of natural populations of echinoderms and thereby to the stability of benthic ecosystems.

Acknowledgements. G. Coteur and B. Danis were holders of doctoral grants from the "Fonds pour la formation à la recherche dans l'industrie et dans l'agriculture" (FRIA) and from the "Fondation David et Alice Van Buuren". G. Coteur and Ph. Dubois are, respectively, Postdoctoral Researcher and Research Associate of the National Fund for Scientific Research (NFSR, Belgium). Research was supported by Belgian federal research programmes ICAS (SSTC contract MN/11/30) and ECOTOX2 (SSTC contract EV/11/23). This is a contribution of the Centre Interuniversitaire de Biologie Marine (CIBIM).

References

Anderson RS, Brubacher LL, Calvo LMR, Burreson EM, Unger MA (1997) Effect of in vitro exposure to tributyltin on generation of oxygen metabolites by oyster hemocytes. Environ Res 74:84–90

Babior M (1984) The respiratory burst of phagocytes. J Clin Invest 73:599–601

Baier-Anderson C, Anderson RS (1997) The effect of pentachlorophenol on pyridine nucleotide production in oyster hemocytes: NADPH and immunomodulation. J Shellfish Res 16:111–114

Baier-Anderson C, Anderson RS (2000) Immunotoxicity of environmental pollutants in marine invertebrates. Rec Adv Mar Biotechnol 5:189–225

Brousseau P, Pellerin J, Morin Y, Cyr D, Blakley B, Boermans H, Fournier M (2000) Flow cytometry as a tool to monitor the disturbance of phagocytosis in the clam *Mya arenaria* hemocytes following in vitro exposure to heavy metals. Toxicology 142:145–156

Buechter DD (1988) Free radicals and oxygen toxicity. Pharmaceut Res 5:253–260

Chia FS, Xing J (1996) Echinoderm coelomocytes. Zool Stud 35:231–254

Chong KY, Lai CC, Lille S, Chang CS, Su CY (1998) Stable overexpression of the constitutive form of heat shock protein 70 confers oxidative protection. J Mol Cell Cardiol 30:599–608

Coteur G, Danis B, Fowler SW, Teyssie J-L, Dubois Ph, Warnau M (2001) Effects of PCBs on reactive oxygen species (ROS) production by the immune cells of *Paracentrotus lividus* (Echinodermata). Mar Pollut Bull 42:667–672

Coteur G, Warnau M, Jangoux M, Dubois Ph (2002) Reactive oxygen species (ROS) production by amoebocytes of *Asterias rubens* (Echinodermata). Fish Shellfish Immunol 12:187–200

Coteur G, Gillan D, Joly G, Pernet Ph, Dubois Ph (2003a) Field contamination of the starfish *Asterias rubens* by metals. Part 2: effects on cellular immunity. Environ Toxicol Chem 22:2145–2151

Coteur G, Gosselin P, Wantier P, Chambost-Manciet Y, Danis B, Pernet Ph, Warnau M, Dubois Ph (2003b) Echinoderms as bioindicators, bioessays, and impact assessment tools of sediment-associated metals and PCBs in the North Sea. Arch Environ Contam Toxicol 45:190–202

Coteur G, Corriere N, Dubois Ph (2004) Environmental factors influencing the immune responses of the common European starfish (*Asterias rubens*). Fish Shellfish Immunol 16:51–63

Danis B, Cotret O, Teyssié JL, Fowler SW, Bustamante P, Warnau M (2003) Delineation of PCB uptake pathways in a benthic sea star using a radiolabelled congener. Mar Ecol Prog Ser 253:155–163

Dyrynda EA, Law RJ, Dyrynda PEJ, Kelly CA, Pipe RK, Graham KL, Ratcliffe NA (1997) Modulations in cell-mediated immunity of *Mytilus edulis* following the 'Sea Empress' oil spill. J Mar Biol Assoc UK 77:281–284

Dyrynda EA, Pipe RK, Burt GR, Ratcliffe NA (1998) Modulations in the immune defences of mussels (*Mytilus edulis*) from contaminated sites in the UK. Aquat Toxicol 42:169–185

Fournier M, Cyr D, Blakley B, Boermans H, Brousseau P (2000) Phagocytosis as a biomarker of immunotoxicity in wildlife species exposed to environmental xenobiotics. Am Zool 40:412–420

Ito T, Matsutani T, Mori K, Nomura T (1992) Phagocytosis and hydrogen peroxide production by phagocytes of the sea urchin *Strongylocentrotus nudus*. Dev Comp Immunol 16:287–294

Kehrer JP (2000) The Haber-Weiss reaction and mechanisms of toxicity. Toxicology 149:43–50

LaPeyre JF, Chu FLE, Vogelbein WK (1995) In vitro interaction of *Perkinsus marinus* merozoites with eastern and Pacific oyster hemocytes. Dev Comp Immunol 19:291–304

Larson KG, Roberson BS, Hetrick FM (1989) Effect of environmental pollutants on the chemiluminescence of hemocytes from the American oyster *Crassostrea virginica*. Dis Aquat Org 6:131–136

Lennartz MR (1999) Phospholipases and phagocytosis: the role of phospholipid-derived second messengers in phagocytosis. Int J Biochem Cell Biol 31:415–430

Matranga V, Toia G, Bonaventura R, Muller WEG (2000) Cellular and biochemical responses to environmental and experimentally induced stress in sea urchin coelomocytes. Cell Stress Chaper 5:113–120

Matranga V, Bonaventura R, Di Bella G (2002) hsp70 as a stress marker of sea urchin coelomocytes in short term cultures. Cell Mol Biol 48:345–359

Noble PB (1970) Coelomocyte aggregation in *Cucumaria frondosa*: effect of ethylenediaminetetraacetate, adenosine, and adenosine nucleotides. Biol Bull 139:549–556

North Sea Task Force (1993) Assessment report 1993. Subregion 6 (Norway). State Pollution Control Authorities. HS-Trykk A/S Press, Oslo

Pipe RK, Coles JA, Thomas ME, Fossato VU, Pulsford AL (1995) Evidence for environmentally derived immunomodulation in mussels from the Venice Lagoon. Aquat Toxicol 32:59–73

Reeves EP, Lu H, Jacobs HL, Messina CGM, Bolsover S, Gabella G, Potma EO, Warley A, Roes J, Segal AW (2002) Killing activity of neutrophils is mediated through activation of proteases by K^+ flux. Nature 416:291–297

Regoli F, Nigro M, Bompadre S, Winston GW (2000) Total oxidant scavenging capacity (TOSC) of microsomal and cytosolic fractions from Antarctic, Arctic and Mediterranean scallops: differentiation between three potent oxidants. Aquat Toxicol 49:13–25

Torreilles J, Guerin MC, Roch P (1996) Reactive oxygen species and defense mechanisms in marine bivalves. C R Acad Sci Paris Life Sci 319:209–218

Torreilles J, Guerin MC, Roch P (1999) Modified Alsever's solution is not a good medium for reactive oxygen metabolite study in bivalves. Fish Shellfish Immunol 8:65–69

Wheatley K, Brown RG, Scheibling RE, Jellett JF (1998) Coelomocyte oxidative activity of the green sea urchin (*Strongylocentrotus droebochiensis*) following challenge by bacterial and amoebic pathogens. In: Mooi R, Telford M (eds) Echinoderms: San Francisco. AA Balkema, Rotterdam, pp 881–886

Monitoring Chemical and Physical Stress Using Sea Urchin Immune Cells

V. Matranga, A. Pinsino, M. Celi, A. Natoli, R. Bonaventura, H.C. Schröder, W.E.G. Müller

Abstract. Coelomocytes are the cells freely circulating in the body fluid contained in echinoderm coelom and constitute the defence system, which, in response to injuries, host invasion, and adverse conditions, is capable of chemotaxis, phagocytosis, and production of cytotoxic metabolites. Red and colourless amoebocytes, petaloid and philopodial phagocytes, and vibratile cells are the cell types that, in different proportions, constitute the mixed coelomocyte cell population found in sea urchins. Advances in cellular and molecular biology have made it possible to identify a number of specific proteins expressed in coelomocytes under resting conditions or when activated by experimentally induced stress. Only recently, coelomocytes have been used for pollution studies with the aim of introducing a new biosensor for detection of stress at both cellular and molecular levels, as sentinel of sea health. In this chapter, we briefly review the important features of these valuable cells and describe studies on their use in the laboratory and in the field for the assessment of chemical and physical pollution of the sea.

1 Introduction

Life on earth originated in the sea. Marine animals were the first animals that had to evolve defence strategies for their survival. To cope with possibly dangerous events, marine organisms have developed over millions of years what

V. Matranga (e-mail: matranga@ibim.cnr.it), A. Pinsino, M. Celi, A. Natoli, R. Bonaventura
Istituto di Biomedicina e Immunologia Molecolare (IBIM) "Alberto Monroy", Consiglio Nazionale delle Ricerche, Via U. La Malfa 153, 90146 Palermo, Italy
H.C. Schröder, W.E.G. Müller
Institut für Physiologische Chemie, Abteilung Angewandte Molekularbiologie, Johannes Gutenberg-Universität, Duesbergweg 6, 55099 Mainz, Germany

Progress in Molecular and Subcellular Biology
Subseries Marine Molecular Biotechnology
V. Matranga (Ed.), Echinodermata

we now call an immune system, as a strong defence mechanism able to resist host attacks in an efficient way. The term "immunis" comes from the Latin meaning "exempt", referring to protection against foreign agents. It is still believed that in all organisms of the animal kingdom the cells of self are virtually marked, so that they are not attacked by their own defence mechanism. Therefore, the immune system must have the capacity to discriminate between self and non-self, to transform itself to deal with future dangers, and, in addition, to change, since the self also evolves with time (e.g. during embryo development). However, the self/non-self model of the immune system has recently been disproved because of many inconsistencies: (1) not all foreign cells need to be destroyed, some in fact must be assimilated for nourishment; (2) in mammals the growing embryo, in principle a host, is not destroyed by the immune system of the mother. This is just to quote two examples. To solve these apparent paradoxes one very attractive theory has been proposed by Polly Matzinger, the present head of the National Institute of Allergy and Infectious Diseases at NIH, Bethesda, Maryland, USA. In her September 2001 lecture entitled, "An Innate Sense of Danger", she proposed that cellular apoptosis signals and directs the immune mechanism (reviewed in Matzinger 2002). In other words, the triggering factor comes from the "sense of danger" originated during damage to body tissues, rather than from the classical recognition of foreign antigens. Her theory has given hints to studies on tumour regression, graft rejection, pregnancy failure and autoimmune diseases. Another puzzling question, which the self/non-self model can neither predict nor explain, is: why is the mere foreignness of a protein not enough to elicit immunity, and why are noxious substances like mineral oil, mycobacteria, and aluminum hydroxide needed in order to get a clear response? Charles Janeway, one of the most brilliant and imaginative immunologists, recently deceased (April 2003), proposed, together with his colleague Medzhitov, the so-called Pathogen-Associated Molecular Pattern theory (Medzhitov and Janeway 2002). They postulate that immune responses cannot occur unless antigen-presenting cells are first activated, and that this activation occurs via pattern recognition receptors. Their prediction of the presence of evolutionarily conserved receptors recognizing molecules from infectious non-self organisms was confirmed when, studying *Drosophila* Toll innate immune sensor, they found the mammalian Toll-like receptors (Medzhitov et al. 1997), which are now studied by laboratories around the world.

The above-mentioned new ways of interpreting the machinery of the immune system bring to the concept that the innate immune system is an evolutionarily ancient form of host defence found in most multicellular organisms. The basic mechanisms operating in sensing and attacking foreign agents are already found in echinoderms. It is stimulating that these hypotheses can be tested in the echinoderm model and that, to a certain extent, studies on their defence system might provide many useful insights into the understanding of the mechanisms involved. In the following text, we briefly review the state of the art on what is known about the echinoderm immune system, with particular emphasis on mechanisms and/or molecules involved in sensing experimental or environmental stress.

2 Echinoderm Coelomocytes as Immune Effector Cells: Morphological Features and Recognized Functions

In all organisms the first defence against infections involves a physical barrier to invasions by pathogens or foreign substances, provided by their body coats which physically prevent the interaction between the host and the invading organism. However, when pathogens accidentally or experimentally penetrate this barrier they first encounter generic or specific molecules that restrict the infection. The second-line defence consists of cells that can engulf (phagocytose) foreign substances or pathogens or can aggregate under formation of large coagulates (clots) in the case of wounds. Other cells secrete in the medium specific anti-host molecules, which will be discussed below. Phagocytosis is a complex process which involves several steps, including: (1) chemotaxis, whereby phagocytic cells are attracted towards chemotactic chemicals like microbial products, complements, or damaged cells; (2) adhesion, whereby phagocytes stick to each other and attach to foreign agents; (3) opsonization, a process that enhances adhesion, whereby specific proteins (opsonins) are coated on the microbial surface; and (4) ingestion, whereby phagocytes extend their cellular projections, engulfing the foreign organism. Finally, the foreign organism can be digested by enzymes in the lysosome. All the steps described above have been studied using echinoderms as a model system. In the past, each of the echinoderm classes has preferentially been used for a certain type of study; in the case of classical immune/defence studies, involving self/non-self recognition, which occurs when grafting cells from one individual to another, scientists have mostly used the classes of Holothuriodea and Echinoidea. For the purpose of our description and to take care of the amount and the heterogeneity of information present in the literature, we will focus our interest on echinoids, specifically sea urchins. In the body of adult sea urchins a cell population lives, which has been identified as the true immune effector because of many proven characteristics. Quite vast, although not so recent, literature is available which describes the morphologies present in this mixed cell population (see Smith 1981; Matranga 1996). However, possibly due to the way in which cells have been collected or preserved before their observation by optical or electron microscopy, a variety of misleading classifications are reported, which have generated confusion. The easiest way to give a nomenclature that reflects the actual morphology of the cells is the immediate observation of fresh and live cells just taken from the sea urchin without any addition of anti-coagulant solution (Fig. 1). Under these conditions it appears clear that some cells are capable of rapid movements, while others show very slow motions. Within the first group, fast-moving cells are red and colourless, each constituting 5–7 % of the total cell population. Peculiar to this cell type is their characteristic locomotion (about 0.5 μm s^{-1}) achieved by rapid changes in their body shape, where a leading edge of the cell is protruded towards their march direction (Fig. 2). Consequently, these cells are referred to as amoebocytes, from the Greek *amoeba*, which means *change*, referring to the shape of the cell (Fig. 1a,b). The red pigment, called echinochrome, is thought to be utilized by echinoderms as an anti-bactericidal agent. This notion comes from the single report present to our knowledge in the literature, describing the release of a red pigment by the sand dollar *Mellita quinquiesperforata* coelomocytes in response to stress (Smith and Smith 1985). In the same study the authors showed the release of histamine on challenge and suggested that the coelomocyte stress response may be an evolutionary precursor to the mammalian allergic response. Due to their fast movement, it seems conceivable to suppose that this cell type is utilized as a first defence mobilization, as in the case of wound healing or prompt attack by foreign agents. Indeed, red amoebocytes have been observed trapped in clotting and encapsulation (Smith 1981).

Another type of very fast cells, constituting 5–6 % of the total cell population, are the so-called vibratile cells (Fig. 1 c); these are round cells which, thanks to a long flagellum, can move in a straight direction along a helicoidal pattern. The function of this less studied cell type is at present unknown.

Finally, the most abundant morphology observed (80–85 %) is constituted by cells with a dendritic-like phenotype, namely "petaloid" or "philopodial." Their conversion from the first to the second morphology is easily observed under the microscope within 5–10 min, and a detailed description, as well as

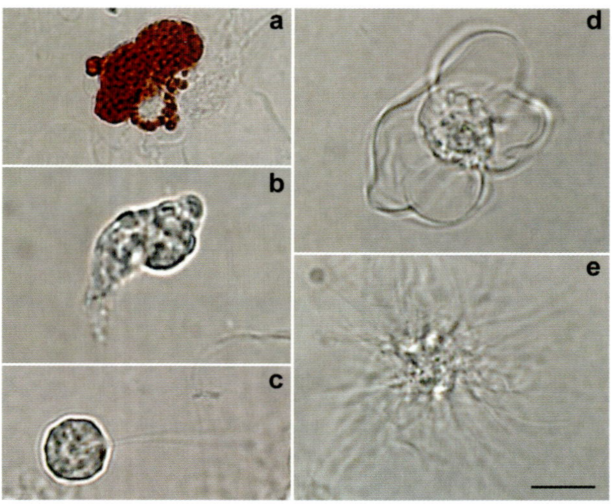

Fig. 1. *Paracentrotus lividus* coelomocytes: **a** red amoebocyte; **b** colourless amoebocyte; **c** vibratile cell; **d** petaloid phagocyte; **e** philopodial phagocyte. *Bar* 10 μm

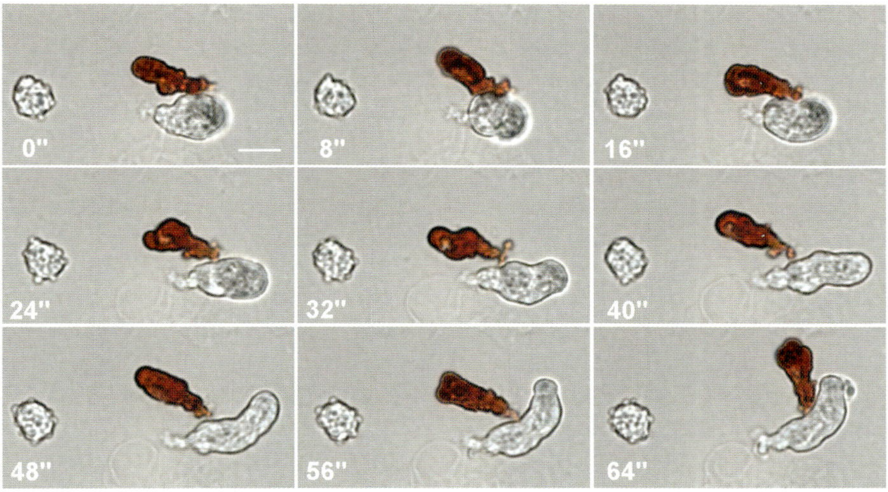

Fig. 2. Red and colourless amoebocytes freshly collected and rapidly moving under the microscope. Bright fields of cells viewed on a Zeiss Axioscope 2 plus using a 63× objective lent. Pictures taken every 8 s. *Bar* 10 μm

different functional states of the same cell type, can be found in several papers (Edds 1985, 1993; Henson et al. 1999).

These four major cell types have been consistently seen and described for at least three different species, i.e. the Pacific Ocean *Strongylocentrotus purpuratus* (Hillier and Vacquier 2003), the North Sea *Strongylocentrotus droebachiensis* (Bertheussen and Seljelid 1978), and the Mediterranean Sea *Paracentrotus lividus* (Matranga and Bonaventura 2002). However, the proportion of each cell type can vary not only among species but also between individuals of the same species, according to their size and physiological conditions, as we will see below. It should be recalled that these cells, within a few minutes (5–10 min) after collection from the sea urchin body, tend to clump; therefore, for their use and/or observation, they need to be placed in an anti-coagulant solution, which, on the other hand, might affect their physiological morphology. However, much work still needs to be done on the function of all these cell types. Most of the immune responses that we expect to be carried out are performed by phagocytes (reviewed by Smith 1981). It seems that the petaloid stage is involved in migration towards the sites of injuries, while the fillopodial stage is involved in clotting. In fact, because of their high numeric representation and the ability to change the shape of their cytoplasmic protrusions, phagocytes are able to form large clots in which other coelomocyte types are trapped or possibly contribute to their formation. However, of course, phagocytes deserve their name from their ability to engulf (phagocyte) foreign particles, such as pathogens, as observed in the clearance of injected bacteria (Yui and Bayne 1983) or even synthetic beads (Bertheussen 1981). In addition, encapsulation, a process by which the host is surrounded by a layer of cells and eventually digested by lytic enzymes released into the cavity, is performed by phagocytes and so-called spherule cells (possibly colorless amoebocytes), the latter being claimed to release bactericidal substances (Johnson 1969; Gerardi et al. 1990).

However, due to the small number of controversial reports present in the literature, the real function of each of these cell types is still not acknowledged.

3 The Origin of Coelomocytes: Terminally Differentiated or Circulating Stem Cells?

Nowadays, the origin of coelomocytes is becoming an extremely interesting field of research for at least two reasons: (1) the access to a new, interesting and easy to use model system to work on and (2) the potential source of stem cells for the production of new therapeutic tools (see, for example, their involvement in regeneration phenomena). The axial organ, a complex circulatory system composed of closed vessels, is believed to be the source of coelomocytes which, in analogy to the vertebrate system, has been regarded as the ancestral primary lymphoid gland. This notion comes from an old study which described the release of coelomocytes from the axial organ after echinoid injury (Millott 1969). Indeed, 10 years later, another report confirmed these results, describing the presence in the axial organ of many coelomocytes rich in pigment and granules, separated into compartments by elastic fibers. The authors, reporting what they called "transformational stages of coelomocytes", suggested the axial organ as the site of their storage and maturation, as well as excretory functions via the water vascular system (Bachmann and Goldschmid 1978).

More recently, two populations of cells, namely adherent (B-like) cells resembling mammalian B lymphocytes and non-adherent (T-like) cells resembling mammalian T lymphocytes, have been fractionated from the axial organ of the sea star *Asteria rubens* (Leclerc and Bajelan 1992). The characterization and identification of molecules produced and released into the coelom by these two cell types would be of extreme interest. An alternative hypothesis explaining the presence of the mixed cell population in the coelomic fluid postulates that these cells represent different maturation stages of the same cell line (Smith 1981). Although attractive, no evidence has been provided confirming this hypothesis. On the contrary, recent studies have shown a rapid increase in the amount of the red amoebocytes (previously referred to as red spherulae cells because of lack of movement caused by the medium used) in response to pollution or experimentally induced stresses (Matranga et al. 2000). This evidence can be explained by either the rapid division of circulating stem cells or the recruitment from the axial organ or similar reservoirs. Both hypotheses need experimental confirmation.

Unfortunately, these lines of investigation have not been followed very much over the years. The development of techniques enabling the culture of coelomocytes as mixed cell population and, more interestingly, as single cell types constitutes a necessary step for studies on the origin and function of coelomocytes; efforts are awaited in this direction.

4 Molecules Expressed by Echinoderm Coelomocytes

Many studies performed in the early 1950s and 1960s demonstrated that, in response to graft reactions or invading agents, sea urchin coelomocytes produce and release into the so-called coelomic fluid a variety of different proteins. In addition, several other stimuli or even the puncture of a needle were shown to produce the activation of a certain number of genes (Smith et al. 1992). The advent of molecular biology techniques, combined with efforts in biochemical purification of molecules, led to the discovery of an extended collection of proteins, expressed in coelomocytes under physiological or adverse/stress conditions (reviewed in Matranga 1996; Matranga and Bonaventura 2002). The repertoire of identified and characterized proteins specifically expressed in sea urchin coelomocytes is growing fast. The list includes most of the immune molecules already described for the vertebrate immune system, such as cytokines, lectins, perforins, lysosomal enzymes, serum proteins, profilins, and others. Below we give a brief overview of these proteins, arbitrarily grouped into five main categories, which we think are of interest for the exploitation of new biotechnological tools.

4.1 Innate Immune Molecules: The Complement System

Aquatic organisms are under constant assault by pathogens, which must be kept in check by the fast-acting innate immune system. Indeed, innate immunity is fighting against pathogens using defence mechanisms that are quickly mobilized and triggered by receptors recognizing a broad spectrum of pathogens. The lower deuterostomes, including echinoderms, possess an innate immune system part of which shares similarities with the vertebrate complement system. The latter is composed of a complex group of serum proteins which are activated in a cascade-like fashion. Only the alternative pathway and the lectin pathway are of interest here, since the classical pathway concerns the adaptive immune system which involves antigen–antibody complexes that, to our best knowledge, are not present in echinoderms. The alternative complement system pathway activates C1, C4, C2, C3, C5 and finally C6 to C9, which form the membrane attack complex. The lectin pathway activates C2, C3, C4 and some C1 homologue calcium-dependent lectin family proteins. There is growing evidence from several studies that describe the identification of genes coding for homologues of the vertebrate complement system proteins. It has been shown, for example, that complement component C3 is specifically expressed in sea urchin coelomocytes (Al-Sharif et al. 1998). While in earlier studies the whole mixed coelomocyte population was used, in very recent reports some authors have separated by gradients the four populations of cells and were able to show that only phagocytes express the *Sp*C3 gene and contain the protein (Gross et al. 2000). Further studies have shown that *Sp*C3 acts as a humoral opsonin, tagging target cells for ingestion by the cell type called, by authors, polygonal phagocytes (Clow et al. 2004), most presumably the petaloid form of phagocytes. Interestingly, the same authors found that *Sp*C3 is already present in the developing sea urchin embryo, with a peak in transcript levels just prior to and during gastrulation (Shah et al. 2003). In addition, continuous exposure to heat killed *Vibrio diazatrophicus*, a marine pathogen of sea urchins, from the hatching blastula to the pluteus stages, highly increased *Sp*C3 expression. These results provided evidence for a complement-based immune system in defence against pathogens, which sea urchin embryos may use in their aquatic environment. In later studies, two cDNAs from coelomocytes, *Sp*CRL (*S. purpuratus* complement related protein, long form) and *Sp*CRS (short form), were characterized. Their deduced protein sequences have domains that are also found in regulatory proteins, such as factor H and factor I, and in the terminal pathway components C6 and C7 (Multerer and Smith 2004).

4.2 Agglutinins, Lectins and Adhesion Molecules

One of the major events that are macroscopically observed in response to pathogens or experimental manipulation is the aggregation of coelomocytes to form clots or to encapsulate hosts. Classically, studies on agglutination of exogenously injected red blood cells into the sea urchin coelomic cavity led to the partial characterization of many molecules. The aggregation of phagocytes at inflammatory sites is a feature that invertebrates and vertebrates have in common. The recruitment of cells, mainly phagocytes and red amoebocytes, is due to chemotaxis and cellular proliferation is triggered by cytokines (Prendergast et al. 1983). In most cases the recognition and binding of non-self material is based on the presence of carbohydrate-specific molecules, which may therefore represent an evolutionary ancient binding principle. Candidates of the last functions are proteins isolated from the coelomic fluid and most probably produced by coelomocytes. Evidence has been reported for a so-called aggregation factor found in *Holothuria polii* (Canicattì and Rizzo 1991) as well as in sea urchins (Canicattì et al. 1992). It is now generally accepted that the invertebrate host defence functions first by immobilizing and then by encapsulating offensive (attacking) microorganisms. During this process, two mechanisms are operating, possibly in cooperation: some molecules are involved in the adhesion between cells surrounding and encapsulating the host, while others, such as perforins (for a review see Glinski and Jarosz 2000), attack the invading agent.

Some years ago we showed that colourless amoebocytes, at the time referred to as colourless spherule cells (Cervello et al. 1994), serve as reservoirs for the precursor of a cell adhesion protein (toposome) which is later used by the embryo (Matranga et al. 1986) and whose synthesis has recently been demonstrated (Unuma et al. 2001; see Zito et al., this Vol.). The protein is released into the medium in response to stress induced by centrifugation, suggesting its involvement in the clotting of coelomocytes after injuries.

Hillier and Vacquier (2003) have recently demonstrated that intercoelomocyte adhesion is mediated by a coelomic plasma protein with a relative molecular mass (Mr) of 75 kDa, named amassin. The protein forms large disulfide-bonded aggregates to which coelomocytes attach. What is interesting is that, by sequence comparison data, it has been shown that amassin shares a protein domain with olfactomedin (OLF), a group of extracellular matrix vertebrate proteins specific to the olfactory neuroepithelium (Yokoe and Anholt 1993). The authors claim that this is the first report assigning a function to a protein of the OLF family and suggest that other as yet undescribed proteins belonging to this family may function in intercellular adhesion.

Among cell–cell and cell–matrix adhesion molecules, integrins, a family of proteins, play a fundamental role. Through the recent use of polymerase chain reaction, it has been possible to identify integrin beta subunits in coelomocytes. Comparative sequence analysis with vertebrate integrins will contribute to the understanding of the evolution and function of this important group of receptors (for a review on the argument see Burke 1999).

Even if not related to the immune response, a few reports have described the presence of globin/hemoglobin in echinoderms. This is the case for four different globin molecules – two of them have been sequenced – found in the sea cucumber *Caudina arenicola* (McDonald et al. 1992). By partial amino acid sequencing (about 40 amino acids) little homology with holothurian globins has been found in recently discovered hemoglobins contained in coelomocytes circulating in the water vascular system of the ophiuroid *Hemipholis elongata* (Christensen et al. 2003). The authors were able to demonstrate a higher affinity for oxygen in juveniles than in adults; the latter exhibiting a heterogeneous mixture composed of three major fractions. Taken together, these findings, confirming once again the role of the "hemal" system carried out by coelomocytes in some of the echinoderm classes (holothurians and crinoids), provide new insights into the understanding of the evolution and function of globin genes.

4.3 Cytoskeleton Proteins

The cytoskeleton is a cellular "scaffolding" or "skeleton," used to maintain and/or alter cellular shape. It appears to be evident that an organized set of proteins must operate in order to accomplish the rapid movements that all coelomocyte populations carry out. The expression of one such specific protein, kinesin, was first shown by domain-specific monoclonal antibodies (McAbs); the protein is highly concentrated in restricted cellular domains, always in association with microtubules which are well spread in fast-moving cells and perinuclear in stationary cells (Henson et al. 1992).

The partial purification of a 110-kDa peptide, possessing K^+EDTA-, Ca^{2+}-, Mg^{2+}-, and F-actin-activated Mg^{2+}-ATPase activities characteristic of myosin-like motor proteins involved in particle/vesicle intracellular trafficking, has been reported (D'Andrea et al. 1994). The authors claim that coelomocytes display particle/vesicle movements even in cells devoid of microtubules (petaloid) and possess an unconventional myosin which may be the motor protein driving particle/vesicle translocation.

The presence of other intermediate filament proteins has been shown in petaloid coelomocytes from two species of echinoderms, the sea urchin *Strongylocentrotus droebachiensis* and the sea cucumber *Cucumaria frondosa* (Holy et al. 2000). Three isoforms have been characterized: a 68-kDa sea urchin lamin, as well as two putative lamin isoforms of approximately 70 and 68 kDa in sea cucumber coelomocytes. In this study, McAbs were produced that recognize lamin isoforms in immunofluorescence and Western blot and were used for tracing studies on their intercellular localization and function. Other authors have studied the actin-based centripetal flow process in sea urchin coelomocytes, demonstrating that it is the result of a two-part mechanism: actin polymerization at the cell edge coupled with actomyosin contraction at the cell centre (Henson et al. 1992). Extended studies suggested a significant cross-talk between the two underlying mechanisms and indicate that changes in actomyosin tension may be translated into alterations of the structural organization of the actin cytoskeleton (Henson et al. 2003).

A molecular approach provided evidence for the increased expression of a gene (*Sp*Coel1) coding for a newly discovered sea urchin profilin, in response to minor injuries, such as puncture by a needle of the peristomal membrane (Smith et al. 1992). Its sequence has striking similarity to the mammalian profilin, a small protein that has both actin and phosphatidyl-inositol-biphosphate binding functions (Goldschmidt-Clermont et al. 1991). It is supposed that sea urchins can utilize this mechanism which links the signal transduction system to the cytoskeleton.

4.4 NK Antigens and Cytokines

Special consideration should be given to these classes of proteins because they could constitute a potential source of new therapeutic drugs, to be derived from coelomocyte exploitation.

The obvious analogies in the features and functions of coelomocytes to human lymphocytes prompted the search for natural killer (NK) cells and their related antigens. It was not surprising that CD56 and CD57 antigens were shown to be expressed in a subpopulation, although not specified, of *Lythechinus variegatus* coelomocytes, by immunocytometry analysis (Koros et al. 2000). On the other hand, only phagocytes from *Arbacia punctulata* were shown to have the strongest cytotoxic activity and the highest binding to human NK markers, namely CD14, CD56 and CD158b (Lin et al. 2001).

Interestingly, using a mouse thymocyte proliferation assay, Burke and Watkins (1991) showed the presence of a cytokine activity extractable from the coelomic fluid of the sea star *Pisaster ochraceus*, which can be blocked with antibodies to mammalian interleukin (IL)-1 alpha. Based on the increase in bacteria clearance after addition of recombinant IL-1 alpha to coelomocytes cultured in vitro, the authors proposed that endogenous interleukins stimulate recruitment of phagocytic cells as part of the non-specific cellular defence mechanism operating in asteroids.

The same line of evidence came from studies reporting that IL-1 and 6 and tumour necrosis factor (TNF) are found in the coelomic fluid of the echinoderm *Asterias forbesi* (Beck and Habich 1986, 1996). However, attempts at isolation and characterization of the true echinoderm TNF proved elusive since its presence has been demonstrated only indirectly by the use of cross-reacting antibodies to the human inflammatory cytokine.

Recently, in the framework of the UVTOX European project, the presence of a stress-inducible TNF-alpha-like activity in *P. lividus* sea urchin coelomocytes has been demonstrated (Matranga and Madaratz, pers. comm.). This evidence came from experiments in which the culture medium obtained from coelomocytes exposed to physical stress (temperature and UV-B radiation) was added to a mouse macrophage cell line, displaying a dose-dependent apoptosis in response to TNF-alpha (Mosman 1983; Hansen et al. 1989). Furthermore, coelomocyte-secreted activity was, at least partly, neutralized by antibodies raised against human or mouse TNF-alpha (not shown). More importantly, by RT-PCR with different oligonucleotide primer pairs designed on the basis of murine TNF-alpha sequence, these authors were able to successfully amplify TNF-alpha fragments from total RNA obtained from sea urchin coelomocytes (not shown). Products of the expected size ranges were cloned into bacterial vectors and amplified for sequencing (not shown). The full length sequence analysis of the putative TNF-alpha mRNA, which is in progress, will show whether sea urchin coelomocytes can produce vertebrate-like TNF-alpha molecules.

4.5 Stress Response Proteins

The "4-D" model postulated by Matzinger for the functioning of the vertebrate immune system, namely Danger, Death, Destruction, and Distress, constitutes an attractive working hypothesis that could be tested using sea urchin coelomocytes. However, very few, if any, systematic and consistent studies testing this hypothesis are provided in the literature. We will give just a few examples of proteins, related to stress response, which have attracted the attention of investigators. It is known that, during an inflammatory insult, changes in the plasma concentration of certain divalent cations, involving, for example, iron-specific binding and transporting proteins, provide a major host defence response in vertebrates. Strikingly, a ferritin molecule was cloned from an echinoderm coelomocyte cDNA library showing a conserved iron-responsive element sequence and sharing a high analogy to the vertebrate homologue (Beck et al. 2002). These authors showed an increased expression of ferritin mRNA after stimulation of coelomocytes with drugs such as LPS or PMA, with a parallel decrease in the amount of iron in "in vitro" culture supernatants over time. These results suggest that echinoderm ferritin is a protein involved in acute phases of toxification and that sequestration of iron is an ancient host defence response in animals.

Recently, acute effects, as opposed to chronic effects, of temperature, chemical, UV-B radiation and pollution stresses have been demonstrated in studies on the expression of a few stress proteins. Briefly, we have demonstrated the overexpression of hsp70 (Matranga et al. 2000, 2002) and acetylcholine esterase (Angelini et al. 2003); these results will be discussed in the following sections.

5 Cell Cultures of Coelomocytes: New Tools to Detect Marine Pollution

As already mentioned, the development of in vitro cultures of coelomocytes has been considered important and recommended in order to better analyze the mechanisms involved in defence/stress response and to overcome the stress of manipulation. Media for keeping sea urchin coelomocytes outside of the animal, for a reasonably sufficient period of time, generally involve the use of anti-coagulant reagents, such as the chelating agents EDTA and EGTA. In our studies, *P. lividus* coelomocytes were kept and separated into subpopulations by gradients in an EDTA-containing buffer (ISO-EDTA). This medium was good enough to show, for the first time, that coelomocytes respond to physical and chemical stresses by elevating the hsp70 expression levels (Matranga et al. 2000). However, it was not preserving the morphological features of freshly collected cells and it was not usable at all for in vitro exposure to chemicals or other stresses. A medium involving EGTA, although utilized with success for the American species *Lythechinus variegatus* (Koros 1993), was, in our hands, not working. We then used successfully a medium modified from that originally described by Henson et al. (1992). We were able, under these conditions, to keep cells for short-term cultures in vitro, but not longer than 4–6 h (Matranga et al. 2002). In fact, by Hoechst staining, we observed that the number of apoptotic cells increased after prolonged time culture: from 5–6 % for culture times up to 6 h, gradually increasing to 10–12 % between 8 and 12 h, and above 20 % after 14 h. The number of apoptotic cells reached 25 and 44 % after 16 and 18 h, respectively (not shown). Therefore, for studies on chronic effects of pollutants on sea urchin coelomocytes, it was necessary to define a culture medium for long-term cultures. For this purpose we took advantage of the technique described for primary cultures of epithelial cells and myocytes obtained from the heart of the mollusc *Pecten maximus* (Le Marrec-Croq et al. 1999), which utilizes a simple medium based on sterile seawater. In fact, mollusc cultures were maintained viable in vitro for at least 2 months and used as a sensitive test to study the effects of pollutants at the cellular level (Pennec et al. 2002).

The medium has been adapted to the *P. lividus* coelomocytes osmolarity, thanks to collaboration with G. Dorange within the framework of the UVTOX European project. Total cells were collected by bleeding with a sterile syringe introduced through the peristomal membrane and plated on tissue culture-treated 24-well plates. Half of the medium was refreshed every other day. Under these conditions, cells attached to the plastic-ware without the need of any special adhesive factor. The number of attaching cells grad-

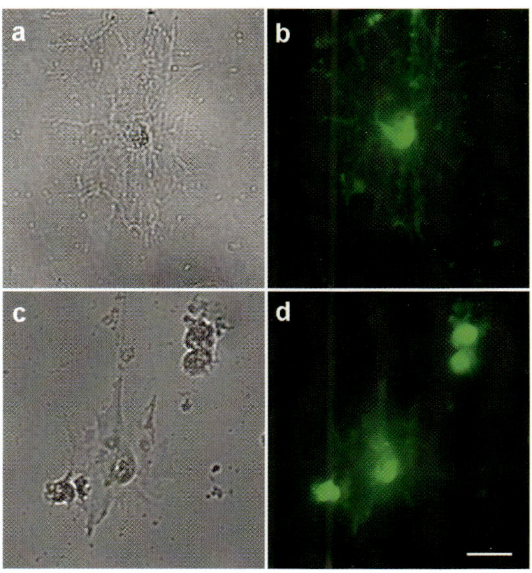

Fig. 3a–d. Surface antigens localized on sea urchin coelomocytes attached to the substrate. Bright field and immunofluorescence images of whole mounts decorated with WGA-FITC. *Bar* 10 µm

ually increased with time, and within 1 h virtually all cells plated were bound to the substratum. The vitality of cells was confirmed by the expression of surface antigens, as assessed by WGA-FITC fluorescence on whole mount specimen (Fig. 3).

When maintained in culture for longer periods of time, cells have the tendency to form bundles and fibres, although maintaining their individuality (Fig. 4); petaloid-shaped phagocytes were no longer visible and possibly philopodial phagocytes were converted into a fibroblastoid-like morphology. Interestingly, both red and colourless amoebocytes, as well as vibratile cells, continued to be observable, constituting an easy criterion for assessing the viability of the culture.

An attempt to keep cells in culture for more than 8 days, which in some cases lasted up to 20 days (Fig. 5), resulted in the formation of a meshwork of cell clumps interconnected by macrofibres made of cellular protrusions (Fig. 5a). Even after such a long time it was possible to distinguish red amoebocytes. Possibly, colourless amoebocytes were passively trapped or adhered to the large phagocyte aggregates present in the culture (Fig. 5b,c), in agreement with earlier reports on hanging drops cultures (Johnson 1969).

The culture media tested, namely the Henson and the Le Marrec-Croq modified coelomocyte media, respectively, referred to as HMCC and LMCC below, were used for the next experiments and constitute good starting points to define an appropriate coelomocyte culture medium for establishing continuous cell cultures.

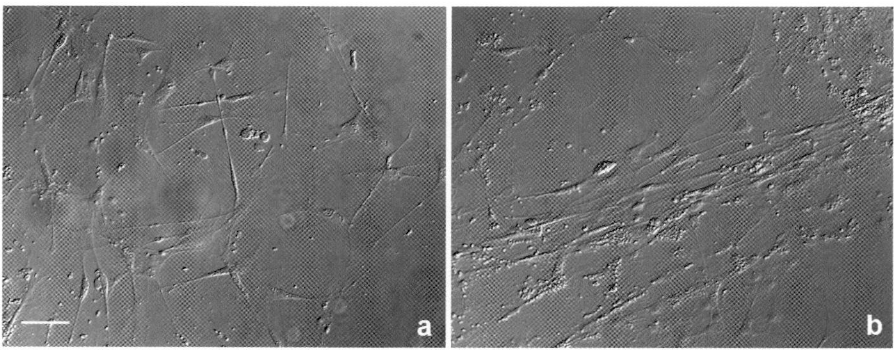

Fig. 4. Sea urchin coelomocyte cultures. Cells, maintained in culture in vitro for 4 (**a**) and 8 (**b**) days tend to form bundles and fibres. *Bar* 10 μm

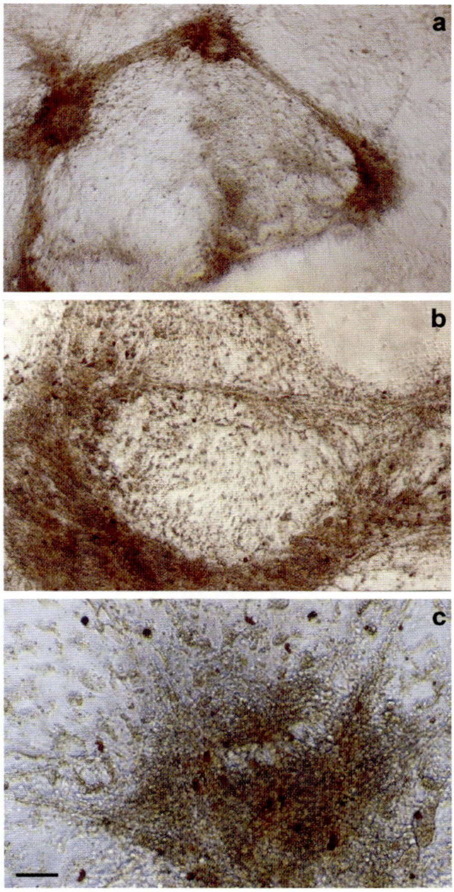

Fig. 5. Sea urchin coelomocyte long-term cultures. Cells were maintained for 20 days in culture after plating. *Bar*: **a** 158 μm; **b** 100 μm; **c** 50 μm

6 Laboratory Experiments

The growing concern over the health state of the sea has attracted the attention of many laboratories to the use of aquatic invertebrates, e.g. echinoderms, as tools for monitoring environmental hazards. However, only few investigations report the effects of pollution in terms of cellular and biochemical modifications on adults of higher marine invertebrates, such as the sea urchin. Most of the reports present in the literature describe in fact the toxic/teratogenic effects of various chemical agents and drugs on the development of sea urchin embryos (Kobayashi 1980; Ozretic and Krajnovic-Ozretic 1985; Sconzo et al. 1995; Morale et al. 1998; Russo et al. 2003; Geraci et al. 2004; Roccheri et al. 2004). The studies described below were aimed at finding a biological indicator to be used as a stress marker and to characterize the response to induced or accidental stresses by the use of molecular markers using sea urchin coelomocytes as biosensors.

It was already known that sea urchin embryos respond to temperature stress by the activation of specific genes (Giudice 1989), whose sequence for the *P. lividus* species has been identified and described (Sconzo et al. 1992). However, there were no studies on the presence of constitutive and/or inducible hsp70 protein in adult sea urchin cells at that time. In a first attempt, we tested the hypothesis that coelomocytes play a role in defence mechanisms activated by adverse external conditions, challenging whole *P. lividus* sea urchin to low and high temperature stress. For our purpose, adult individuals were kept at 16 °C, the control temperature (CT), with subsequent 35 °C heat stress (HS) or 4 °C cold stress (CS) for 4 h. After a 1-h recovery at 16 °C, coelomocytes were gently collected at 4 °C to prevent handling stress as much as possible, lysed, and analyzed by Western blotting. The anti-hsp70 McAb used for immunodetection, previously shown to cross-react with other invertebrate hsp70s (Koziol et al. 1997), recognized both the constitutive and the inducible forms. We found an overexpression of the hsp70 protein in coelomocytes obtained from temperature-stressed urchins, with a two- and a five-fold increase in the protein levels for heat and cold stress respectively (Matranga et al. 2000). The inductive response was time-dependent, i.e. individuals kept for 60 min at the non permissive temperature had twice the hsp70 levels as those exposed to the stress for 30 min. No increases in hsp70 expression were found for periods between 1 and 4 h, and periods longer than 4 h under temperature stress led to death of the animals (Matranga et al. 2000).

The presence and function of acetylcholinesterase (AChE) in vertebrate blood cells and plasma have been elucidated, suggesting its pivotal role in ion exchange (for a review of AchE activity in sea urchin embryos see Angelini et al., this Vol.); in addition, its activity has been found to be associated with the perinuclear region of leucocytes (Falugi 1985). Bearing in mind the similarity between coelomocytes and vertebrate blood cells, we investigated the presence of AChE activity in coelomocytes exposed to cold stress, with the aim of assessing the possibility of using the AChE activity present in these cells as a biomarker for both field and laboratory studies on environmental stress. Our results showed that cholinesterase activity increased with the time of exposure to cold stress and it was detected in petaloid phagocytes and colourless amoebocytes (Angelini et al. 2003). In both cases, the function might be linked to cytoskeletal dynamics, which follows changes in intracellular ion concentrations. This has been observed in sea urchin zygotes, as a result of the action of molecules related to the cholinergic system (Harrison et al. 2002).

In our experiments on whole sea urchins, we found a discrete variability in the basal levels of hsp70 as well as a different inductive response when comparing coelomocytes that were obtained from different individuals. This can be explained by the fact that different sea urchins possess a different personal history, thus reflecting different physiological conditions. We cannot in fact avoid the possibility that individuals utilized for the experiment were already under stress pressure caused, for example, by pathogens or other foreign agents (drugs, pollutants). We now know that pollution can be one of such factors increasing the hsp70 expression (Matranga et al. 2000; see following section).

It was then important to repeat these experiments using the same cell population taken from one individual and confirm the results obtained with whole adults as object of study. Then, using in vitro short-term cultures (1–4 h) of sea urchin coelomocytes, an identical number of cells, having an identical cell type composition, could be exposed to the same stress conditions, giving us also the possibility of monitoring their morphology. When their response to temperature, acidic pH, and heavy metals was tested, using again the hsp70 protein as a stress marker, results recapitulated those obtained with whole sea urchins (Matranga et al. 2002). This was an important achievement since, for the first time, coelomocyte cultures were used to "sense" heavy metal pollution, induced in our case experimentally by the addition of cadmium chloride to the medium. Moreover, since a dose-dependent increase in the hsp70 levels was shown, this could have been conveniently used for studies applied to environmental monitoring.

However, a concern for the applicability of the measurements was given by the fact that in order to elicit a response in a short period of time (4 h), high concentrations of cadmium were used, in the micromolar range. We know that such high cadmium levels are never reached in the marine environment, even in highly polluted areas. Then it was important to keep cells in culture for longer periods of time and to test lower cadmium concentrations, in the nanomolar range. This was possible thanks to the definition of a coelomocyte culture medium, LMCC medium, by which cells could be easily maintained for 3–8 days in culture (see previous section). The hsp70 expression was then monitored in total coelomocytes continuously exposed to $CdCl_2$ for 3 days, after plating the cells for 5 days in LMCC medium in 24-well plates. Under these conditions cells were responsive to the heavy metal and showed an increase in hsp70 levels. However, higher expression levels were observed at the lowest $CdCl_2$ doses, probably due to damage of cells exposed to the heavy metal for long periods of time. Figure 6 shows the results of a representative experiment.

It is well known that UV radiation causes damage to DNA and/or protein in a variety of organisms (for more information see Schröder et al., this Vol.). Many invertebrates, including sea urchins, spawn and/or develop in the 1-m zone of the sea and thus are susceptible to genetic damage. The effects on sea urchin embryo development have already been recently described by a few authors (Epel et al. 1999; Lesser et al. 2003; Bonaventura et al. 2005).

To test whether sea urchin coelomocytes respond to UV-B radiation (312 nm) (1,000 J/m^2) alone, or in combination with different concentrations of $CdCl_2$, by activating the expression of hsp70, in vitro cultures have been used as follows. Sea urchins were bled through a cut in the peristomal mem-

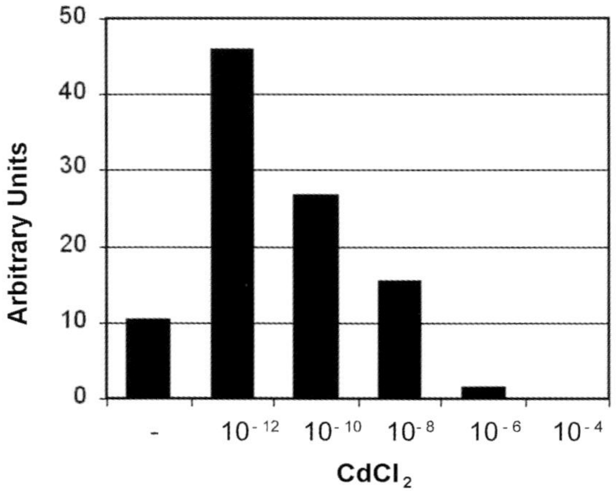

Fig. 6. Densitometric analysis of Western blot with anti-hsp70 antibody of lysates from coelomocytes exposed for 3 days in culture to different $CdCl_2$ concentrations

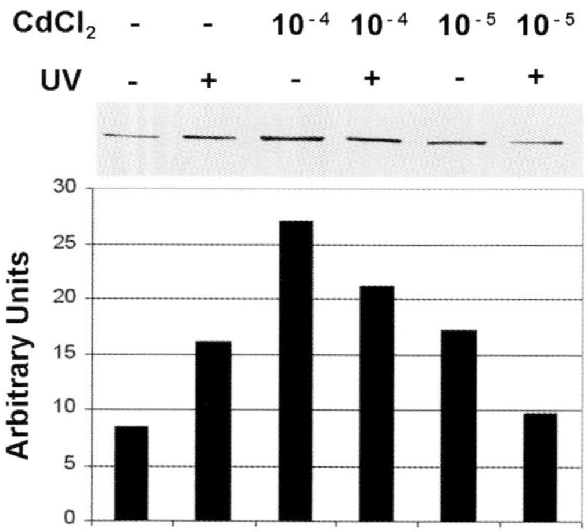

Fig. 7. Western blot analysis of coelomocytes exposed to different $CdCl_2$ concentrations for 2 h, with (+) or without (–) UV-B irradiation (1,000 J/m^2), probed with anti-hsp70 antibody. Histogram shows the densitometric scanning of the filter

brane. The fluid was poured onto ISO-EDTA and the cell suspension was divided into a certain number of Petri dishes. Cultures of coelomocytes were kept under different stress (cadmium) conditions for 2 h; then the cells were or were not exposed to UV-B, collected, centrifuged and the pellet lysed as previously described (Matranga et al. 2002). Equal amounts of protein were loaded onto SDS-PAGE and analysed by Western blotting for the expression of hsp70 using a commercially available antiserum (Fig. 7). A two-fold increase in the expression of hsp70 is found in UV-B-stressed coelomocytes as compared to the control. Similarly, a three-fold increase in the hsp70 level is observed in coelomocytes exposed to 10^{-4} M $CdCl_2$. When coelomocytes were exposed to both 10^{-4} M $CdCl_2$ and UV-B radiation, we did not observe an additive effect, the increase in hsp70 level being about 2.5-fold. We observed the same trend for coelomocytes exposed to both 10^{-5} M $CdCl_2$ and UV-B. This apparent paradox could be explained by the experimental conditions used: i.e., cells were first exposed to cadmium for 2 h and then UV-B treated. We can hypothesize that, in agreement with what is known about the protective effect on apoptosis produced by hsp70 expression (Samali and Cotter 1996), under these conditions the first stress is somehow protecting cells from the second one (UV-B).

7 Field Studies

Only recently, we have tried to establish the use of echinoderm coelomocytes, obtained from sea urchins and sea stars, as a cell laboratory wherein to test different environmental hazards. Below we report a few examples of analyses of samples collected in oceanographic campaigns held in the northern and southern Adriatic Sea as well as in the North Sea.

7.1 The Northern Adriatic Sea: First Example of Monitoring Metal Pollution

Back in 1995, during an EU-sponsored summer course on “Monitoring of environmental stress using modern techniques”, we assessed pollution by means of analysing cellular and molecular markers of stress, using coelomocytes from sea urchins collected in the seawaters around Croatia, during the war. Some tracts of this coast had been previously designated as polluted areas, as a result of urban runoff and/or industrial waste from a nearby fish cannery, and others as non-polluted (Müller et al. 1998; Schröder et al. 1999). Animals were collected from three sites: (1) along the coast near Rovinj (Istria), known as the “urban contaminated area”, (2) in front of Ruder Boskovic Marine Station, the “industrial contaminated site” and (3) from the uncontaminated Limski Canal (north of Rovinj), the “unpolluted controls”. The first difference observed between polluted and non-polluted coelomocytes was found at the cellular level. In fact, in those samples obtained from sea urchins collected from polluted areas, regardless of urban runoff or industrial waste contamination, a consistently greater number of red amoebocytes was observed (Matranga et al. 2000). We do not have an explanation yet to account for this increase in cell number. A few hypotheses can be put forward: (1) conversion of pre-existing phenotypes, for example from colourless to red amoebocytes; (2) rapid cell duplication activated by a mitogenic signal, possibly linked to the production of trophic factors, such as inflammatory cytokines IL-1 and 6 and/or TNF; and (3) recruitment from the so-called hematopoietic areas present in the adult urchin (axial organ). Although all are attractive and supported by a few reports present in the literature, these hypotheses need experimental confirmation.

At the molecular level, the differential expression of stress markers in coelomocytes obtained from sea urchins collected from polluted seawaters was demonstrated by a two-fold increase in hsp70 levels in comparison with controls (Matranga et al. 2000). In addition, we found that both phagocytes and red amoebocytes, isolated by gradients, expressed hsp70 at high levels.

7.2 The Southern Adriatic Sea: A Case Study for the Assessment of TNT Exposure

The studies described above were the first example of the use of sea urchin coelomocytes for monitoring environmental stress and opened the way to other field studies. This was the case for a campaign that took place in June 2003 around Pianosa Island (Tremiti Islands Marine Protected Area, Adriatic Sea, Italy), sponsored by the Istituto Centrale per la Ricerca Scientifica e Tecnologica Applicata al Mare (ICRAM). This site was chosen because of the large number of TNT-containing bombs dumped at sea at the end of World War II. Obviously, the disposal of the explosive 2,4,6-trinitrotoluene (TNT) represents a serious hazard to marine ecosystems, and several agencies strongly recommend their monitoring. Analyses of *P. lividus* specimens, collected around Pianosa Island, and compared with an unimpacted site in the same archipelago near Caprara Island (Tremiti Islands Marine Protected Area, Adriatic Sea, Italy), were aimed at detecting the hsp70 expression levels and the differences in the subpopulation ratio of coelomocytes. In Western blotting experiments using total cell lysates, we found a significant increase in the expression of hsp70 in samples from the TNT-impacted site around Pianosa Island with respect to the Tremiti Islands (not shown). In addition, when separated into subpopulations, coelomocytes obtained from sea urchins collected around Pianosa Island showed a high proportion of red amoebocytes, which, in contrast, were scarcely visible in specimens from control sites (not shown). These results support the suitability of sea urchin coelomocytes as valid tools for monitoring the effects of TNT exposure on aquatic fauna and emphasize their important role as stress bioindicators for future ecotoxicological studies in the marine environment.

7.3
The North Sea: A Norwegian Fiord as a Natural Gradient of Metal Contamination

An interesting confirmation of the results obtained in field studies came from the analysis of hsp70 levels present in coelomocytes from the sea star *Asteria rubens* collected in the Norwegian fiords during the campaign held in May–June 2000, thanks to collaboration with Drs. Coteur and Dubois. The fiord, located in southwestern Norway, is about 38 km long, 1–3 km wide, and at maximum 390 m deep (see for details Coteur et al. 2003). The study area and sampling sites are extremely interesting because they constitute a sort of natural dose-dependent open-sea laboratory where the heavy metal concentration decreases going from the head of the Sørfjord, where first zinc then titanium/iron smelters are located, to the open sea. In fact, studies on the accumulation of Cd, Pb, Zn, and Cu in the sea star *Asteria rubens*, living along the natural gradient, demonstrated a direct relationship with the level of contamination of the environment, at least for Cd and Pb (Coteur et al. 2003). We then tested hsp70 expression in the same samples and found again an increase in levels of the protein, which also correlated with the levels of heavy metals measured in animal tissues (not shown). This result, demonstrating hsp70 as a marker of stress for another class of echinoderms, reinforces its efficiency in pollution monitoring.

8
Concluding Remarks

Our current understanding of inflammatory and immune responses in humans probably has its roots in the comparative approach to immunology. For example, in the late 1900s, research on echinoderms provided initial evidence of the importance of phagocytic cells in reactions to foreign material. Over the last 40 years it has been widely documented that sea urchin coelomocytes, the echinoderm immune effector cells, mediate cellular responses to danger through chemotaxis, phagocytosis, encapsulation and cytotoxicity. More recently, the production of a variety of proteins, whose genes are activated by an efficient responsive system, has been reported. However, despite recent advances in molecular and cellular biology, a clear understanding of the molecular mechanisms operating in echinoderms' response to damage/stress is still lacking. This is partially due to the small number of reports, often contradictory, which makes interpretation of results a very difficult task. The possibility of using the sea urchin coelomocyte as a stress model under controlled laboratory conditions was first proposed by Koros (1993) and recently described by Matranga (Matranga et al. 2000; Matranga 2002). Their use has been extended to in vitro short-term cultures and field studies, which made coelomocytes from *P. lividus*, and possibly from other echinoderm species, a good tool for the assessment of molecular and cellular alterations in response to chemical and physical stresses. The introduction of novel techniques for the assessment of contamination in the marine environment is one of the top priorities within the European Community. Efforts to identify and apply "early" stress markers, long before pathological conditions such as diseases, mortality, or population changes occur, are greatly awaited. Moreover, recent evidence has suggested that the use of a battery of biomarkers, being more advantageous than the use of a single biomarker, offers an effective early warning system in biomonitoring aquatic environments (Ringwood et al. 1999). To this end, it would be interesting to develop a panel of sensible molecular stress markers expressed by sea urchin coelomocytes maintained in large plants or large-scale facilities where different stress conditions might be tested.

Marine invertebrates rely on innate immunity, the study of which has recently grown in higher vertebrates. It is now believed that the Toll-like receptors, discovered by comparative immunological studies, are responsible for a large proportion of the innate immune recognition of pathogens, sensing the "pathogen-associated molecular patterns" and/or providing the "danger signal", as speculated by Janeway and Matzinger, respectively. It is stimulating that the most abundant class of sea urchin coelomocyte, namely phagocytes, has a striking similarity in morphology, and probably in function, to dendritic cells, the immune cells present in the mammalian system in those tissues that are in contact with the environment (skin, nose, lungs, stomach and intestine). Recently described endogenous "danger signals", released by tissues undergoing stress, damage or abnormal death, and necessary for the activation of dendritic cells (a necessary step for the initiation of primary and secondary immune responses in humans), are: heat-shock proteins, nucleotides, reactive oxygen intermediates, extracellular-matrix breakdown products, neuromediators and cytokines like the IFNs (Gallucci and Matzinger 2001). The fact that most of these molecules have been found in sea urchin coelomocytes in response to physical and chemical stress once again demonstrates the suitability of this model system to sense pollution and the interest in this field for the exploitation of new biotechnological tools.

Acknowledgements. The authors wish to thank Dr. F. Zito for stimulating discussions and critical reading of the manuscript and Mr. Davide Comacchio for technical assistance and help in maintenance of sea urchins in aquaria. This work was fully supported by the UVTOX EU project, contract no. EVK3-CT-1999-00005. Partial support from ASI project no. I/R/073/01 and ICRAM project REDCOD no. B4-3070/2003/368585/SUB/D3 is also acknowledged. A.N. and R.B. were supported by the Italian Ministry for Education, University and Research.

References

Al-Sharif WZ, Suyer JO, Lambris JD, Smith LC (1998) Sea urchin coelomocytes specifically express a homologue of the complement component C3. J Immunol 160:2983–2997

Angelini C, Amaroli A, Falugi C, Di Bella G, Matranga V (2003) Acetylcholinesterase activity is affected by stress conditions in *Paracentrotus lividus* coelomocytes. Mar Biol 143:623–628

Bachmann S, Goldschmid A (1978) Fine structure of the axial complex of *Sphaerechinus granularis* (Lam.) (Echinodermata: Echinoidea). Cell Tissue Res 193:107–123

Beck G, Habich GS (1986) Isolation and characterization of a primitive interleukin-1-like protein from an invertebrate, *Asterias forbesi*. Proc Natl Acad Sci USA 83:7429–7433

Beck G, Habich GS (1996) Characterization of an IL-6-like molecule from an echinoderm (*Asterias forbesi*). Cytokine 8:507–512

Beck G, Ellis TW, Habich GS, Schluter SF, Marchalonis JJ (2002) Evolution of the acute phase response: iron release by echinoderm (*Asterias forbesi*) coelomocytes, and cloning of an echinoderm ferritin molecule. Dev Comp Immunol 26:11–26

Bertheussen K (1981) Endocytosis by echinoid phagocytosis in vitro. I. Recognition of foreign matter. Dev Comp Immunol 5:241–250

Bertheussen K, Seljelid R (1978) Echinoid phagocyte in vitro. Exp Cell Res 111:401–412

Bonaventura R, Poma V, Costa C, Matranga V (2005) UVB prevents skeleton growth and simulates the expression of stress markers in sea urchin embryo. Biophys Biochem Res Comm 328:150–157

Burke RD (1999) Invertebrate integrins: structure, function, and evolution. Int Rev Cytol 191:257–284

Burke RD, Watkins RF (1991) Stimulation of starfish coelomocytes by interleukin-1. Biochem Biophys Res Commun 180:579–584

Canicattì C, Rizzo A (1991) A 220 kDa coelomocyte aggregating factor involved in *Holothuria polii* cellular clotting. Eur J Cell Biol 56:79–83

Canicattì C, Pagliara P, Stabili L (1992) Sea urchin coelomic fluid agglutinin mediates coelomocyte adhesion. Eur J Cell Biol 58:291–295

Cervello M, Arizza V, Lattuca G, Parrinello N, Matranga V (1994) Detection of vitellogenin in a subpopulation of sea urchin coelomocytes. Eur J Cell Biol 64:314–319

Christensen AB, Colacino JM, Bonaventura C (2003) Functional and biochemical properties of the hemoglobins of the burrowing brittle star *Hemipholis elongata* say (Echinodermata, Ophiuroidea). Biol Bull 205:54–65

Clow LA, Raftos DA, Gross PS, Smith LC (2004) The sea urchin complement homologue, SpC3, functions as an opsonin. J Exp Biol 207:2147–2155

Coteur G, Pernet P, Gillan D, Joly G, Maage A, Dubois P (2003) Field contamination of the starfish *Asterias rubens* by metals. Part 1: short- and long-term accumulation along a pollution gradient. Environ Toxicol Chem 22:2136–2144

D'Andrea L, Danon MA, Sgourdas GP, Bonder EM (1994) Identification of coelomocyte unconventional myosin and its association with in vivo particle/vesicle motility. J Cell Sci 107:2081–2094

Edds KT (1985) Morphological and cytoskeletal transformation in sea urchin coelomocytes. In: Cohen WD (ed) Blood cells of marine invertebrates: experimental systems in cell biology and comparative physiology. AR Liss, New York, pp 53–74

Edds KT (1993) Effects of cytochalasin and colcemid on cortical flow in coelomocytes. Cell Motil Cytoskeleton 26:262–273

Epel D, Hemela K, Schick M, Patton C (1999) Development in the floating world: defenses of eggs and embryos against damage from UV radiation. Am Zool 39:271–278

Falugi C (1985) Histochemical localization of acetylcholinesterase in blood cells. Basic Appl Histochem 29:105–113

Gallucci S, Matzinger P (2001) Danger signals: SOS to the immune system. Curr Opin Immunol 13:114–119

Geraci F, Pinsino A, Turturici G, Savona R, Giudice G, Sconzo G (2004) Nickel, lead, and cadmium induce differential cellular responses in sea urchin embryos by activating the synthesis of different HSP70s. Biochem Biophys Res Commun 322:873–877

Gerardi P, Lassegues M, Canicattì C (1990) Cellular distribution of sea urchin antibacterial activity. Biol Cell 70:153–157

Giudice G (1989) Heat shock proteins in sea urchin embryos. Dev Growth Differ 31:103–106

Glinski Z, Jarosz J (2000) Immune phenomena in echinoderms. Arch Immunol Ther Exp 48:189–193

Goldschmidt-Clermont P, Machesky L, Doberstein S, Pollard T (1991) Mechanism of the interaction of human platelet profilin with actin. J Cell Biol 113:1081–1089

Gross PS, Clow LA, Smith LC (2000) SpC3, the complement homologue from the purple sea urchin, *Strongylocentrotus purpuratus*, is expressed in two subpopulations of the phagocytic coelomocytes. Immunogenetics 51:1034–1044

Hansen MB, Nielsen SE, Berg K (1989) Re-examination and further development of a precise and rapid dye method for measuring cell growth/cell kill. J Immunol Methods 119:203–210

Harrison PK, Falugi C, Angelini C, Whitaker MJ (2002) Muscarinic signalling affects intracellular calcium concentration during the first cell cycle of sea urchin embryos. Cell Calcium 31:289–297

Henson JH, Nesbitt D, Wright BD, Scholey JS (1992) Immunolocalisation of kinesin in sea urchin coelomocytes: association of kinesis with intracellular organelles. J Cell Sci 103:309–320

Henson JH, Svitkina TM, Burns AR, Hughes HE, MacPartland KJ, Nazarian R, Borisy GG (1999) Two components of actin-based retrograde flow in sea urchin coelomocytes. Mol Biol Cell 10:4075–4090

Henson JH, Kolnik SE, Fried CA, Nazarian R, McGreevy J, Schulberg KL, Detweiler M, Trabosh VA (2003) Actin-based centripetal flow: phosphatase inhibition by calyculin-A alters flow pattern, actin organization, and actomyosin distribution. Cell Motil Cytoskeleton 56:252–266

Hillier BJ, Vacquier VD (2003) Amassin, an olfactomedin protein, mediates the massive intercellular adhesion of sea urchin coelomocytes. J Cell Biol 160:597–604

Holy J (2000) Intermediate filament proteins in echinoderm coelomocytes. Comp Biochem Physiol B Biochem Mol Biol 127:491–504

Johnson PT (1969) The coelomic elements of sea urchins (*Strongylocentrotus*). 3. In vitro reaction to bacteria. J Invertebr Pathol 13:42–62

Kobayashi N (1980) Comparative sensitivity of various developmental stages of sea urchins to some chemicals. Mar Biol 58:163–171

Koros AMC (1993) Neuroendocrine marker expression on sea urchin coelomocytes and other immunoregulatory parameters may be used to monitor environmental changes. Mar Environ Res 35:137–140

Koros AMC, Goodwin DG, Siderits RH, Malavasi F (2000) "Natural Killer" (NK) cell antigens, CD 56, CD57, and others are expressed on breast and lung tumor cells as well as sea urchin coelomocytes. Tissue Antigens 55:77–78

Koziol C, Batel R, Arinc E, Schröder HC, Müller WEG (1997) Expression of the potential biomarker heat shock protein 70 and its regulator, the metazoan DnaJ homolog, by temperature stress in the sponge *Geodia cydonium*. Mar Ecol Prog Ser 154:261–268

Leclerc M, Bajelan M (1992) Homologous antigen for T cell receptor in axial organ cells from the asterid *Asterias rubens*. Cell Biol Int Rep 16:487–490

Le Marrec-Croq F, Glaise D, Guguen-Guillouzo C, Chesne C, Guillouzo A, Boulo V, Dorange G (1999) Primary cultures of heart cells from the scallop *Pecten maximus* (mollusca-bivalvia). In Vitro Cell Dev Biol Anim 35:289–295

Lesser MP, Kruse VA, Barry TM (2003) Exposure to ultraviolet radiation causes apoptosis in developing sea urchin embryos. J Exp Biol 206:4097–4103

Lin W, Zhang H, Beck G (2001) Phylogeny of natural cytotoxicity: cytotoxic activity of coelomocytes of the purple sea urchin, *Arbacia punctulata*. J Exp Zool 290:741–750

Matranga V (1996) Molecular aspects of immune reactions in Echinodermata. In: Müller WEG, Rinkevich B (eds) Invertebrate immunology, vol 15. PMSB Series. Springer, Berlin Heidelberg New York, pp 235–247

Matranga V, Bonaventura R (2002) Sea urchin coelomocytes, the progenitors of vertebrate immune effectors, as bio-indicators of stress and pollution. In: Yokota Y, Matranga V, Smolenicka Z (eds) The sea urchin: from basic biology to aquaculture. Swets and Zeitlinger, Lisse, The Netherlands, pp 161–176

Matranga V, Kuwasaki B, Noll H (1986) Functional characterization of toposomes from sea urchin blastula embryos by a morphogenetic cell aggregation assay. EMBO J 5:3125–3132

Matranga V, Toia G, Bonaventura R, Muller WEG (2000) Cellular and biochemical responses to environmental and experimentally induced stress in sea urchin coelomocytes. Cell Stress Chaperones 5:158–165

Matranga V, Bonaventura R, Di Bella G (2002) hsp70 as a stress marker of sea urchin coelomocytes in short term cultures. Cell Mol Biol 48:345–359

Matzinger P (2002) The danger model: a renewed sense of self. Science 296:301–305

McDonald GD, Davidson L, Kitto GB (1992) Amino acid sequence of the coelomic C globin from the sea cucumber *Caudina* (*Molpadia*) *arenicola*. J Protein Chem 11:29–37

Medzhitov R, Janeway CA Jr (2002) Decoding the patterns of self and nonself by the innate immune system. Science 12:298–300

Medzhitov R, Preston-Hurlburt P, Janeway CA Jr (1997) A human homologue of the *Drosophila* Toll protein signals activation of adaptive immunity. Nature 388:394–397

Millott N (1969) Injury and the axial organ of echinoids. Experientia 25:756

Morale A, Coniglio L, Angelini C, Cimoli G, Bolla A, Alleteo D, Russo P, Falugi C (1998) Biological effects of a neurotoxic pesticide at low concentrations on sea urchin early development. A teratogenic assay. Chemosphere 37:3001–3010

Mosman T (1983) Rapid colorimetric assay for cellular growth and cytotoxicity assays. J Immunol Methods 65:55

Müller WEG, Batel R, Lacorn M, Steinhart H, Simat T, Lauenroth S, Hassanein HMA, Schröder HC (1998) Accumulation of cadmium and zinc in the marine sponge *Suberites domuncula* and its potential consequences on single-strand breaks and on expression of heat shock protein: a natural field study. Mar Ecol Prog Ser 167:127–135

Multerer KA, Smith LC (2004) Two cDNAs from the purple sea urchin, *Strongylocentrotus purpuratus*, encoding mosaic proteins with domains found in factor H, factor I, and complement components C6 and C7. Immunogenetics 56:89–106

Ozretic B, Krajnovic-Ozretic M (1985) Morphological and biochemical evidence of the toxic effect of pentachlorophenol on the developing embryos of the sea urchin. Aquat Toxicol 7:255–263

Pennec JP, Gallet M, Gioux M, Dorange G (2002) Cell culture of bivalves: tool for the study of the effects of environmental stressors. Cell Mol Biol (Noisy-le-grand) 48:351–358

Prendergast RA, Lutty GA, Scott AL (1983) Directed inflammation: the phylogeny of lymphokines. Dev Comp Immunol 7:629–632

Ringwood AH, Hameedi MJ, Lee RF, Brouwer M, Peters EC, Scott GI, Luoma SN, Digiulio RT (1999) Bivalve biomarker workshop: overview and discussion summaries. Biomarkers 4:391–399

Roccheri MC, Agnello M, Bonaventura R, Matranga V (2004) Cadmium induces the expression of specific stress proteins in sea urchin embryos. Biophys Biochem Res Commun 321:80–87

Russo R, Bonaventura R, Zito F, Schröder HC, Müller I, Müller WEG, Matranga V (2003) Stress to cadmium monitored by metallothionein gene induction in *Paracentrotus lividus* embryos. Cell Stress Chaperones 8:232–241

Samali A, Cotter TG (1996) Heat shock proteins increase resistance to apoptosis. Exp Cell Res 223:163–170

Schröder HC, Hassanein HMA, Lauenroth S, Koziol C, Mohamed TAAA, Lacorn M, Steinhart H, Batel R, Müller WEG (1999) Induction of DNA strand breaks and expression of hsp70 and GRP78 homolog by cadmium in the marine sponge *Suberites domuncula*. Arch Environ Contam Toxicol 36:47–55

Sconzo G, Scardina G, Ferraro MG (1992) Characterization of a new member of the sea urchin *Paracentrotus lividus* hsp 70 gene family and its expression. Gene 121:353–358

Sconzo G, Romancino D, Fasulo G, Cascino D, Giudice G (1995) Effect of doxorubicin and phenytoin on sea urchin development. Pharmacie 50:616–619

Shah M, Brown KM, Smith LC (2003) The gene encoding the sea urchin complement protein, SpC3, is expressed in embryos and can be upregulated by bacteria. Dev Comp Immunol 27:529–538

Smith LC, Britten RJ, Davidson EH (1992) SpCoel1: a sea urchin profilin gene expressed specifically in coelomocytes in response to injury. .Mol Biol Cell 3:403–414

Smith SL, Smith AC (1985) Sensitization and histamine release by cells of the sand dollar, *Mellita quinquiesperforata*. Dev Comp Immunol 9:597–603

Smith VJ (1981) The echinoderms. In: Ratcliffe NA, Rowley AT (eds) Invertebrate blood cells, vol 2. Academic Press, London

Unuma T, Okamoto H, Konishi K, Ohta H, Mori K (2001) Cloning of cDNA encoding vitellogenin and its expression in red sea urchin, *Pseudocentrotus depressus*. Zool Sci 18:559–565

Yokoe H, Anholt RR (1993) Molecular cloning of olfactomedin, an extracellular matrix protein specific to olfactory neuroepithelium. Proc Natl Acad Sci USA 90:4655–4659

Yui MA, Bayne CJ (1983) Echinoderm immunology: bacterial clearance by the sea urchin *Strongylocentrotus purpuratus*. Biol Bull 165:473–486

DNA Damage and Developmental Defects After Exposure to UV and Heavy Metals in Sea Urchin Cells and Embryos Compared to Other Invertebrates

H.C. Schröder, G. Di Bella, N. Janipour, R. Bonaventura, R. Russo, W.E.G. Müller, V. Matranga

Abstract. The depletion of the stratospheric ozone layer and the resulting increase in hazardous ultraviolet-B (UV-B) radiation reaching the Earth are of major concern not only for terrestrial but also for aquatic organisms. UV-B is able to penetrate clear water to ecologically significant depths. This chapter deals with the effects of UV radiation on DNA integrity in marine benthic organisms, in particular sea urchins in comparison to other marine invertebrates (sponges and corals). These animals cannot escape the damaging effects of UV-B radiation and may be additionally exposed to pollution from natural or anthropogenic sources. Besides eggs and larvae that lack a protective epidermal layer and are particularly prone to the damaging effects of UV radiation, coelomocytes from the sea urchin *Paracentrotus lividus* were used as a "cellular sensor" to analyse the effects on DNA caused by UV-B, heavy metals (cadmium), and their combined actions. From our data we conclude that sea urchin coelomocytes as well as cells from other marine invertebrates are useful bioindicators of UV-B and heavy metal stress, responding to these stressors with different extents of DNA damage.

H.C. Schröder (e-mail: hschroed@uni-mainz.de), N. Janipour, W.E.G. Müller
Institut für Physiologische Chemie, Abteilung Angewandte Molekularbiologie, Johannes Gutenberg-Universität, Duesbergweg 6, 55099 Mainz, Germany
G. Di Bella, R. Bonaventura, R. Russo, V. Matranga
Istituto di Biomedicina e Immunologia Molecolare (IBIM) "Alberto Monroy", Consiglio Nazionale delle Ricerche, Via U. La Malfa 153, 90146 Palermo, Italy

Progress in Molecular and Subcellular Biology
Subseries Marine Molecular Biotechnology
V. Matranga (Ed.), Echinodermata

1 Introduction

It is now well established that the thinning of the stratospheric ozone (O_3) layer results in higher levels of mid-ultraviolet (UV-B) radiation reaching the Earth's surface (Smith et al. 1992; El-Sayed et al. 1996). The biologically damaging effects of decreased UV-B absorption in the stratosphere are especially pronounced in the Antarctic where the decline in ozone levels (the so-called ozone hole) amounts to more than 50 % (Gleason et al. 1993). However, reduction of ozone concentration is also occurring at temperate and tropical latitudes (Stolarski et al. 1992).

1.1 Atmospheric Ozone

The UV waveband of the electromagnetic spectrum is subdivided into three regions: UV-A (320–400 nm), UV-B (280–320 nm), and UV-C (200–280 nm). The biological effects of UV radiation strongly depend on wavelength. In general, shorter-wavelength UV radiation (<320 nm) is more active than UV radiation at longer wavelength.

Atmospheric O_3 is created in the stratosphere at altitudes between about 25 and 100 km; it selectively absorbs UV radiation with different efficiency depending on wavelength (Molina and Molina 1986). The most damaging UV-C radiation is completely absorbed before reaching the Earth's surface, while UV-A and visible light (spectral range 400–700 nm) are not significantly absorbed by the ozone layer. The release of chlorofluorocarbons and other anthropogenic chemicals, including nitrous oxide and bromine-containing halons, is causing a degradation of the ozone layer, resulting in an increase in the intensity, in particular of the short-wavelength UV-B radiation penetrating the Earth's atmosphere. In contrast to UV-B, the solar spectrum in the UV-A and visible range is only little or unchanged by ozone depletion (Cadwell et al. 1986).

The intensity of biologically damaging UV radiation at the Earth's surface depends on a variety of factors including cloud cover, sun angle, atmospheric aerosols and seasonal and geographical factors (Frederick et al. 1989). In general, a strong seasonal variation is observed in temperate regions, while seasonal dependence is much less near the equator.

1.2
Penetration of Solar UV Radiation in Marine Waters

In seawater, sunlight is attenuated by absorption and scattering with increasing depth (Jerlov 1968; Dustan 1982). In parallel, a change in the spectral composition of the light is observed. The attenuation of visible light is higher at the wavelengths at either end of its spectral range. In the UV range, shorter wavelengths (UV-B and UV-C) are more attenuated than longer wavelengths (UV-A). Nevertheless, UV-B light still penetrates in clear seawater up to 30 m (Smith et al. 1992) or more (about 50 m in the Antarctica; Smith et al. 1992) and harmful effects can still be measured at depths up to 20–25 m (Smith and Baker 1979, 1981; Karentz and Lutze 1990; Regan et al. 1992; Smith et al. 1992; Prézelin et al. 1994). The penetration of UV-B is, however, influenced by several constituents of the seawater, in particular by dissolved organic carbon (Baker and Smith 1982; Scully and Lean 1994; Crump et al. 1999), the amount of which is higher in coastal waters than the open sea.

1.3
Effects of Solar UV Radiation on Aquatic Organisms

In humans, increased UV-B exposure may cause skin cancer, eye damage, and changes in the immune system (Griffiths et al. 1998). However, deleterious effects of UV-B radiation are also observed in aquatic organisms and ecosystems (Häder et al. 1995). There are many studies concerning the effects of UV-B irradiation on phytoplanktonic organisms (Helbling et al. 1992). However, zooplankton and benthic invertebrates are also prone to UV-B damage (Hunter et al. 1979; Damkaer et al. 1980; Karanas et al. 1981; Malloy et al. 1997). In particular, sessile species such as sponges and corals are sensitive to the adverse effects caused by UV-B radiation (reviewed in Acevedo and Nolan 1993). These animals do not possess a UV-B-adsorbing epidermal layer like higher plants and animals and cannot avoid the harmful effects of solar radiation.

The most serious effects of UV-B radiation are observed in those organisms that are exposed to sunlight during their developmental stages. Eggs and larvae residing in surface waters may receive high levels of UV-B, causing impaired development (Hunter et al. 1979; Damkaer et al. 1981; Damkaer and Dey 1983; Ringelberg et al. 1984; El-Sayed 1988; Jeffery 1990; Shick et al. 1991; Bothwell et al. 1993; Blaustein et al. 1994; Little and Fabacher 1994; Häder et al. 1995).

Protection from harmful UV-B radiation can be achieved by behavioural (e.g. the daily vertical migration of planktonic organisms) or physiological (e.g. development of shells, enhanced melanin production, or induction of antioxidant defense and repair processes) adaptation mechanisms (El-Sayed 1988; Vincent and Roy 1993; Hessen 1994).

1.4 DNA Damage

The deleterious effects of UV-B on marine organisms are caused by photochemical absorption of radiation by macromolecules. UV-B induces direct damage to DNA, proteins, and some other molecules that have absorbance maxima in the UV-B region. The most important macromolecule affected by UV-B radiation is DNA (Friedberg et al. 1995). The absorbing centers within the DNA are the purine and pyrimidine bases, which have absorption maxima at 260–265 nm. The absorption maxima of proteins in the UV-B and UV-C regions are around 280 nm. UV-A radiation causes indirect damage to DNA via formation of oxygen and hydroxyl radicals (Fridovich 1986; Peak and Peak 1989; Karentz et al. 1994). These reactive oxygen species (ROS) interact with DNA to form strand breaks, alkali-labile sites and DNA protein cross-links. The major lesions induced by UV-B are DNA photoproducts [cyclobutane pyrimidine dimers and pyrimidine(6–4)pyrimidone dimers]. The most common lesion caused by UV-A is the formation of 8-hydroxydeoxyguanosine.

1.5 DNA Repair

DNA lesions could be cytotoxic, by inhibiting DNA replication and transcription, or cause mutations, which are possibly harmful for living organisms. Therefore several mechanisms have evolved to repair damaged DNA. There are two major DNA repair mechanisms: photoreactivation and nucleotide excision repair. Photoreactivation repair of UV-induced cyclobutane pyrimidine dimers is a photoenzymatic process consisting of two steps: binding of a photoreactivating enzyme (photolyase) to the dimers and subsequent repair and dissociation of the complex after irradiation with visible light. Aquatic organisms may be highly sensitive to exposure to UV-B in the absence of visible light (Hirosawa and Miyachi 1983; Buhlmann et al. 1987). Nucleotide excision repair involves a series of steps, including recognition of the defective site, incision of the damaged DNA strand at or near the lesion, subsequent excision and synthesis of the DNA around the damaged site using the complementary intact strand as a matrix, and, finally, ligation of the remaining nick in the repaired strand (Sancar 1996a). DNA excision repair does not depend on light.

1.6 Solar UV Radiation and Evolution

The formation of the stratospheric ozone layer was a result of the oxygenation of the atmosphere due to evolution of photosynthesis. In the period before the

"Cambrian explosion", there were only low levels of oxygen in the atmosphere (Cloud 1968; Canfield and Teske 1996). Consequently, during this period, a less dense ozone shield protected living organisms from UV radiation than exists at present (Loomis 1988; Towe 1994; Brasier 2000). This suggests that living organisms were exposed to high levels of UV radiation. Therefore, it can be deduced that the oldest still extant Metazoa, the sponges (Porifera), which developed about 800 million years ago, must have acquired efficient protection systems to resist those environmental effects (Müller and Müller 1998); this assumption has been confirmed by experimental results (see Sect. 2.9.). The absence of a protective ozone layer is also considered to be one major reason for the confinement of life to the ocean during that time period (Fisher 1965).

1.7 Heavy Metals

Cadmium and some other heavy metals are natural constituents of seawater but can also be considered as conservative pollutants. Increased concentrations of these metals may be present in marine organisms due to bioaccumulation. Therefore, marine contamination by even low concentrations of heavy metals can disturb marine ecosystems (Radenac et al. 2001). The toxic as well as the potential mutagenic and teratogenic effects caused by cadmium have been described in many organisms (Flick et al. 1971), including marine invertebrates (Schröder et al. 1999b). They involve the production of ROS, DNA strand breaks, and inhibition of DNA synthesis and repair enzymes (Hartwig 1994; Schröder et al. 1999b) Studies on ROS production in response to environmental factors and contaminants, such as heavy metals and polychlorinated biphenyls (PCBs), by echinoderm immune effector cells are reviewed by Coteur et al. (this Vol.). In particular, marine sponges are able to accumulate high amounts of cadmium (Müller et al. 1998). Also, organic derivatives of heavy metals often found in the marine environment, e.g. tributyltin (TBT), may cause damage to marine invertebrates (Batel et al. 1993). Similarly, studies on the accumulation of cadmium (Radenac et al. 2001; Coteur et al. 2003) and PCBs (Noblet et al. 2003) have been reported for asteroids and echinoids.

2 Marine Invertebrates as Bioindicators

Until recently, research on biological effects of UV-B radiation in the marine environment primarily focused on phytoplankton. However, marine invertebrates are highly sensitive to physical and/or chemical environmental stressors and, consequently, the analysis of their molecular responses to UV-B has gained increasing attention. In addition, these animals developed various protective mechanisms against such stressors; e.g. by induction of heat shock proteins (Koziol et al. 1996; Müller et al. 1998; Matranga et al. 2000). A few stress proteins have been used as tools to detect marine environmental pollution using marine invertebrates as bioindicators, including HSP70 and metallothioneins (Lyons-Alcantara et al. 1998; Müller et al. 1995, 1998; Schröder et al. 1999a,b; Cajaraville et al. 2000; Matranga et al. 2000, 2002; Matranga and Bonaventura 2002; Roccheri et al. 2004).

2.1 Sea Urchins

Sea urchins have been proposed to represent valuable bioindicator organisms to detect environmental perturbations (Dinnel et al. 1988; Sconzo et al. 1995; Morale et al. 1998; Matranga et al. 2000). These animals are exposed to solar UV radiation during all stages of their life cycle. In particular, their (transparent) embryos are poorly protected against irradiance by UV light.

Although DNA repair enzymes (e.g. photolyases) protect sea urchins, UV radiation may cause direct damage to DNA by formation of pyrimidine dimers or cross-links between DNA and protein. As a consequence, delayed cell divisions, developmental delays, and abnormalities may occur. In a European Union project ("UVTOX"), sea urchin coelomocytes have recently been used as a novel biosensor of the effects caused by UV-B and heavy metals (cadmium) and their combinations (Schröder et al. 2000a).

2.2 Sea Urchin Coelomocytes

The coelomocytes are immune effector cells of sea urchins, which are contained in the coelomic cavity of the animals. These cells were found to strongly respond to a number of environmental and experimentally induced stresses (Matranga et al. 2000, 2002; Matranga and Bonaventura 2002; see also Matranga et al., this Vol.).

In the experiments described below, the sea urchin *P. lividus* was investigated. The coelomocytes were obtained as a total cell population by bleeding the animals through a cut in the peristomal membrane. The coelomic fluid was poured into an ISO-EDTA anti-coagulant solution. The resulting cell suspension was divided into several Petri dishes. The coelomocytes were harvested in CCM (coelomocytes culture medium) (Henson et al. 1992), kept in Petri dishes, and then exposed to UV or metal stress. Under long-term culture conditions using a modified medium (Le Marrec et al. 1999), the cells tend to form bundles and fibres (see Matranga et al., this Vol., Figs. 4, 5).

2.3 Assay for DNA Integrity

One achievement of the European Union UVTOX project was the introduction of novel techniques for the assessment of the effects of increased UV-B exposure, due to ozone depletion, in marine invertebrates without or in combination with stress by pollution (Schröder et al. 2000a). Analysis of DNA damage and repair was carried out using a quick and sensitive fluorometric microplate assay, Fast Micromethod (Batel et al. 1999; Schröder et al. 2004). This technique measures the rate of unwinding of cellular DNA upon exposure to alkaline conditions using a fluorescent dye, which preferentially binds to double-stranded DNA. The sensitivity of this assay is similar or even superior to Comet Assay (Bihari et al. 2002). The Fast Micromethod has major advantages compared to other tests for DNA damage: the assay is performed in 96-well microplates within 3 h or less; it requires only minimal amounts of material (30 ng of DNA per well; $\sim 3\times 10^3$ cells or ~25 μg of tissue). The assay can also be applied to frozen tissue samples.

The procedure applied in the experiments described below was as follows. The cells (coelomocytes cultures, sponge primmorphs, or other cells) or tissues under study were exposed to solar UV irradiation, UV-B light or heavy metal stress (or combinations of UV and heavy metals). After stress exposure, the cells were applied to a black microplate; then lysing solution containing PicoGreen was added. The lysing solution consisted of 4.5 M (for coelomocytes) or 9 M (for sponge cells) urea, 0.1 % SDS, 0.2 M EDTA, pH 10. Lysis of cells was allowed for 1 h in the dark. DNA unwinding was started by addition of 250 μl of 0.025 M NaOH to each well. Measurements were performed for up to 1 h at 480 nm excitation and 520 nm emission using a Fluoroskan II fluorescence ELISA plate reader. Effects were expressed as strand scission factor (SSF), which was calculated after 20 min of denaturation, as follows: SSF=log (% dsDNA in sample/% dsDNA in control). Negative values for SSFs are indicative of increased frequencies of strand breaks.

2.4 Radiation Source

2.4.1 Full Solar Spectrum Lamp

A full solar spectrum lamp was used, which allows sunlight simulation (SOL 500; Dr. Hönle AG, Planegg, Germany) with filters H1 (wavelengths $\geq$ 320 nm; UV-A and visible light) or H2 (wavelengths $\geq$ 295 nm; UV-B, UV-A, and visible light). The cell concentration was 2×10^4 cells/ml. The energy output of the lamp was monitored with a UVM-CP radiometer (with A and B sensor; UV-Consulting Peschl, Mainz, Germany).

2.4.2
UV-B Lamp

Irradiation was performed using a monochromatic UV-B lamp (VL215.M from Vilber Lourmat with a peak emission at 312 nm, L+F-Labortechnik, Wasserburg). For elimination of UV-C (<290 nm) the light was filtered through cellulose diacetate (Kodacel; Eastman Kodak, Rochester).

2.5
DNA Damage in Coelomocytes

Sea urchin coelomocytes were used as a bioindicator to study the effects of exposure to (solar) UV-A and UV-B radiation, heavy metals (cadmium chloride), and combinations on the occurrence of DNA strand breakage using the Fast Micromethod.

2.5.1
UV Radiation

Irradiation of sea urchin coelomocytes using the full solar spectrum lamp (SOL 500) at wavelengths ≥320 nm (filter H1) or ≥295 nm (filter H2) resulted in a dose-dependent increase in DNA damage, as determined by Fast Micromethod (Fig. 1). The increase in frequency of DNA single-strand breaks (expressed as SSF) was much higher if a filter was used that excludes only UV-C but not UV-B (filter H2). In the experiment shown in Fig. 1, the cells were irradiated for the same time periods (10 and 60 s) with filters H1 and H2, and the UV-A (filter H1) and UV-B (filter H2) doses were determined (UV-radiometer with a UV-A and UV-B sensor respectively). Interestingly, the SSF

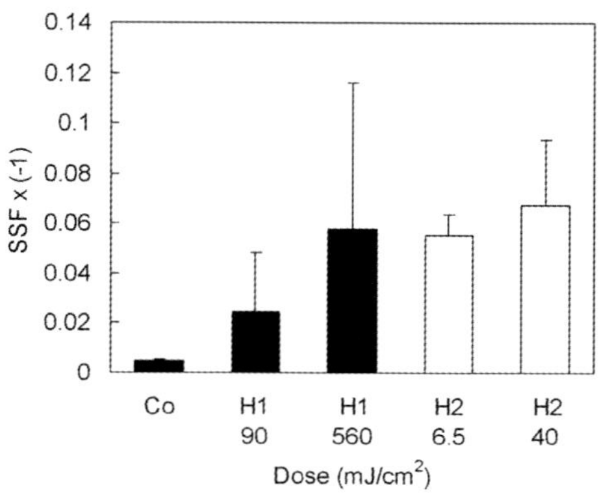

Fig. 1. DNA damage induced by irradiation of sea urchin coelomocytes using a SOL 500 lamp with filter H1 (UV-A and visible light; wavelengths ≥320 nm) or filter H2 (UV-B, UV-A, and visible light; wavelengths ≥ 295 nm). DNA single-strand breaks were determined using Fast Micromethod and are expressed as strand scission factor (*SSF*) value. Control (*Co*): non-irradiated cells

values were found to increase after incubation of the cells following irradiation due to the formation of transient breaks during repair of DNA damage (see Figs. 3 and 4).

Similar results were obtained after irradiation of coelomocytes using a monochromatic (312 nm) UV-B lamp (results not shown).

2.5.2 *Cadmium*

The effects of mixed pollution on marine invertebrates have been intensively studied in the past (Müller et al. 2000), but less attention has been paid to the combined effects of exposure to UV radiation and environmental pollutants. This problem has become increasingly important due to depletion of the UV-B protective ozone layer. Therefore, we determined the effects of cadmium (as a model pollutant) and its combined effects with UV-B (see Sect. 2.5.3) on DNA integrity in sea urchin coelomocytes.

Analysis of DNA damage in coelomocytes treated with various concentrations of cadmium chloride (10^{-6} to 10^{-12} M) for 2 h revealed a dose-dependent effect of the heavy metal on frequency of single-strand breaks (Fig. 2). Even at the low concentration of cadmium chloride of 10^{-12} M, an induction of DNA strand breaks in coelomocytes was found (Fig. 2).

2.5.3 *Combined Effects*

In the following, we determined the effects of different combinations of UV and cadmium exposure on DNA integrity in sea urchin coelomocytes. Unexpectedly, we found that after exposure of the cells to radiation emitted by full

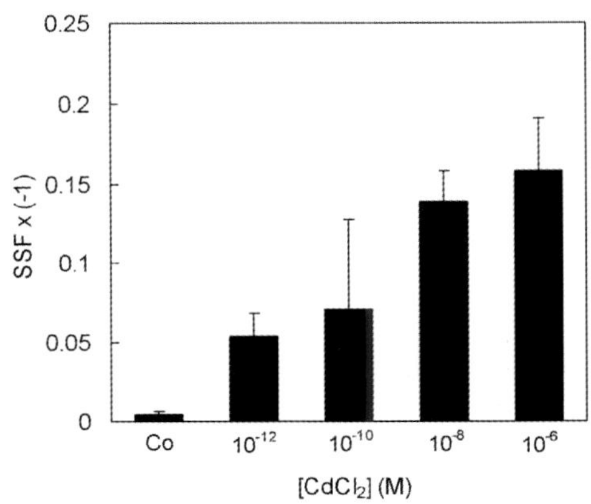

Fig. 2. Effect of sublethal concentrations of cadmium chloride on DNA integrity in sea urchin coelomocytes. The coelomocytes (4.8×10^5 cells/ml) were exposed to various concentrations of cadmium chloride (10^{-6} to 10^{-12} M) for 2 h at 16 °C. For further details, see Fig. 1

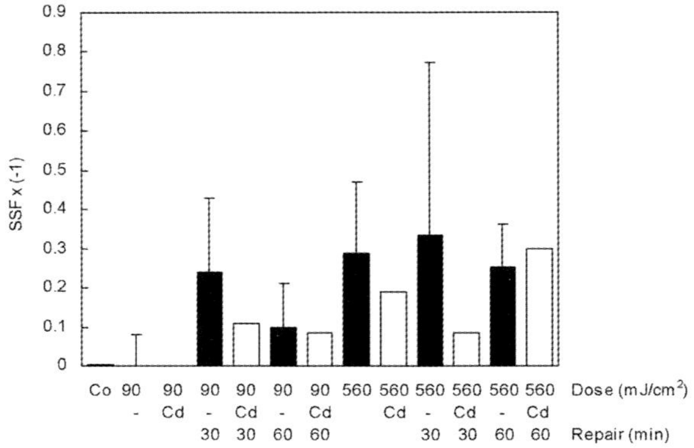

Fig. 3. Effect of cadmium on DNA repair in sea urchin coelomocytes after irradiation using a SOL 500 lamp with filter H1 (UV-A and visible light; wavelengths ≥320 nm). Coelomocytes (4.8×10^5 cells/ml) were irradiated with 90 or 560 mJ/cm² UV-A in the absence or presence of 10^{-6} M cadmium chloride; cadmium chloride was present throughout the repair period of 30 or 60 min at 16 °C. For further details, see Fig. 1

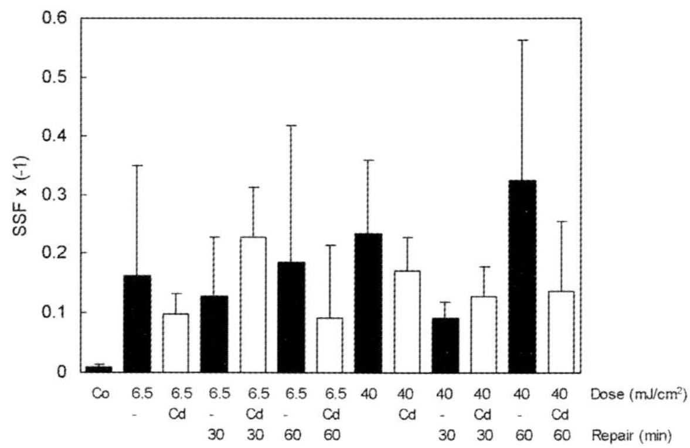

Fig. 4. Effect of cadmium on DNA repair in sea urchin coelomocytes after irradiation using a SOL 500 lamp with filter H2 (UV-B, UV-A, and visible light; wavelengths ≥295 nm). Coelomocytes (4.8×10^5 cells/ml) were irradiated with 6.5 or 40 mJ/cm² UV-B in the absence or presence of 10^{-6} M cadmium chloride; cadmium chloride was present throughout the repair periods of 30 or 60 min at 16 °C. For further details, see Fig. 1

solar spectrum lamp at wavelengths ≥320 nm (UV-A and visible light; filter H1), lower levels of DNA single-strand breaks [decrease in SSF × (–1) value] were present in coelomocytes if cadmium chloride was added during a post-irradiation repair period of 30 or 60 min at 16 °C (Fig. 3), even at a concentration (10^{-6} M) that has a strong DNA damaging effect. The same result was observed after exposure of the cells to wavelengths ≥295 nm (UV-B, UV-A, and visible light; filter H2) and subsequent recovery in the presence of 10^{-6} M cadmium chloride (Fig. 4). These effects are most likely caused by a cadmium-induced inhibition of the enzymes (endonucleases) involved in DNA excision repair occurring after UV damage of DNA (Hartwig 1994).

2.5.4
Expression of HSP70

Sea urchin coelomocytes also respond to UV-B radiation by an induction of expression of the heat shock protein HSP70 (see Matranga et al., this Vol.). Similarly, an increase in HSP70 level was observed in coelomocytes exposed to 10^{-4} M cadmium chloride (Matranga et al. 2002).

2.6
Experimental Approach

2.6.1
Effect of UV-B Radiation on P. lividus Embryo Development

Sea urchin embryos occurring in the superficial layers of seawater and lacking a protective epidermal layer are particularly prone to UV-B damage. The effects of UV radiation have been studied on larvae and early juveniles of the sea urchin *Strongylocentrotus intermedius* (Yabe et al. 1998). We used embryos of the sea urchin *P. lividus* as a cellular biosensor for monitoring the effects of UV-B and its combined effects with pollutants (cadmium).

The results showed that blastula and pluteus larvae are seriously damaged by UV-B when exposed to 500 and 1,000 J/m^2. When embryos were exposed at the mesenchyme blastula stage to different UV-B doses and observed at 68 h of development, we found an increase in the number of defective embryos, which paralleled the increase in the UV dose used. Table 1 summarizes the results obtained, taking into consideration the two major perturbed categories, i.e. gut- and skeleton-defective embryos. The typical morphologies of aberrant embryos are shown in Fig. 5, which illustrates a representative experiment; while control embryos were well-developed plutei (Fig. 5a, b), UV-exposed embryos were partially inhibited in the differentiation of their endoderm derivatives, as the three parts of the digestive apparatus failed to be properly organized (Fig. 5 c, g and d, h). Similarly, a severe inhibition of skele-

Table 1. Abnormalities in sea urchin embryos exposed to UV-B radiation. Numbers refer to the percentage of 100 embryos scored, having the following morphologies: *N* normal; *G* gut defects; *S* skeleton defects

Morphology	UV-B radiation (J/m^2)				
	0	50	300	500	1000
N	92.9	77.6	48.0	19.4	4.8
G	7.1	22.4	33.3	66.7	72.6
S	–	–	18.7	13.9	22.6

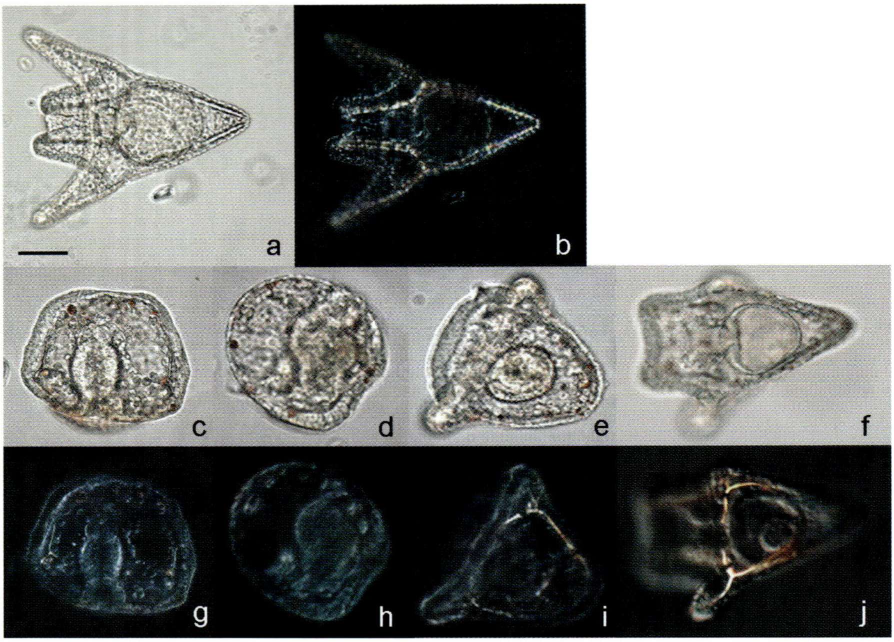

Fig. 5a–j. Morphological observation of sea urchin embryos exposed to UV-B radiation at mesenchyme blastula stage. Embryos were continuously cultured for 66 h (48 h after irradiation). Control embryos (**a**, **b**), embryos with gut defects (**c**, **d**, **g**, **h**), and embryos with skeleton defects (**e**, **f**, **i**, **j**). *Bar* 50 mm

tal patterning was observed in UV-B-exposed embryos, which showed either a failure in proper elongation or incorrect patterning of skeletal rods (Fig. 5e, i and f, j). As a consequence, UV-B-treated embryos had poorly or asymmetrically developed arms.

It has already been mentioned that UV radiation, causing DNA damage, induces the expression of genes involved in the DNA repair machinery. The human *XPB/ERCC-3* DNA excision repair gene, coding for a helicase that unwinds the DNA in 3′ to 5′ direction, is highly conserved and related genes were found in different organisms: *Drosophila melanogaster* (Mounkes et al. 1992), *Dictyostelium discoideum* (Lee et al. 1997), *Saccharomyces cerevisiae* (Park et al. 1992), and *Geodia cydonium* (Batel et al. 1998).

Since it was known that *XPB/ERCC-3* gene expression increases after UV treatment in sponges (Batel et al. 1998), it was conceivable to hypothesize its activation after UV stress also in other marine invertebrate organisms, like the sea urchin. Therefore, it was interesting to isolate the homologous cDNA in *P. lividus* embryos. To this end, we used as source total RNA from 32-cell stage embryos irradiated with UV-B, at a dose of 200 J/m^2, and kept in the dark for 2 h. The degenerated oligonucleotide primers were designed on the

```
  1 - GGTCCAGTGTGCCGAGGTATGGTGCCCAATGGCTCCAGAGTTCTTCAGAGAGTATCTGGC  - 60
       V  Q  C  A  E  V  W  C  P  M  A  P  E  F  F  R  E  Y  L  A
 61 - AATTAGGACTAGAAAGAGATTATTGCTGTATGTAATGAATCCCAATAAGTTCCGGGCATG  - 120
       I  R  T  R  K  R  L  L  L  Y  V  M  N  P  N  K  F  R  A  C
121 - TCAGTTTCTTGTGTGGTTCCACGAGCAGCGGAACGACAAGGTCATCGTCTTCTCAGATAA  - 180
       Q  F  L  V  W  F  H  E  Q  R  N  D  K  V  I  V  F  S  D  N
181 - CGTCTTTGCTCTAAAGCATTATGCAATAGCTATGGGCAGACCGTATATCTATGGGCCTAC  - 240
       V  F  A  L  K  H  Y  A  I  A  M  G  R  P  Y  I  Y  G  P  T
241 - AAGTCAAGGAGAGAGGATGCAGATCTTACAGAACTTTCAACACAACCCTGCCGTCAATAC  - 300
       S  Q  G  E  R  M  Q  I  L  Q  N  F  Q  H  N  P  A  V  N  T
301 - AATCTTCATTTCCAAGGTCGGTGATAATTCCTTTGATCTTCCCGAGGCTAATGTTCTCAT  - 360
       I  F  I  S  K  V  G  D  N  S  F  D  L  P  E  A  N  V  L  I
361 - CCAGGTTTCATCCCATGGTGGATCAAGAAGACAAGAAGCTCAACGTCTAGGTCGTATCCT  - 420
       Q  V  S  S  H  G  G  S  R  R  Q  E  A  Q  R  L  G  R  I  L
421 - CAGAGCTAAGAAAGGTGCTGCAGCGGAGGAGTATAACGCCTTCTTCTAC- 469
       R  A  K  K  G  A  A  A  E  E  Y  N  A  F  F  Y
```

Fig. 6. Nucleotide sequence of *Paracentrotus lividus* cDNA encoding the DNA repair helicase XPB/ERCC3 (*Pl*-ercc3), and its deduced amino acid sequence. Forward and reverse oligonucleotide primers used for RT-PCR amplification are *underlined*

basis of a multiple sequence alignment of the three *XPB/ERCC-3* nucleotide sequences from *Drosophila* (Mounkes et al. 1992), mouse (Weeda et al. 1991), and *G. cydonium* (Batel et al. 1998). The partial cDNA sequence (469 bp) and its deduced amino acid sequence [deposited in the National Center for Biotechnology Information (NCBI) databank; accession number AJ439717] are shown in Fig. 6. The comparative analysis revealed an 87 % amino acid identity with *Ciona intestinalis* (NCBI accession no. T31655), 83 % with *Mus musculus* and *Homo sapiens* (Weeda et al. 1990), 80 % with *D. melanogaster*, 73 % with *G. cydonium* and *Caenorhabditis elegans* (Hartman et al. 1989), 69 % with *S. cerevisiae* (Park et al. 1992), and 67 % with *D. discoideum* (Lee et al. 1997). The *Pl*-ercc3 probe is currently used in Northern blotting and RT-PCR experiments with embryos exposed to UV-B radiation (unpubl. results).

2.6.2
Effect of Cadmium on P. lividus Embryo Development

A study of the effects of sublethal concentrations of cadmium chloride on sea urchin embryo development revealed a decrease in the percentage of normal embryos with increasing concentrations of cadmium chloride (10^{-6} to 10^{-3} M) 24 and 48 h after fertilization (Russo et al. 2003). Embryos exposed to 10^{-5} to 10^{-3} M cadmium chloride developed no significant differences compared to controls until the swimming blastula stage (about 12 h after fertilization). However, these embryos showed developmental defects when control embryos reached the late gastrula stage (about 24 h of culture). The number of delayed and abnormal embryos increased with increasing concentrations of cadmium chloride. At 10^{-3} M cadmium chloride, 51–67 % of the embryos were delayed at the early gastrula stage. Concentrations of cadmium chloride higher than 10^{-3} M were lethal to the embryos. These results confirm previous studies (Pagano et al. 1982; Fernandez and Beiras 2001; Radenac et al. 2001) and are in agreement with the results from studies on vertebrates (Chernoff 1973; Layton and Layton 1979; Samarawickrama and Webb 1979). Our recent studies on time-dependent continuous exposure of *P. lividus* sea urchin embryos reveal the synthesis of a specific set of stress proteins (90, 72–70, 56, 28 and 25 kDa) which was dependent on the duration of the treatment (Roccheri et al. 2004).

Metallothioneins are possibly involved in detoxification processes in marine organisms occurring after exposure to heavy metals such as cadmium, zinc and copper (Bonham et al. 1987; Lyons-Alcantara et al. 1998; Cajaraville et al. 2000). Therefore, we also investigated the effects of exposure to sublethal concentrations of cadmium chloride on the expression of the metallothionein gene during the development of *P. lividus* sea urchin embryos (Russo et al. 2003). Northern blot analysis and RT-PCR experiments revealed that the metallothionein gene is constitutively expressed at low levels in control embryos at cleavage, swimming blastula, late gastrula and pluteus stages (6, 12, 24 and 48 h after fertilization; Russo et al. 2003). The levels of metallothionein transcripts increase with the developmental stage, in agreement with results reported by others (Wilkinson and Nemer 1987). However, when embryos were cultured in the presence of sublethal concentrations of cadmium chloride and harvested at cleavage, swimming blastula, late gastrula and pluteus stages, a time- and dose-dependent increase in the levels of transcription of metallothionein gene was observed (Russo et al. 2003). Embryos exposed to 10^{-5} M cadmium chloride showed, if at all, only a very small increase in metallothionein mRNA, in agreement with the absence of morphological abnormalities; at 10^{-4} M cadmium chloride a modest increase in metallothionein expression was observed, while embryos exposed to 10^{-3} M cadmium chloride showed a strong increase in the levels of metallothionein transcripts at the blastula (12 h; three-fold) and gastrula (24 h; two-fold) stages. Quantitative analysis of sea urchin embryos using a relative RT-PCR showed a two-fold increase in the levels of metallothionein transcripts even after 6 h following cadmium exposure, and a three-fold (two-fold) increase after 12 h (24 h), confirming results obtained by Northern blotting. The levels of metallothionein transcripts decreased in embryos treated for 48 h. Interestingly, morphological abnormalities were observed only after 24 h of exposure to the pollutant. An increase in expression of metallothionein gene after cadmium exposure was also found in the marine sponge *Suberites domuncula* (Schröder et al. 2000b).

Besides development of malformations, DNA damage and disturbances of mitotic spindle formation have been described in sea urchin gametes and embryos following cadmium exposure (Pagano et al. 1982).

2.7 Porifera

Sponges (Porifera) represent the phylogenetically oldest metazoan phylum; they separated from the common ancestor of the other metazoan phyla at the Proterozoic–Phanerozoic boundary, close to 800 million years ago, before the "Cambrian explosion", which began about 540 million years ago (reviewed in Müller 1998). The increased UV-B radiation reaching the Earth during that period (see above) restricted the evolution of those pro- and eukaryotes that were provided with only a weak UV-light-protection system. Therefore, it can be assumed that sponges must have acquired efficient protection systems to resist these environmental effects (Müller and Müller 1998).

2.8 Sponge Primmorphs

In the experiments described below, the marine sponge *S. domuncula* was used to monitor the response to UV-A and UV-B radiation and visible light. A special cell culture system was used to study the effects of exposure to these radiations on DNA integrity of sponge cells (Fig. 7). The sponge cell culture

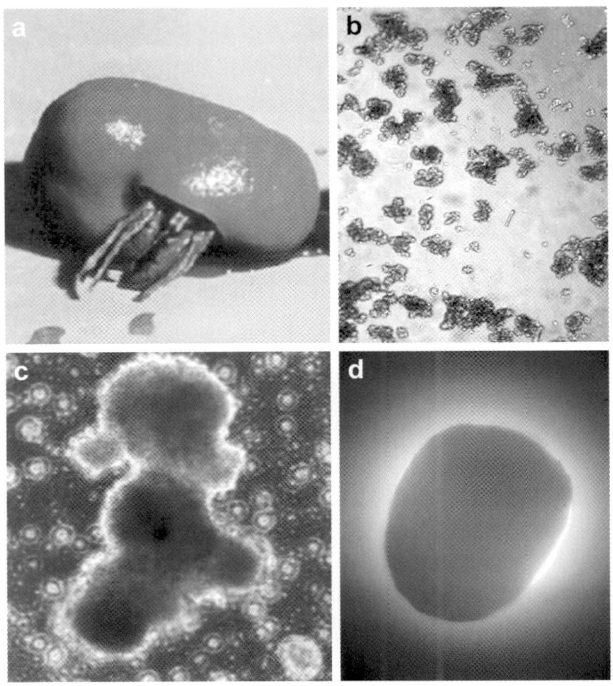

Fig. 7a–d. Formation of primmorphs from single cell suspensions of the sponge *S. domuncula*. Sponge tissue (**a**) (×0.7) was dissociated into single cells (**b**) using Ca^{2+}- and Mg^{2+}-free artificial seawater containing EDTA (Rottmann et al. 1987) (×100); after transfer into seawater/antibiotics, aggregates are formed (**c**) (×100), which finally resulted in the formation of primmorphs after an incubation period of 3–4 days (**d**) (×20)

starts from dissociated cells of *S. domuncula*, which subsequently form aggregates; in those aggregates the cells start DNA synthesis and proliferate. The aggregates, termed primmorphs, show a tissue-like appearance and can be cultured for more than 5 months.

2.8.1 UV Radiation

The primmorphs from *S. domuncula* cells were used as a model to analyze the effects of UV-A and UV-B on DNA integrity in sponge cells in vitro. The Fast Micromethod was adapted to work with low-integrity sponge DNA; unwinding was performed at pH 11.6. At this pH only strand breaks and cross-links could be measured. At higher pH (>12.0), alkali-labile sites (enzyme binding sites, apurinic and apyrimidinic sites) are the predominant effects of DNA denaturation.

Sponge cells (primmorphs) were irradiated with a SOL 500 lamp, which covered the whole solar spectrum using either H1 (wavelengths ≥320 nm; UV-A and visible light) or H2 (wavelengths ≥295 nm; UV-B, UV-A and visible light) filters. Subsequently, aliquots were analyzed for DNA single-strand breaks using Fast Micromethod. As in sea urchin coelomocytes, a dose-dependent increase in DNA damage was observed following exposure of sponge cells to 80–940 J/m^2 of UV-A or 5–60 J/m^2 of UV-B (Fig. 8). The UV-A and UV-B doses were monitored using a UV radiometer with an A or B sensor respectively. Again, the effects of UV-B (plus UV-A and visible light; filter H2) were much (several-fold) stronger than those observed with UV-A (plus visible light; filter H1). The extent of UV-A-induced DNA single-strand breaks

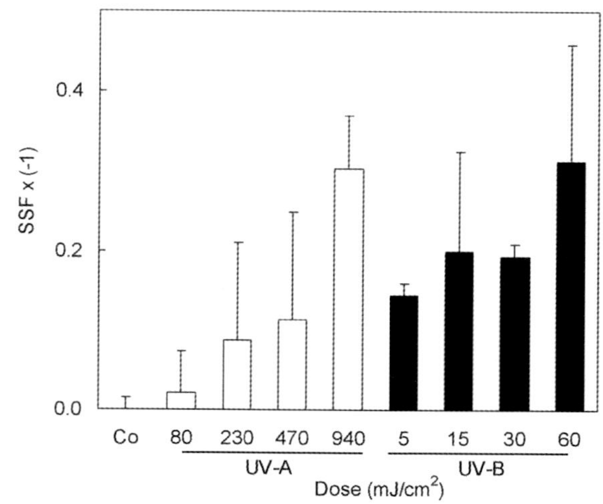

Fig. 8. UV-induced DNA damage in *S. domuncula* cells. Cells (1.5×10^5 cells/ml) were exposed to the indicated doses of UV-A and UV-B using a SOL 500 lamp with filter H1 (UV-A and visible light; wavelengths ≥320 nm) or filter H2 (UV-B, UV-A, and visible light; wavelengths ≥295 nm). Control (*Co*): non-irradiated cells

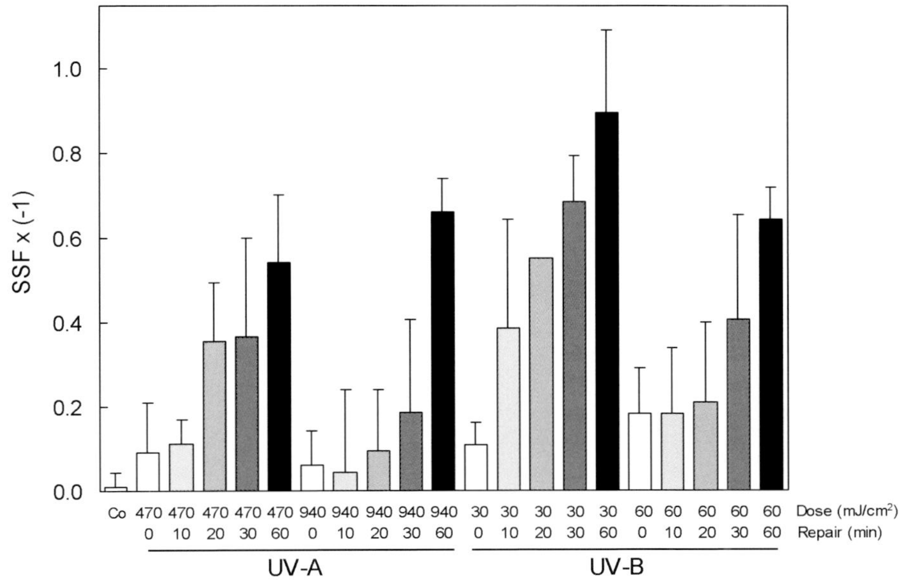

Fig. 9. DNA damage and repair in *S. domuncula* cells after exposure of cells (1.5×10^5 cells/ml) to the indicated doses of UV-A and UV-B using a full solar spectrum SOL 500 lamp with filter H1 (UV-A and visible light; wavelengths ≥320 nm) or filter H2 (UV-B, UV-A, and visible light; wavelengths ≥295 nm). DNA repair was allowed to occur for 0, 10, 20, 30 and 60 min at 16 °C. Control (*Co*): non-irradiated cells

increased during a post-irradiation repair period at 16 °C for 1.5 h (Fig. 9), due to the formation of transient breaks by the action of repair enzymes. The apparent decrease in strand breaks observed at higher UV doses (Fig. 9) may be caused by UV-induced formation of DNA cross-links resulting in lower SSF × (−1) values.

Induction of DNA single-strand breaks was also measured following irradiation of sponge cells (primmorphs) using a monochromatic UV-B lamp (peak at 312 nm; results not shown). DNA repair was accompanied by a time-dependent increase in the number of single-strand breaks; after a repair period of 2 h, the extent of DNA single-strand breaks decreased, reaching similar levels as in non-irradiated control cells after an incubation period of 18 h. The production of DNA single-strand breaks strongly depended on the medium used; the SSF × (−1) values were lower (negative) in Ca^{2+}-/Mg^{2+}-free seawater containing EDTA, but increased in Ca^{2+}-/Mg^{2+}-containing seawater. The decrease in SSF × (−1) value in EDTA-containing, Ca^{2+}-/Mg^{2+}-free seawater may be caused by a higher radiosensitivity (induction of formation of DNA cross-links) of the sponge single cells present under these conditions, compared to sponge primmorphs formed in the presence of calcium (Ca-/Mg-containing seawater).

The Fast Micromethod has also been applied for measuring the extent of DNA single-strand breaks in whole tissue samples of the marine sponge *Geodia cydonium*, irradiated with UV light under controlled experimental conditions, or from specimens collected in the field (Batel et al. 1998). The tissue samples were irradiated with 0–300 mJ/cm² UV-B (monochromatic UV-B lamp). One hour later, samples were analyzed for DNA strand breaks. In the untreated controls, which remained in the dark, the SSF was set to zero (0.00±0.04; n=5). Irradiation at a dose of 10 mJ/cm² significantly increased the number of strand breaks [SSF × (−1)] to 0.14±0.02. Higher doses of UV-B further increased the extent of DNA damage; at 300 mJ/cm² the SSF × (−1) reached a value of 0.32±0.08.

2.8.2 Cadmium

DNA damage in sponges is also induced by cadmium chloride. Exposure of whole specimen of marine sponges, e.g., *S. domuncula*, to cadmium resulted in a strong accumulation of the metal in the sponge (Schröder et al. 1999b). As a consequence a strong increase in the extent of DNA single-strand breaks was found using Fast Micromethod. The maximal increase was observed after an incubation period of 12 h in the presence of 1 mg/l of cadmium chloride and after an incubation period of 1–3 days in the presence of 10 or 100 µg/l of cadmium chloride; after a prolonged incubation period, the number of damages decreased, most likely due to DNA repair (Schröder et al. 1996).

Field experiments using *S. domuncula* collected from five stations in the northern Adriatic Sea, characterized by a gradient of pollution, revealed significant differences in the cadmium levels between these stations (Müller et al. 1998). The frequency of DNA strand breaks roughly paralleled the gradient of pollution (cadmium levels) at these sites.

2.8.3 Combined Effects

Results of the effect of cadmium on DNA repair in *S. domuncula* cells after irradiation with full solar spectrum lamp with filter H1 (wavelengths ≥320 nm) or filter H2 (wavelengths ≥295 nm) did not indicate an inhibition of DNA repair processes as found in sea urchin coelomocytes (results not shown). Possibly some mechanism(s) preventing cadmium toxicity despite its high accumulation in sponge tissue (a 17,500 enrichment has been determined under field conditions; Müller et al. 1998), such as expression of metallothionein-like proteins (Schröder et al. 2000b), are responsible for the finding that an inhibitory effect of cadmium on DNA repair could not be detected in the sponge system.

2.9 DNA Repair

Demosponges have an efficient DNA excision repair system. They are able to repair DNA damage caused by genotoxins, e.g. benzo[a]pyrene (Zahn et al. 1983). The mechanisms of repair in sponges are similar to those found in higher Metazoa. We found that in *G. cydonium* the expression of a homologue of the human *XPB/ERCC-3* excision repair gene is induced after exposure to UV radiation (Batel et al. 1998). In humans, the product of the *XPB/ERCC-3* gene is involved in the early step of DNA excision repair (Sancar 1996a); it corrects the repair defect in xeroderma pigmentosum and in Cockayne's syndrome. The human *XPB/ERCC-3* gene product is a helicase unwinding the DNA in the 3′→5′ direction from the site of the lesion (Weeda et al. 1991). Laboratory studies revealed that after irradiation of *G. cydonium* with 30 or 100 mJ/cm^2 UV-B light, a dose-dependent increase in the steady-state level of *GCXPB* occurs; values of up to 29-fold with respect to the controls which were kept in the dark have been determined. In parallel, the DNA integrity in the sponge samples was measured using Fast Micromethod. The data revealed that the degree of DNA strand breaks paralleled the increase in expression of the *GCXPB* gene (Batel et al. 1998).

2.9.1 (6–4) Photolyase

Sponges have not only an excision DNA repair system, but also a photolyase-based photoreactivating system. Two types of DNA photolyases have been found that cleave cyclobutane pyrimidine dimers (CPD): CPD photolyase and (6–4) photolyase (Sancar 1996b). The photolyases are part of the cryptochrome family and associated with the blue light photoreceptors (Kanai et al. 1997; Kobayashi et al. 2000). The photolyase of the hexactinellid sponge *Aphrocallistes vastus* was studied (Schröder et al. 2003). A cDNA was isolated from *A. vastus*, which comprises high sequence similarity to genes encoding the protostomian and deuterostomian (6–4) photolyases. The *A. vastus* sequence was assigned to the class I photolyases based on its high sequence similarity, especially within the N-terminal α/β-domain and C-terminal helical domain. Using functional studies, we demonstrated that this gene codes for a photolyase-related protein. After transfection into the *Escherichia coli* SY2 strain, which is deficient in photoreversal activity for pyrimidine cyclobutane dimers (Kim and Sundin 2001), the sponge gene caused resistance of the bacteria to UV light (Fig. 10). Irradiation was performed with a UV-B lamp (peak at 312 nm). The UV-induced damage in the bacteria was almost completely repaired during the light-repair phase; at a dose of 0.2 J/m^2 only a small reduction in the survival rate was seen (Fig. 10). A strong UV-B sensitivity was observed if the cells were transformed with the empty vector,

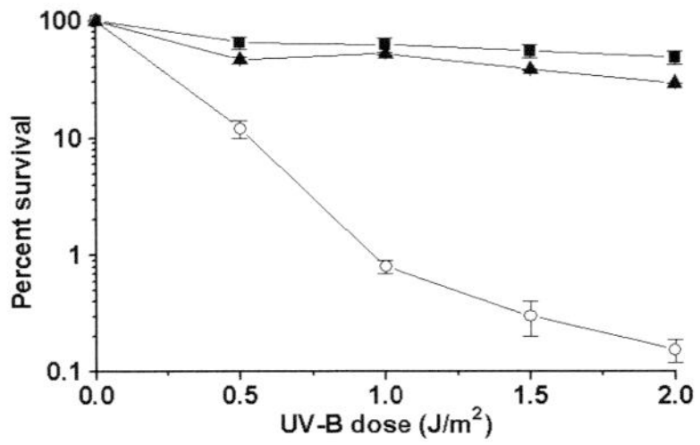

Fig. 10. Survival of *E. coli* strain SY2/pGEX (which does not have photolyase activity; *open circles*), *E. coli* SY2/pMS969 [complemented with the *ph* (6–4) photolyase gene from *E. coli*; *solid squares*), and *E. coli* SY2 carrying the *A. vastus APHVAPH* gene (*solid triangles*) after UV-B irradiation at doses between 0.5 and 2 J/m^2. Following UV irradiation, photoreactivation was allowed to occur by irradiation of the bacteria for 1 h with a lamp emitting visible light (Translux EC halogen photocuring unit, 400 and 520 nm with a maximum at 480 nm; Heraeus Kulzer, Wehheim, Germany). Subsequently, the bacteria were incubated overnight at 26 °C. Survival rate, which is given in percent, was calculated on the basis of number of colonies formed. Mean ± SE (n=5)

irrespective of a post-treatment with light. The recombinant sponge protein bound to UV-modified DNA that contained thymine dimers, while it failed to bind to non-treated DNA, suggesting that the sponge gene displays thymidine dimer-repairing enzyme activity (Schröder et al. 2003).

2.10 Corals

The detrimental effects of increasing solar UV radiation on corals have been recognized for a long time (Jokiel 1980; Shick et al. 1996). Exposure to UV light has been implicated in the process of coral bleaching. In addition, UV-induced DNA damage in coral-reef microbial communities has been observed (Lyons et al. 1998). As a result of the increased UV exposure, a decrease in skeletal growth and an increase in larval production have been described (Jokiel and York 1982). Most of these studies did not distinguish between the effects of UV-A and UV-B, although the latter wavelengths are more affected by stratospheric ozone depletion.

Primmorphs from *Dendronephthya klunzingeri* were used to determine the effect of increasing exposure to UV-B and visible light (480 nm). The modified procedure was used to dissociate the coral cells and form primmorphs (Custodio et al. 1998; Müller et al. 1999). The results revealed that primmorphs irradiated with UV-B responded with an increased expression of HSP90 only if exposed to low levels of UV-B (Wiens et al. 2000; Ammar et al. 2001). After irradiation with 30 J/cm^2 of UV-B, the primmorphs reacted with a 5.5-fold increase in HSP90 protein, compared with the controls, as revealed in Western blot experiments. Upregulation was observed following an increase in UV-B irradiation to 100 J/cm^2, while at levels above 100 J/cm^2 no

HSP90 was detected. In contrast to the studies with UV-B, visible light causes an up-regulation of the expression of the gene for HSP90 in a wider exposure range, between 20 and 1,000 J/cm^2. The increase in HSP90 expression following exposure to visible light was less pronounced and reached a maximum at 300 J/cm^2, with a 3.2-fold increase (Wiens et al. 2000).

Interestingly, in the absence of UV-B no transcripts of UVS-related protein could be visualized, while after treatment of the primmorphs with 30–300 J/cm^2 a strong up-regulation was seen in the expression of this gene (Wiens et al. 2000).

3 Concluding Remarks

In summary, the results show that marine invertebrates are highly sensitive to solar UV-B radiation without or in combination with heavy metal (cadmium) exposure. However, the degree of DNA damage and the course of subsequent repair processes somewhat differ between different species (sea urchins compared to sponges) and different developmental stages (sea urchin embryos). The results show that marine invertebrates may be endangered by the increase in UV-B radiation due to ozone depletion, especially during early stages of development. In addition, the results show that cell cultures of marine invertebrates (sea urchin coelomocytes and sponge primmorphs) might be useful tools for monitoring stress caused by physical (UV-B radiation) and chemical (cadmium) factors in the aquatic environment.

Acknowledgements. This work was supported by grants from the European Commission (EVK3-CT-1999-00005 "UVTOX") and the Bundesministerium für Bildung und Forschung (project Center of Competence BIOTECmarin).

References

Acevedo J, Nolan C (eds) (1993) Environmental UV radiation: causes – effects – consequences. Commission of the European Communities, Directorate-General XII for Science, Research and Development, Environment Programme, Brussels

Ammar MSA, Schröder HC, Wiens M, Müller WEG (2001) Induction of heat-shock gene expression by selected natural and anthropogenic disturbances in the octocoral (*Dendronephthya klunzingeri*). ACP-EU Fish Res Rep 10:85–95

Baker KS, Smith RC (1982) Spectral irradiance penetration in natural waters. In: Calkins J (ed) The role of solar ultraviolet radiation in marine ecosystems. Plenum, New York, pp 233–246

Batel R, Bihari N, Rinkevich B, Dapper J, Schäcke H, Schröder HC, Müller WEG (1993) Modulation of organotin-induced apoptosis by the water pollutant methyl mercury in a human lymphoblastoid tumor cell line and a marine sponge. Mar Ecol Prog Ser 93:245–251

Batel R, Fafandjel M, Blumbach B, Schröder HC, Hassanein HM, Müller IM, Müller WEG (1998) Expression of the human *XPB/ERCC-3* excision repair gene-homolog in the sponge *Geodia cydonium* after exposure to ultraviolet radiation. Mutat Res 409:123–133

Batel R, Jak×iæ Z, Bihari N, Hamer B, Fafan×jel M, Chauvin C, Schröder HC, Zahn RK, Müller WEG (1999) A microplate assay for DNA damage determination (Fast Micromethod) in cell suspensions and solid tissues. Anal Biochem 270:195–200

Bihari N, Batel R, Jak×iæ Z, Müller WEG, Waldmann P, Zahn RK (2002) Comparison between the Comet assay and Fast Micromethod for measuring DNA damage in HeLa cells. Croatia Chem Acta 75:793–804

Blaustein AR, Hoffman PD, Hokit DG, Kiesecker JM, Walls SC, Hays JB (1994) UV repair and resistance to solar UV-B in amphibian eggs: link to population declines? Proc Natl Acad Sci USA 19:1791–1795

Bonham K, Zafarullah M, Gedamu L (1987) The rainbow trout metallothioneins: molecular cloning and characterization of two distinct cDNA sequences. DNA 6:519–528

Bothwell ML, Sherbot D, Roberge AC, Daley RJ (1993) Influence of natural ultraviolet radiation on lotic periphytic diatom community growth, biomass accrual, and species composition: short-term versus long-term effects. J Phycol 29:24–35

Brasier MD (2000) The Cambrian explosion and the slow burning fuse. Sci Progr 83:77–92

Buhlmann B, Bossard P, Uehlinger U (1987) The influence of longwave ultraviolet radiation (UV-A) on the photosynthetic activity of phytoplankton. J Plankton Res 9:935–943

Cadwell MM, Camp LB, Warner CW, Flint SD (1986) Action spectra and their key role in assessing biological consequences of solar UV-B radiation change. In: Worrest RC, Caldwell MM (eds) Stratospheric ozone reduction, solar ultraviolet radiation and plant life. Springer, Berlin Heidelberg New York, pp 87–111

Cajaraville MP, Bebianno MJ, Blasco J, Porte C, Sarasquete C, Viarengo A (2000) The use of biomarkers to assess the impact of pollution in coastal environments of the Iberian Peninsula: a practical approach. Sci Total Environ 257:295–311

Canfield DE, Teske A (1996) Late Proterozoic rise in atmospheric oxygen concentration inferred from phylogenetic and sulphur-isotope studies. Nature 382:127–132

Chernoff N (1973) Teratogenic effects of cadmium in rats. Teratology 8:29–32

Cloud PE (1968) Atmospheric and hydrospheric evolution of primitive earth. Science 160:729–736

Coteur G, Pernet P, Gillan D, Joly G, Maage A, Dubois P (2003) Field contamination of the starfish *Asterias rubens* by metals. Part 1: Short- and long-term accumulation along a pollution gradient. Environ Toxicol Chem 22:2136–2144

Crump D, Lean D, Berrill M, Coulson D, Toy L (1999) Spectral irradiance in pond water: influence of water chemistry. Photochem Photobiol 70:893–901

Custodio MR, Prokic I, Steffen R, Koziol C, Borojevic R, Brümmer F, Nickel M, Müller WEG (1998) Primmorphs generated from dissociated cells of the sponge *Suberites domuncula*: a model system for studies of cell proliferation and cell death. Mech Ageing Dev 105:45–59

Damkaer DM, Dey DB (1983) UV damage and photoreactivation potentials of larval shrimp, *Pandalus platyceros*, and adult euphausiids, *Thysanoessa raschii*. Oecologia 60:169–175

Damkaer DM, Dey DB, Heron GA, Prentice EF (1980) Effects of UV-B radiation on near-surface zooplankton of Puget Sound. Oecologia 44:149–158

Damkaer DM, Dey DB, Heron GA (1981) Dose/dose-rate responses of shimp larvae to UV-B radiation. Oecologia 48:178–182

Dinnel PA, Pagano GG, Oshida PS (1988) A sea urchin test system for marine environmental monitoring. In: Burke RD, Mladenov PV, Lambert P, Parsley RL (eds) Echinoderm biology. AA Balkema, Rotterdam, pp 611–619

Dustan P (1982) Depth-dependent photoadaptation by zooxanthellae of the reef coral *Montastrea annularis*. Mar Biol 68:253–264

El-Sayed SZ (1988) Fragile life under the ozone hole. Nat Hist 97:73–80

El-Sayed SZ, Van Dijken G, Gonzales-Rhodas G (1996) Effects of ultraviolet radiation on marine ecosystems. Int J Environ Stud 51:199–216

Fernandez N, Beiras R (2001) Combined toxicity of dissolved mercury with copper, lead and cadmium on embryogenesis and early larval growth of the *Paracentrotus lividus* sea urchin. Ecotoxicology 10:263–271

Fisher AG (1965) Fossils, early life, and atmospheric history. Proc Natl Acad Sci USA 53:1205–1215

Flick DF, Krayville HF, Dimitroff JM (1971) Toxic effects of cadmium: a review. Environ Res 4:71–85

Frederick JE, Snell HE, Haywood EK (1989) Solar ultraviolet radiation at the earth's surface. Photochem Photobiol 50:443–450

Fridovich I (1986) Biological effects of the superoxide radical. Arch Biochem Biophys 247:1–11

Friedberg EC, Walker GC, Siede W (eds) (1995) DNA repair and mutagenesis. ASM Press, Washington

Gleason JF, Bhartia PK, Herman JR, McPeters R, Newman P, Stolarski RS, Flynn L, Labow G, Larko D, Seftor C, Wellemeyer C, Komhyr WD, Miller AJ, Planet W (1993) Record low global ozone in 1992. Science 260:523–526

Griffiths H, Mistry P, Herbert KE, Lunec J (1998) Molecular and cellular effects of ultraviolet light-induced genotoxicity. Crit Rev Clin Lab Sci 35:189–237

Häder DP, Worrest RC, Kumar HD, Smith RC (1995) Effects of increased solar ultraviolet radiation on aquatic ecosystems. Ambio 24:174–180

Hartman PS, Hevelone J, Dwarakanath V, Mitchell DL (1989) Excision repair of UV radiation-induced DNA damage in *Caenorhabditis elegans*. Genetics 122:379–385

Hartwig A (1994) Role of DNA repair inhibition in lead- and cadmium-induced genotoxicity: a review. Environ Health Perspect 102 (Suppl 3):45–50

Helbling EW, Villafañe V, Ferrario M, Holm-Hansen O (1992) Impact of natural ultraviolet radiation on rates of photosynthesis and on specific marine phytoplankton species. Mar Ecol Prog Ser 80:89–100

Henson JH, Nesbitt D, Wright BD, Sholey JM (1992) Immunolocalization of kinesin in sea urchin coelomocytes. J Cell Sci 103:309–320

Hessen DO (1994) *Daphnia* responses to UV-light. Arch Hydrobiol Beih 43:185–195

Hirosawa T, Miyachi S (1983) Inactivation of Hill reaction by long-wavelength ultraviolet radiation (UV-A) and its photoreactivation by visible light in the cyanobacterium, *Anacystis nidulans*. Arch Microbiol 135:98–102

Hunter JR, Taylor JH, Moser HG (1979) The effect of ultraviolet irradiation on eggs and larvae of the northern anchovy, *Engraulis mordax* and the Pacific mackerel, *Scomber japonicus* during the embryonic stage. Photochem Photobiol 29:325–338

Jeffery WR (1990) Ultraviolet irradiation during ooplasmic segregation prevents gastrulation, sensory cell induction, and axis formation in the ascidian embryo. Dev Biol 140:388–400

Jerlov NG (1968) Optical oceanography. Elsevier, Amsterdam

Jokiel PL (1980) Solar ultraviolet radiation and coral reef epifauna. Science 207:1069–1071

Jokiel PL, York RH (1982) Solar ultraviolet photobiology of the reef coral *Pocillopora damicornis* and symbiotic zooxanthellae. Bull Mar Sci 32:301–315

Kanai S, Kikuno R, Toh H, Ryo H, Todo T (1997) Molecular evolution of the photolyase-blue-light photoreceptor family. J Mol Evol 45:535–548

Karanas JK, Worrest RC, Van Dyke H (1981) Impact of UV-B radiation on the fecundity of the copepod *Acartia clausii*. Mar Biol 65:125–133

Karentz D, Lutze LH (1990) Evaluation of biologically harmful ultraviolet radiation in Antarctica with a biological dosimeter designed for aquatic environments. Limnol Oceanogr 35:549–561

Karentz D, Bothwell ML, Coffin RB, Hanson A, Herndl GJ, Kilham SS, Lesser MP, Lindell M, Moeller RE, Morris DP, Neale PJ, Sanders RW, Weiler CS, Wetzel RG (1994) Impact of UV-B radiation on pelagic freshwater ecosystems. Report of working group on bacteria and phytoplankton. Arch Hydrobiol Beih Ergebn Limnol 43:31–69

Kim JJ, Sundin GW (2001) Construction and analysis of photolyase mutants of *Pseudomonas aeruginosa* and *Pseudomonas syringae*: contribution of photoreactivation, nucleotide excision repair, and mutagenic DNA repair to cell survival and mutability following exposure to UV-B radiation. Appl Environ Microbiol 67:1405–1411

Kobayashi T, Ishikawa T, Hirayama J, Daiyasu H, Kanai S, Toh H, Fukuda I, Tsujimura T, Terada N, Kamei Y, Yuba S, Iwai S, Todo T (2000) Molecular analysis of zebrafish photolyase/cryptochrome family: two types of cryptochromes present in zebrafish. Genes Cells 5:725–738

Koziol C, Wagner-Hülsmann C, Cetkovic H, Gamulin V, Kruse M, Pancer Z, Schäcke H, Müller WEG (1996) Cloning of the heat-inducible biomarker, the cDNA encoding the 70-kDa heat shock protein, from the marine sponge *Geodia cydonium*: response to natural stressors. Mar Ecol Prog Ser 136:156–161

Layton WM Jr, Layton WM (1979) Cadmium-induced limb defects in mice: strain associated differences in sensitivity. Teratology 19:229–235

Lee SK, Yu SL, Garcia MX, Alexander H, Alexander S (1997) Differential developmental expression of the rep B and rep D xeroderma pigmentosum related DNA helicase genes from *Dictyostelium discoideum*. Nucleic Acids Res 25:2365–2374

Le Marrec F, Glaise D, Guguen-Guillouzo C, Chesnè C, Guillouzo A, Boulo V, Dorange G (1999) Primary cultures of heart cells from the scallop *Pecten maximus* (Mollusca-Bivalvia). In Vitro Cell Dev Biol 35:289–295

Little EE, Fabacher DL (1994) Comparative sensitivity of rainbow trout and two threatened salmonids, Appache Trout and Lahontan Cutthout Trout, to UV-B radiation. Arch Hydrobiol Beih 43:217–226

Loomis WF (1988) Four billion years. Sinauer Associates, Sunderland

Lyons MM, Aas P, Pakulski JD, Van Waasbergen L, Miller RV, Mitchell DL, Jeffrey WH (1998) DNA damage induced by ultraviolet radiation in coral-reef microbial communities. Mar Biol 130:537–543

Lyons-Alcantara M, Mooney R, Lyng F, Cottell D, Mothersill C (1998) The effects of cadmium exposure on the cytology and function of primary cultures from rainbow trout. Cell Biochem Funct 16:1–13

Malloy KD, Holman MA, Mitchell D, Detrich HW (1997) Solar UV-B-induced DNA damage and photoenzymatic DNA repair in Antarctic zooplankton. Proc Natl Acad Sci USA 94:1258–1263

Matranga V, Bonaventura R (2002) Sea urchin coelomocytes, the progenitors of vertebrate immune effectors, as bio-indicators of stress and pollution. In: Yokota Y, Matranga V, Smolenicka Z (eds) The sea urchin: from basic biology to aquaculture. Swet and Zeitlinger, Lisse, The Netherlands, pp 161–176

Matranga V, Toia G, Bonaventura R, Müller WEG (2000) Cellular and biochemical responses to environmental and experimentally induced stress in sea urchin coelomocytes. Cell Stress Chaperones 5:113–120

Matranga V, Bonaventura R, Di Bella G (2002) Hsp70 as a stress marker of sea urchin coelomocytes in short term cultures. Cell Mol Biol 48:345–359

Molina LT, Molina MJ (1986) Absolute absorption cross sections of ozone in the 185- to 350-nm wavelength range. J Geophys Res 91:14501–14508

Morale A, Coniglio L, Angelini C, Cimoli G, Bolla A, Alleteo D, Russo P, Falugi C (1998) Biological effects of a neurotoxic pesticide at low concentrations on sea urchin early development. A teratogenic assay. Chemosphere 37:3001–3010

Mounkes LC, Jones RS, Liang BC, Gelbart W, Fuller MT (1992) A *Drosophila* model for xeroderma pigmentosum and Cockayne's syndrome: haywire encodes the fly homolog of ERCC3, a human excision repair gene. Cell 71:925–937

Müller WEG (1998) Origin of Metazoa: sponges as living fossils. Naturwissenschaften 85:11–25

Müller WEG, Müller I (1998) Sponge cells and tissue as biological monitors of aquatic pollution. In: Wells PG, Lee K, Blaise C (eds) Microscale testing in aquatic toxicology – advances, techniques and practice. CRC Press, Boca Raton, pp 97–112

Müller WEG, Koziol C, Kurelec B, Dapper J, Batel R, Rinkevich B (1995) Combinatory effects of temperature stress and nonionic organic pollutants on stress protein (hsp70) gene expression in the fresh water sponge *Ephydatia fluviatilis*. Environ Toxicol Chem 14:1203-1208

Müller WEG, Batel R, Lacorn M, Steinhart H, Simat T, Lauenroth S, Hassanein H, Schröder HC (1998) Accumulation of cadmium and zinc in the marine sponge *Suberites domuncula* and its potential consequences on single-strand breaks and on expression of heat-shock protein: a natural field study. Mar Ecol Prog Ser 167:127–135

Müller WEG, Wiens M, Batel R, Steffen R, Schröder HC, Borojevic R, Custodio MR (1999) Establishment of a primary cell culture from a sponge: primmorphs from *Suberites domuncula*. Mar Ecol Prog Ser 178:205–219

Müller WEG, Koziol C, Wiens M, Schröder HC (2000) Stress response in marine sponges: genes and molecules involved and their use as biomarkers. In: Storey KB, Storey J (eds) Environmental stressors and gene responses. Elsevier, Amsterdam, pp 193–208

Noblet JA, Zeng EY, Ritter KJ (2003) Evaluation of polychlorinated biphenyl bioaccumulation patterns in white sea urchins (*Lytechinus pictus*) using multiple approaches. Environ Toxicol Chem 22:2719–2226

Pagano G, Esposito A, Giordano GG (1982) Fertilization and larval development in sea urchins following exposure of gametes and embryos to cadmium. Arch Environ Contam Toxicol 11:47–55

Park E, Guzder SN, Koken MH, Jaspers-Dekker I, Weeda G, Hoeijmakers JH, Prakash S, Prakash L (1992) RAD25 (SSL2), the yeast homolog of the human xeroderma pigmentosum group B DNA repair gene, is essential for viability. Proc Natl Acad Sci USA 89:11416–11420

Prézelin BB, Boucher NP, Smith RC (1994) Marine primary production under the influence of the Antarctic ozone hole: Icecolors '90. In: Weiler CS, Penhale PA (eds) Ultraviolet radiation in Antarctica: measurements and biological effects. Antarctic Res Ser 62. American Geophysical Union, Washington, DC, pp 159–186

Radenac G, Fichet D, Miramand P (2001) Bioaccumulation and toxicity of four dissolved metals in *Paracentrotus lividus* sea-urchin embryo. Mar Environ Res 51:151–166

Regan JD, Carrier WL, Gucinski H, Olla BL, Yoshida H, Fujimura RK, Wicklund RI (1992) DNA as a solar dosimeter in the ocean. Photochem Photobiol 56:35–42

Ringelberg J, Keyser AL, Flik BJG (1984) The mortality effect of ultraviolet radiation in a translucent and in a red morph of *Acanthodiaptomus denticornis* (Crustacea, Copepoda) and its possible ecological relevance. Hydrobiologia 112:217–222

Roccheri MC, Agnello M, Bonaventura R, Matranga V (2004) Cadmium induces the expression of specific stress proteins in sea urchin embryos. Biophys Biochem Res Commun 321:80–87

Russo R, Bonaventura R, Zito F, Schröder HC, Müller I, Müller WEG, Matranga V (2003) Stress to cadmium monitored by metallothionein gene induction in *Paracentrotus lividus* embryos. Cell Stress Chaperones 8:232–241

Samarawickrama GP, Webb M (1979) Acute effects of cadmium on the pregnant rat and embryo-fetal development. Environ Health Perspect 28:245–249

Sancar A (1996a) DNA excision repair. Annu Rev Biochem 65:43–81

Sancar A (1996b) No "end of history" for photolyases. Science 272:48–49

Schröder HC, Batel R, Lauenroth S, Hassanein HMA, Lacorn M, Steinhart H, Müller WEG (1999a) Induction of DNA damage and expression of heat shock protein HSP70 by polychlorinated biphenyls in the marine sponge *Suberites domuncula* Olivi. J Exp Mar Biol Ecol 233:285–300

Schröder HC, Hassanein HMA, Lauenroth S, Koziol C, Mohamed TA-AA, Lacorn M, Steinhart H, Batel R, Müller WEG (1999b) Induction of DNA strand breaks and expression of HSP70 and GRP78 homolog by cadmium in the marine sponge *Suberites domuncula*. Arch Environ Contam Toxicol 36:47–55

Schröder HC, Batel R, Müller WEG (2000a) Novel cellular and molecular tools to assess the effects of pollution and solar UV radiation on marine invertebrates. In: Proc EurOCEAN 2000, Eur Conf on Marine Science and Ocean Technology, Hamburg, 29 Aug–2 Sept. Project synopses, vol I: marine processes, ecosystems and interactions. Office for Official Publications of the European Communities, Luxembourg, pp 128–134

Schröder HC, Shostak K, Gamulin V, Lacorn M, Skorokhod A, Kavsan V, Müller WEG (2000b) Purification, cDNA cloning and expression of a cadmium-inducible cysteine-rich metallothionein-like protein from the marine sponge *Suberites domuncula*. Mar Ecol Prog Ser 200:149–157

Schröder HC, Krasko A, Gundacker D, Leys SP, Müller IM, Müller WEG (2003) Molecular and functional analysis of the (6–4) photolyase from the hexactinellid *Aphocallistes vastus*. Biochim Biophys Acta 1651:41–49

Schröder HC, Batel R, Schwertner H, Boreiko O, Müller WEG (2004) Fast Micromethod DNA single-strand break assay. In: Henderson DS (ed) DNA repair protocols: eukaryotic systems. Humana Press, Totowa, New Jersey (in press)

Sconzo G, Romancino D, Fasulo G, Cascino D, Giudice G (1995) Effect of doxorubicin and phenytoin on sea urchin development. Pharmacie 50:616–619

Scully NM, Lean DRS (1994) The attenuation of ultraviolet radiation in temperate lakes. Arch Hydrobiol Beih Ergebn Limnol 43:135–144

Shick JM, Lesser MP, Stochaj WR (1991) Ultraviolet radiation and photooxidative stress in zooxanthellate Anthozoa: the sea anemone *Phyllodiscus semoni* and the octocoral *Clavularia* sp. Symbiosis 10:145–173

Shick JM, Lesser MP, Jokiel PL (1996) Effects of ultraviolet radiation on corals and other coral reef organisms. Global Change Biol 2:527–545

Smith RC, Baker KS (1979) Penetration of UV-B and biologically effective dose-rates in natural waters. Photochem Photobiol 29:311–323

Smith RC, Baker KS (1981) Optical properties of the clearest natural waters (200–800 nm). Appl Opt 20:177–184

Smith RC, Prézelin BB, Baker KS, Bidigare RR, Boucher NP, Coley T, Karentz D, Maclntyre S, Matlick HA, Menzies D, Ondrusek M, Wan Z, Waters KJ (1992) Ozone depletion: ultraviolet radiation and phytoplankton biology in Antarctic waters. Science 255:952–959

Stolarski RS, Bojkov R, Bishop L, Zerefos C, Staehelin J, Zawodny J (1992) Measured trends in stratospheric ozone. Science 256:342–349

Towe KM (1994) Earth's early atmosphere: constraints and opportunities for early evolution. In: Bengtson S (ed) Early life on earth. Columbia University Press, New York, pp 36–47

Vincent WF, Roy S (1993) Solar ultraviolet-B radiation and aquatic primary production: damage, protection, and recovery. Environ Rev 1:1–12

Weeda G, van Ham RC, Masurel R, Westerveld A, Odijk H, de Wit J, Bootsma D, van der Eb AJ, Hoeijmakers JH (1990) Molecular cloning and biological characterization of the human excision repair gene ERCC-3. Mol Cell Biol 10:2570–2581

Weeda G, Ma L, van Ham RC, Bootsma D, van der Eb AJ, Hoeijmakers JH (1991) Characterization of the mouse homolog of the XPBC/ERCC-3 gene implicated in xeroderma pigmentosum and Cockayne's syndrome. Carcinogenesis 12:2361–2368

Wiens M, Shokry M, Nawar AH, Koziol C, Hassanein HMA, Eisinger M, Müller IM, Müller WEG (2000) Induction of heat-shock (stress) protein gene expression by selected natural and anthropogenic disturbances in the octocoral *Dendronephthya klunzingeri*. Aquat Toxicol 245:265–276

Wilkinson DG, Nemer M (1987) Metallothionein genes *Mta* and *MTb* expressed under distinct quantitative tissue-specific regulation in sea urchin embryos. Mol Cell Biol 7:48–58

Yabe K, Onodera S, Makino M, Suzuki M, Kagoshima K (1998) The effects of ultraviolet irradiation on larvae and early juveniles of the sea urchin, *Strongylocentrotus intermedius*. Nippon Suisan Gakkaishi 64:606–609

Zahn RK, Zahn G, Müller WEG, Michaelis ML, Kurelec B, Rijavec M, Batel R, Bihari N (1983) DNA damage by PAH and repair in a marine sponge. Sci Total Environ 26:137–156

Echinoderms: Their Culture and Bioactive Compounds

M.S. Kelly

Abstract. Of the five extant classes of echinoderms, it is the sea urchins (Echinoidea) and the sea cucumbers (Holothuroidea) that are both commercially fished and heavily overexploited. In sea urchins, it is the gonad of both males and females, normally referred to as 'roe', that is a sought-after food. In the sea cucumber, the principal product is the boiled and dried body-wall or '*bêche-de-mer*' for which there is an increasing demand. Many sea urchin and sea cucumber fisheries still have no management system or restrictions in place, and for those that do, the prognosis for catches to continue even at a reduced level is poor. Cultivation of these species increasingly becomes a necessity, both for stock enhancement programs and as a means to meet market demand. Sea urchin culture has been practised on a large scale in Japan for many decades, and effective methods for the culture and reseeding of species in these waters have been long established. Juvenile urchins are produced in their millions in state-sponsored hatcheries, for release to managed areas of seafloor. Outside of Japan, sea urchin cultivation is still a fairly recent practice, less than 10 years old, and largely still at a research level, although a range of species are now being produced in a variety of different culture systems. It is essential that the culture systems are adapted to be species-specific and meet with local environmental constraints. Sea cucumber cultivation originated in Japan in the 1930s and is now well established there and in China. Methods for mass cultivation of the tropical *Holothuria scabra* are now well established and practised in India, Australia, Indonesia, the Maldives and the Solomon Islands, with the focus of the research effort for both temperate and tropical species being centred on the production of juveniles in hatcheries for the restoration and enhancement of wild stocks. Like many other marine organisms, echinoderms have been, and continue to be, examined as a source of biologically active compounds with biomedical applications. Sea cucumber has been valued in Chinese medicine for hundreds of years as a cure for a

M.S. Kelly (e-mail: maeve.kelly@sams.ac.uk)
Scottish Association for Marine Science, Oban, Argyll, Scotland, PA37 1QA, UK

Progress in Molecular and Subcellular Biology
Subseries Marine Molecular Biotechnology
V. Matranga (Ed.), Echinodermata

wide variety of ailments. Some more recently isolated compounds, mainly from sea cucumbers and starfish, and including those with antitumour, antiviral, anticoagulant and antimicrobial activity, are summarised below. When wild stocks decline, the demand created in the market place raises the price of the product and, consequently, culturing is more likely to become viable economically. As this review shows, there have been dramatic advances in the culture methods of sea urchins and sea cucumbers in the last 10–15 years, to the extent that one can conclude that currently the major obstacles to successful cultivation are indeed economic rather than biological. Hence the future of the echinoculture industry is closely linked to that of the fisheries, whose fate will ultimately determine the market forces that will shape this growing industry.

1 Why Cultivate?

Of the five extant classes of echinoderms, it is the sea urchins (Echinoidea) and the sea cucumbers (Holothuroidea) that are both commercially fished and heavily overexploited. In sea urchins, it is the gonad of both males and females, normally referred to as 'roe', that is a sought-after food. The largest market is Japan, but there are also undersupplied markets in Europe. In the sea cucumber, the principal product is the boiled and dried body-wall or '*bêche-de-mer*', for which there is an increasing demand from China, where it is valued as food but also for its curative properties. There is also a trade in sea cucumbers for home aquaria and biomedical products (Bruckner et al. 2003), and at one time there was fishing activity for starfish (Asteroidea), which were processed for fish or poultry meal or for fertiliser (Sloan 1984). The extent and demise of the world's sea urchin and sea cucumber fisheries have been reviewed by Sloan (1984), Conand and Byrne (1993) and Keesing and Hall (1998), and more comprehensively for sea urchins by Andrew et al. (2002) and for sea cucumbers by Bruckner et al. (2003).

Most, if not all, sea urchin fisheries have followed the same pattern of rapid expansion to an unsustainable peak, followed by an equally rapid decline. World landings of sea urchin, having peaked at 120,000 t (metric tonnes) in 1995, are now in the region of 90,000 t/year. However, over half this catch comes from the recently expanded Chilean fishery for *Loxechinus albus*; which also appears to have peaked, the catch now being sustained by the discovery of new fishing grounds. The other major sea urchin fisheries, in terms of tonnage landed, are in Japan, Maine (USA), British Columbia (Canada) and California (USA) (Andrew et al. 2002). In Europe, the sea urchin stocks (*Paracentrotus lividus*) of first France and then Ireland were overfished in the 1980s to supply the French markets and those stocks have never recovered (Barnes and Crook 2001; Barnes et al. 2002). There are large populations of edible urchins in Scotland (*Echinus esculentus* and *Psammechinus miliaris*) and Norway (*Strongylocentrotus droebachiensis*), but these stocks are unsuited to commercial fishing as their roe content is either too low or too variable (Hagen 2000; Kelly 2000; Kelly et al. 2001; Sivertsen 2004).

The global landing of sea cucumbers was estimated as 13,000 t dried weight (130,000 t live weight) in 1995. The largest fisheries today are in Indonesia and the Philippines; however, in the last decade, the number of countries and species involved in the trade have increased in both tropical and temperate regions and the fishery has spread to non-traditional areas such as Mexico, the Galapagos and North America. Bruckner et al. (2003) list 29 species from 50 countries that are fished for import into Hong Kong, many of which will be re-exported to China and other trade centres in Singapore and Chinese Taipei.

Regulations are now imposed on some sea urchin and sea cucumber fisheries, although few are based on formal stock assessments. For example, in the state of Maine, on the east coast of the USA, the sea urchin fishery is 'zoned', in part based on the reproductive/spawning cycle along the coast. Regulation limits fishing effort; there is a limited entry scheme to the fishery, restrictions on gear size, fishing times and minimum and maximum landing sizes. However, researchers believe even this diminished catch is likely to prove unsustainable (L. Harris, University of New Hampshire, USA, pers. comm.).

In contrast, the Japanese sea urchin fishery has endured for more than 50 years, yielding more than 13,000 t in 1998. Six species account for the bulk of the commercial landings, but it is important to note that there is government subsidy for urchin stock enhancement and this is used as a management tool to conserve and rebuild stocks (Andrew et al. 2002).

Many sea cucumber fisheries show evidence of overexploitation or have already collapsed and some still have no known management or restrictions, including the world's largest fishery in Indonesia (Bruckner et al. 2003). Possible management measures suggested for sea cucumber fisheries include the imposition of minimum landing sizes, closed seasons, no-take marine protected areas, bag limits, prohibition of night fishing for nocturnal species and restrictions on the use of SCUBA for harvesting. However, the artisanal nature of the fishery makes implementation of such measures difficult. Bruckner et al. (2003) propose that listing sea cucumbers in the Convention on International Trade in Endangered Species (CITES) Appendix II might be an appropriate way to ensure that harvest to international markets is conducted in a sustainable manner, without detriment to the target species or their ecosystem. Compounding the impact of overexploitation of wild echinoderm populations is the increasing conviction among researchers that recruitment of juvenile sea urchins is both sporadic and unpredictable (see Lawrence 2001 for reviews; also Kelly 2000; Harris et al. 2001), and the fact that some species of sea urchin are extremely long-lived (Ebert 1998). Sea cucumbers are a valuable source of income for many coastal communities, particularly among the developing nations of the Indo-Pacific, but as with sea urchins, their high value and sedentary, shallow-water habit have left them vulnerable to overexploitation through unregulated and illegal fishing activity. Once their density is reduced below a critical mass, populations may take as long as 50 years to recover (Dalzell et al. 1996; Battaglene 1999; Bruckner et al. 2003).

As the prognosis for catches to continue at the current level is poor, cultivation of these species increasingly becomes a necessity, both for stock enhancement programs and as a means to meet market demand. In addition, as wild stocks decline, cultivation is more likely to become viable economically. The need for effective culture methods is now reflected in increased research effort. Culture of sea urchins for reseeding is practised on a large scale in Japan and to a lesser extent in South Korea and the Philippines (Andrew et al. 2002). Sea cucumbers are cultured for reseeding on a large scale in China. Outside these countries most echinoculture is at a research or semi-commercial scale. There is no commercial-scale cultivation of starfish, brittle stars (ophiuroids) or feather stars (crinoids).

2 Sea Urchin Aquaculture

2.1 Life History and State of the Art

Edible sea urchins are among the orders of regular Echinoidea (Lawrence 2001), dioecious (separate sexes) and broadcast spawners; mature individuals shed gametes to seawater where fertilisation occurs. The eggs develop to form pluteus larvae, which, after a period of planktonic development, feeding on microalgae, settle to a substrate and undergo metamorphosis to form tiny juvenile sea urchins. The estimated time for these individuals to reach market size (40–50 mm horizontal test diameter) is commonly in the order of 1–3 years, again varying according to species.

Sea urchin culture has been practised on a large scale in Japan for many decades, and effective methods for the culture and reseeding of species in these waters have been long established. Juvenile urchins are produced in their millions in state-sponsored hatcheries, for release to managed areas of seafloor. The nationally co-ordinated reseeding program initiated in the 1960s has now developed to the extent that over 66 million juveniles were released onto reefs in 2000; of these, over 80 % were *Strongylocentrotus intermedius* (Agatsuma et al. 2004). The effectiveness of the reseeding program has not been easy to assess, in part because of the difficulty in discriminating between reseeded and wild individuals. In Hokkaido prefecture, reseeding of *S. intermedius* began in 1985 with the release of 1 million juveniles, increasing to over 60 million in 1996. However, for the first 8 years of this program, the total harvest of urchins from this area continued to decline and it is only since 1992 that the catch has begun to stabilise. The contribution of released sea urchins to the overall catch has been estimated to be between 62 and 80% (Agatsuma et al. 2004). There are also much smaller-scale reseeding programs operating in South Korea and on Luzon Island in the Philippines (Andrew et al. 2002).

Outside of Japan, sea urchin cultivation is still a fairly recent practice, less than 10 years old. There have been researchers and companies developing methods for sea urchin (*Paracentrotus lividus*) cultivation in southern Ireland for over 20 years (Leighton 1995), and until relatively recently a concentrated effort into developing 'closed-cycle' land-based systems in France, also with *P. lividus* (Grosjean et al. 1998). Echinoculture (*Psammechinus miliaris*, *E. esculentus*, *P. lividus*) has been conducted in Scotland since 1995 and there are also established research teams on the east coast of North America – Florida, Alabama, Maine, New Hampshire, New Brunswick and Newfoundland – working on *S. droebachiensis* and *Lytechinus variegatus*; on the west coast of North America, including California and British Columbia (*S. droebachiensis*, *S. franciscanus*, *S. purpuratus*); in Chile (*Loxechinus albus*); Norway (*S. droebachiensis*); Israel (*P. lividus*); and in New Zealand (*Evechinus chloroticus*). This list is not intended to be exclusive or exhaustive but to illustrate the extent of the research base. The biology and ecology of these and other edible echinoids have recently been reviewed (Lawrence 2001).

While including discussion of the basic methodologies employed, both in and outside of Japan, this chapter will concentrate on examining some of the more recent advances and innovations in sea urchin culture, outside of Japan, with reference to the particular challenges remaining for the culturist at each stage of the sea urchin life cycle.

2.2 Sea Urchin Larviculture

Echinoids have been successfully raised in the laboratory for over 100 years (MacBride 1903). The reproductive periodicity for many echinoid species is well described; temperate water species in culture tend to have one spawning period per year (Himmelman 1977; Byrne 1993; Kelly 2000; Fig. 1) and brood stock are collected locally. Gravid individuals are induced to spawn either by temperature shock or, commonly, by injection of 0.5 M KCl to the coelom via the peristomal membrane. The concentration of sperm allowed to mix with the eggs must be controlled to optimise fertilisation and development success rates. The fertilised eggs hatch in approximately 10–15 h, depending on the species, to release a ciliated blastula which develops to the four-armed, then six-armed, then eight-armed pluteus larvae (Fig. 2). To raise large numbers of larvae in a commercial context the culture techniques must be refined in terms of food quality and quantity, larval density and water quality; and then shown to be effective once scaled up to large batches of larvae (>100,000). Static (no through-flow) aerated systems with a variable number of complete or

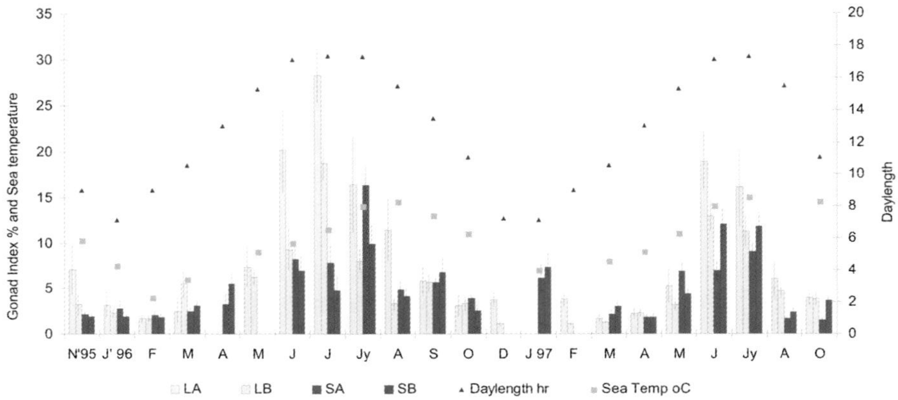

Fig. 1. Annual reproductive cycles (Nov 1995 to Oct 1997) of the sea urchin *Psammechinus miliaris* in a Scottish sea loch as described by the gonad index (GI). GI was calculated as wet weight of gonad divided by drained wet weight of sea urchin expressed as a percentage (n=10) for two littoral (*LA* and *LB*) and two subtidal (*SA* and *SB*) populations. Seawater temperature (°C) and day length (h) ▲ are illustrated. *Error bars* represent 95 % confidence limits. (Kelly 2000)

partial water changes throughout the larval life have been widely used (Fenaux et al. 1988; Leighton 1995; Grosjean et al. 1998; Kelly et al. 2000). In large-scale culture in Japan, partial exchange systems (Sakai et al. 2004) and continuous flow systems are used, the water flow being increased as the larvae develop (Hagen 1996). Upwelling silos, of the type used for fragile halibut yolk-sac larvae, have been tested on a small scale (M. Russell, Villanova University, USA, pers. comm.) and may also prove suitable for the large-scale culture of sea urchin larvae.

The planktonic diatom *Chaetoceros gracilis* is widely used as larval food in sea urchin hatcheries in Japan (Sakai et al. 2004). Many studies have compared different species or combinations of species of microalgae as larval foods for other species of sea urchin. Some species of microalgae used regularly and with success include *Isocrysis galbana* (Gonzalez et al. 1987), *Cricosphaera* (*Hymenomonas*) *elongata* (Fenaux et al. 1988) and *C. carterae* (Leighton 1995), the diatom *Phaeodactylum tricornutum* (Grosjean et al. 1998) and *Dunaliella tertiolecta* (Kelly et al. 2000; Jimmy et al. 2003). One noteworthy observation from these studies is there is no one optimal larval diet; different sea urchin species are reported to respond best to different algae. Of course, the biochemical and therefore nutritional value of the same species of microalgae, grown in different laboratories, may not be identical. However, it seems likely that there are true species-specific differences in echinoid larval dietary requirements.

Echinoid larvae demonstrate considerable plasticity in their morphology in response to varying food ration and quality. Growing larvae must increase

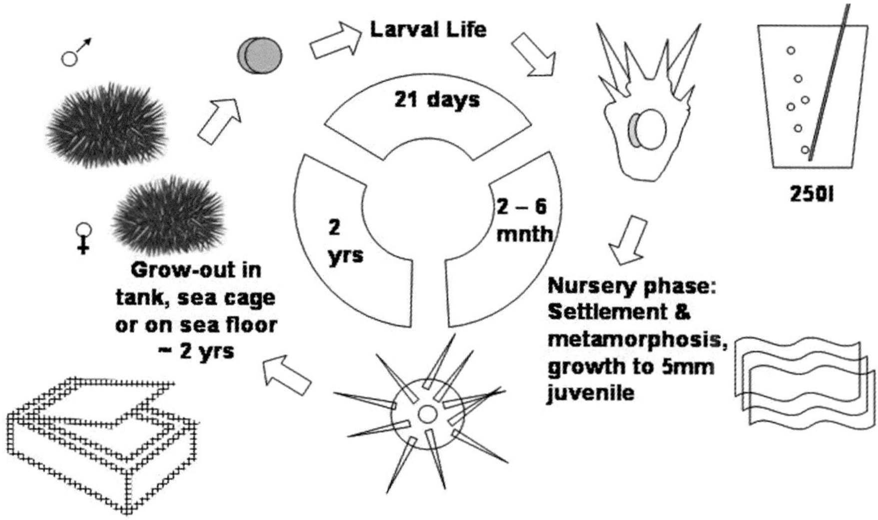

Fig. 2. Life cycle of the regular echinoid *Psammechinus miliaris* in culture. Gravid adults shed their gametes to seawater where fertilisation and first cleavage of the developing embryo occur within hours. Over the next 21 days, the planktonic larvae, maintained in aerated 250-l containers of seawater and fed microalgae, develop to a point where they are competent to settle. Newly metamorphosed juveniles are maintained on PVC wave plates coated with diatoms. At approximately 5-mm test diameter they are weaned to other foods (soft macroalgae or artificial diets) and transferred to a grow-out system where they mature to market size (40–50 mm test diameter)

the ciliated band length in order to increase feeding capability (McEdward 1984; Strathmann et al. 1992). Ciliated band length is increased by increasing arm length and by developing additional pairs of larval arms. The relative proportions of the larval body, e.g. post-oral arm length to larval body length, can therefore be a useful indicator of the nutritional status of larvae in culture (Fig. 3). Underfeeding will increase arm length relative to body length and overfed larvae may show a reduction in the length of the larval and in particular post-oral arms (Kelly et al. 2000; Jimmy et al. 2003).

One labour-intensive aspect of larval culture is the need for the simultaneous production of microalgae as live feed. However, sea urchin larvae may prove suited to culture using artificial diets, as research on *Lytechinus variegatus* (J.M. Lawrence, University of South Florida, USA, pers. comm.) has shown. It is the lipid or fatty acid component that is lost or destroyed in some forms of preserved algae. For example, their loss renders spray-dried microalgae a relatively poor food source for bivalve larvae which require poly- and highly unsaturated fatty acids (PUFAs and HUFAs) (Caers et al. 1998). Some species of sea urchin larvae have been shown to grow well when fed the green microalga *Dunaliella tertiolecta* (Kelly et al. 2000, Jimmy et al. 2003), which is

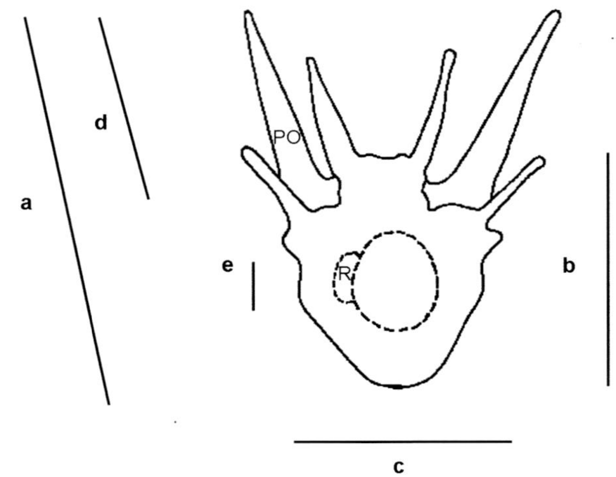

Fig. 3. Relative proportions of the echinoid larva that can be used as a measure of its nutritional status. *a* Larval length; *b* larval body length; *c* larval width; *d* post-oral arm length; *e* rudiment length; *R* echinorudiment; *PO* post-oral arm

known to be deficient in eicosapentaenoic acid (EPA) and docosahexaenoic acid (DHA). Therefore, sea urchins may prove suited to culture using spray-dried or other preserved algal preparations.

2.3 Metamorphosis and the Post-Larval Stage

Sea urchin juveniles have been produced on a commercial or semi-commercial scale by hatcheries in Japan, South Korea, Ireland, Norway, Scotland and in British Columbia, Canada. When deemed competent to settle (judged by the size and state of development of the echinorudiment; Fig. 3), sea urchin larvae are presented with a substrate likely to induce metamorphosis, but which will subsequently serve as a food source for the early juvenile. Most culturists use a natural biofilm or a specially seeded diatom substrate created from species isolated locally and grown on a PVC wave plate. Optimising diets for the early juveniles and/or the replacement of diatom biofilms with artificial diets is probably one of the most challenging areas left to research. The variation in size and subsequent variation in growth rates of post-larvae remain a bottleneck in the supply of hatchery-reared juveniles. Hatchery-reared juveniles are robust enough to survive transfer to sea cages or other grow-out systems from a small size (5-mm test diameter) (Kelly 2002; Sakai et al. 2004). At this point they are weaned onto other diets, soft macroalgae or artificial diets, depending on the grow-out system.

2.4
Sea Urchin Grow-Out Systems

In contrast to the Japanese systems where hatchery-reared juveniles are mainly released to managed areas of seafloor (Hagen 1996; Sakai et al. 2004), researchers in other countries newer to echinoculture have experimented with a wide range of grow-out systems for juvenile and adult urchins, ranging from relocation from poor to good feeding grounds (Moylan 1997) to the ranching of urchins caged on the seafloor (Cuthbert et al. 1995). Wild-collected adults of many species have been held in a variety of tank and sea-cage systems for roe conditioning (Lawrence et al. 1997; Cook et al. 1998; McBride et al. 1998; Fernandez and Boudouresque 2000; Hammer et al. 2000; Kelly et al. 2001; Kennedy 2002; Pearce et al. 2002a). Hatchery-reared juveniles have been grown in suspended culture (Kelly 2002, and also at Instituto de Fomento, Pesquero, Hueihue, Chile) in closed recirculation systems (Grosjean et al. 1998) and in dammed rock pools in southern Ireland (J. Chamberlain, Dunmanus Seafoods Ltd., pers. comm.). A sea-cage cultivation system of stacking baskets suspended from a ladder-like structure over which a work barge or raft can operate is being developed by Norwegian researchers (Aas 2004). The time taken for juveniles of most species to reach market size is in the order of 1–3 years.

Systems that accelerate growth to market size while producing a uniform size class would give an economic advantage. One possible route to obtaining sustainable and environmentally friendly systems for urchin culture is to further examine their potential in integrated systems. They have already been shown to thrive in polyculture with the Atlantic salmon (Kelly et al. 1998) and to have a role in land-based integrated systems (Shpigel et al. 2004). However, many species are true omnivores, so the potential for their integration into systems where natural prey items, for example, mussels, are already produced should be explored.

2.5
Juvenile and Adult Somatic Versus Gonadal Growth

Sea urchins can produce gonads with viable gametes years before they reach the typical 'adult' size for their species (Jensen 1969; Kelly 2001). Such small/young urchins are frequently termed 'juvenile' despite the fact they may have gonads, and young *Psammechinus miliaris* can successfully reproduce in their first year (Kelly 2001). However, it is likely once gametes start to form that urchins begin to partition ingested energy differently (Hagen 1998; Guillou and Lumingas 1999; Otero-Villanueva et al. 2004), and ultimately diets should be designed that promote somatic growth in urchins that are below market size, rather than encourage the sequestering of nutrients as gonad biomass.

2.6 Artificial Diets

The use of artificial diets has been widely adopted by researchers outside of Japan as they generally produce better growth rates than seaweeds. Seaweeds, potentially a cheap source of feed, have the disadvantage of being of variable quality and in variable supply over a season. In some regions, the large-scale harvesting of seaweeds would be regarded as an environmentally unsound practice.

Artificial diets are required to raise sea urchins in monoculture from juveniles to market size, for use as a finishing diet to perfect roe quality in urchins from polyculture systems and for use in enhancing the roe of fished urchins with unmarketable roe content. There is a large body of literature on the formulation of artificial diets for juvenile and adult sea urchins, many of which have been tested in comparison to seaweed as a reference diet. Most artificial diets contain a selection of soybean meal and cereals, either with or without animal-origin proteins and lipids (Cook et al. 1998; Fernandez and Boudouresque 2000; Spirlet et al. 2001), and range from simple moist or agar-bound diets (Klinger et al. 1994; Goebel and Barker 1998) to pellets extruded in commercial processing equipment (Lawrence et al. 1997; Pantazis et al. 2000; Olave et al. 2001). The impact of differing protein levels (de Jong-Westman et al. 1995; McBride et al. 1998; Hammer et al. 2000), the relative value of different protein sources (Pearce et al. 2002a), the necessity of minerals (Kennedy 2002) and effect of binder type (Pearce et al. 2002b) have all been examined. There now appears to be a consensus emerging from the literature that there is little advantage to feeding protein levels in excess of 30 %, and that lipid levels between 4 and 8 % are satisfactory. Although relatively little is known of the lipid biosynthetic pathways in sea urchins, they appear to have some capability for the elongation and desaturation of fatty acids (Bell et al. 2001; Kennedy 2002), and the inclusion of more expensive animal-origin oils is not essential for growth in some species (Pantazis et al. 2000; Kennedy 2002), although Floreto et al. (1996) suggested they may benefit juvenile growth in *Tripneustes gratilla*. It is also very likely there are species-specific differences in dietary needs in the juvenile and adult stages.

An inherent outcome of feeding an artificial formulation is a change in the biochemical composition of the urchin and its gonads (Fernandez 1997; Liyana-Pathirana et al. 2002), which will affect both flavour and colour. Free amino acids are the major factors influencing taste (Murata et al. 2002), and fatty acids and carotenoids are important in the development of 'off flavours', post-mortem (Liyana-Pathirana and Shahidi 2003).

Further research is required to better elucidate how each dietary component influences gonad biochemistry and the relation to gonad flavour. Some diets have been tested on more than one species, for example the 'Wenger' diet (Watts et al. 1998; Olave et al. 2001); however, further trials of one pre-defined diet on a range of sea urchin species would amplify species-specific differences in nutritional needs and assist researchers in optimising artificial diets for each species in culture.

2.7 Carotenoids in Sea Urchin Diets

Roe colour is a critical factor in the commercial product; poor or variable gonad colour at point of sale has a detrimental effect on the value of all species. Therefore cost-effective diets that positively enhance roe colour in adult urchins are key to the success of the industry. Sea urchins do not synthesise carotenoid pigments de novo, so the coloration of their gonad is the result of selective accumulation and modification of pigments from their diet. However, relatively little information is available on the way the primary dietary sources of carotenoids, be they of vegetable or animal origin, influence roe colour in the echinoids of commercial importance. For a review of the occurrence and distribution of carotenoids in echinoids, see Matsuno and Tsushima (2004).

In addition to several studies on the efficiency of pigment transfer from diets to gonads (Havardsson and Imsland 1999; McLaughlin and Kelly 2001; Robinson et al. 2002), the effect of carotenoids from natural and artificial diets on gonad development has been researched (Plank et al. 2002). As well as influencing colour, carotenoids are thought to have a role in biological defence (Kawakami et al. 1998) and reproduction (George et al. 2001). Studies using the same diet formulations and with the same pigment sources (β-carotene from a spray-dried microalgal preparation) (Robinson et al. 2002; Kelly 2004) have a different effect in different urchin species, an indication of species-specific pathways of carotenoid metabolism and expression.

Artificial diets do alter gonad carotenoid composition, but as carotenoid composition changes with sex, season, nutritive state and reproductive stage (Griffiths and Perrott 1976; Borisovets et al. 2002; Young et al. 2004), further research is required to unravel the complexities of pigment metabolism and expression in echinoids.

2.8 Sea Urchin Harvest Protocol, Spoilage and Shelf Life

Japan remains the world's largest consumer of sea urchin roe, and roe exported to Japan is usually delivered processed as chilled, frozen or canned produce. However, there is another undersupplied market in Europe; here, and in France particularly, the market demands a whole urchin. Therefore, harvest protocols should be developed that guarantee the shelf life and quality of sea urchins that are marketed intact. In Europe, sea urchins must conform to the EC Directive on Shellfish Hygiene (statutory instrument 994) and

Food Safety (fishery products and live shellfish). Although there is no requirement for classification of growing water, as with bivalves, other shellfish produce for human consumption must meet the End Product Standard for shellfish toxins and bacterial contamination by *Escherichia coli*. There have been comparatively few published studies (Cook 1999) on the impact of handling and packing protocols on the viability of whole, harvested sea urchins. Spoilage will begin as soon as the physical condition of the sea urchin begins to deteriorate; bacteria are the most important cause of seafood spoilage and spoilage rates are temperature dependent (Dalgaard et al. 2002). For many seafood species, increasing the temperature from 0 to 4 °C doubles the rate of spoilage and cuts the shelf life in half. Sanitation in the handling process is also important. Information on the spoilage rate of sea urchin gonads in situ would enable growers to guarantee the shelf life of urchins, when appropriately packed.

2.9 Disease in Cultured Sea Urchins

There are reports of catastrophic sea urchin die-offs attributable to pathogenic water-borne microorganisms (Lessios et al. 1984; Scheibling and Hennigar 1997), and of heavy infestations by a parasitic nematode in Norwegian populations of *S. droebachiensis* (Sivertsen 1996). The appearance of contagious disease typically accompanies the intensification of culture effort. In Japan, where sea urchins have been in culture the longest, there are reports of bacterial diseases affecting juveniles maintained in tanks (Tajima and Lawrence 2001), the outbreaks related to high summer and low spring seawater temperatures. The symptoms include green or black lesions on the body surfaces, spine loss, discoloration of the peristomal membrane and tube feet that are limp or unable to attach to surfaces. Several bacterial strains have been isolated as the causative agents and methods for their control reported. However, as yet, there is no substantial reporting of contagious sea urchin diseases in cultures in other countries.

3 Sea Cucumber Aquaculture

3.1 Life History and State of the Art

The sea cucumber species targeted for culture belong to two families, the deposit-feeding Aspidochirotida, which includes the Holothuriidae and the Stichopodidae, and the suspension-feeding Dendrochirotida, which includes the genus *Cucumaria*. The sea cucumber species in cultivation are dioecious, broadcast spawners, the fertilised eggs developing into planktonic larvae before settling and undergoing metamorphosis to the juvenile sea cucumber. The average life span of a sea cucumber is thought to be 5–10 years and most species first reproduce at 2–6 years. A number of species are reported to reproduce asexually by fission, and this has been examined as a technique to propagate commercially important species (Reichenbach et al. 1996). They also have the capability to eviscerate part or all of their internal organs as a defence against predation, the shed organs being rapidly regenerated.

Sea cucumber cultivation originated in Japan in the 1930s and juveniles of the temperate species *Stichopus japonicus* were first produced in 1950 (Battaglene 1999). In the last 15 years, commercial production in Japan has accelerated, where annually an estimated 2.5 million juveniles are released. In China, cultured rather than fished *S. japonicus* now account for around 50% of the country's estimated annual production of 2375 t of dry sea cucumber. Methods for mass cultivation of the tropical *Holothuria scabra* are now well established and practised in India, Australia, Indonesia, the Maldives and the Solomon Islands (Battaglene 1999). Other tropical species in culture include *Actinopyga mauritania* (Ramofafia et al. 1996) and *H. fuscogilva* (Ramofafia et al. 2000), with the focus of the research effort centred on the production of juveniles in hatcheries for the restoration and enhancement of wild stocks.

3.2 Sea Cucumber Larviculture

Brood stock, collected from the wild, is most commonly induced to spawn through thermal stimulation, by raising the seawater temperature in holding tanks by 3–5 °C for 1 h. *S. japonicus* broodstock is collected in spring, when mature (Hagen 1996). In general, *H. scabra* has a bi-annual peak in gonadosomatic index, indicating two spawning periods a year, but closer to the equator a proportion of the population spawns year-round (Battaglene 1999). Fertilisation occurs spontaneously once the gametes are allowed to mix in seawater; the fertilised eggs are held in suspension by aeration and egg development is rapid. In *H. scabra* the larval life cycle is around 14 days at 28 °C, including the feeding or auricularia stage, the doliolaria or non-feeding stage and settling pentacula stage (Fig. 4). As with larval sea urchins, holothurian larvae are fed a mixture of microalgal species, with the number of algal cells provided gradually being increased over the larval life. *H. scabra* larvae grow well on a diet of the red microalgae *Rhodomonas salina* and the diatom *Chaetoceros calcitrans* (Battaglene 1999). Further refinement of larval culture techniques is required to improve larval survivorship in this and other species of sea cucumbers.

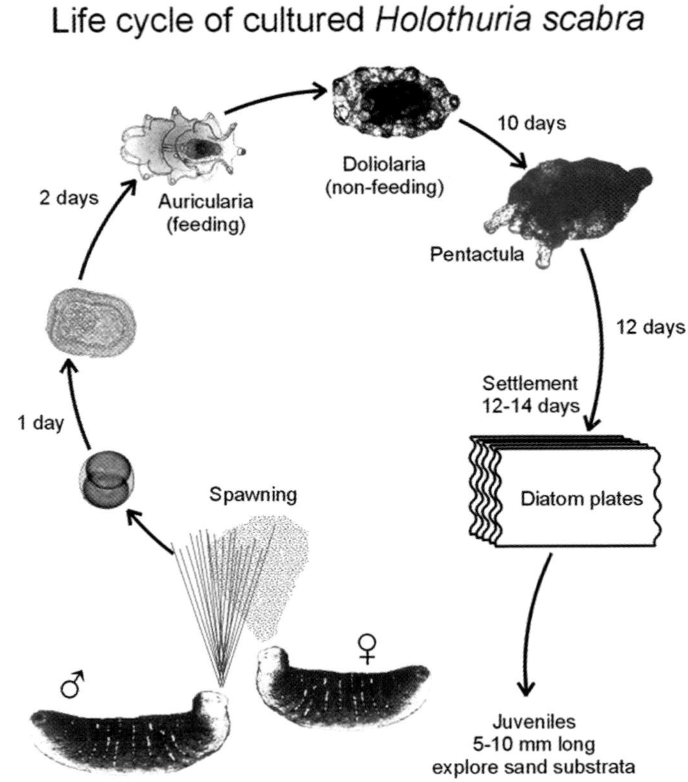

Fig. 4. The 14-day larval cycle of cultured sandfish (*Holothuria scabra*) at a water temperature of 28 °C. (Reproduced with permission from Battaglene 1999)

3.3 Metamorphosis and the Post-Larval Stage

Metamorphosis and settlement are challenging stages in the culture of sea cucumber juveniles. Competent pentacula larvae are provided with a substrate of bacteria and diatoms, which provide the appropriate settlement cues, and to which they adhere with their buccal podia. Typically, *S. japonicus* is settled on PVC plates coated with small periphytic diatoms such as *Navicula*, *Amphora*, *Achnanthes* and *Nitzchia* sp. The plates are coated in outdoor tanks in direct sunlight, although the light intensity, nutrient enrichment and copepod levels must be controlled to produce suitable plates (Ito and Kitamura 1997). Leaves of the sea grass (*Thallassia hemprichii*) are the preferred settlement substrate of *H. scabra* and soluble extracts of the leaves have been shown to induce settlement onto clean plastic surfaces (Mercier et al. 2000).

Post-settlement juvenile sea cucumbers are grown either on diatom-coated plates, held in fine mesh bags in tanks (method used for *S. japonicus* in Japan and China) or on the bottom of tanks, where juveniles of 10–20 mm are transferred to a fine sand substrate and fed a diet supplemented by algal extracts or powdered algae (methods for *H. scabra* developed in India; Battaglene 1999). Throughout the juvenile stage it is necessary to periodically detach the juveniles from the substrate for grading, transfer between tanks or to supply fresh substrates. KCl (1 %) in seawater is an effective agent for detaching *H. scabra* from settlement surfaces (Battaglene and Seymour 1998).

3.4 Sea Cucumber Growth to Maturity

After a 6-month, on-growing nursery phase, and at a length of 2–8 cm, juvenile *S. japonicus* are released to managed areas of the seafloor. They are recovered after 1 year when they measure approximately 20 cm (Hagen 1996). There is a lack of information on growth rates and survivorship in tropical species, and, as with all *Holothuria*, measurements of growth are complicated by their ability to change shape, eviscerate and retain water and sediment in the gut and coelomic cavity. However, Battaglene et al. (1999) suggest there should be no impediment to the large-scale production of juvenile *H. scabra* for stock enhancement programs provided they can be released at a size of 6 cm and with a weight of 20 g.

3.5 Disease in Cultured Holothurians

There is little published information on parasites and diseases of cultured holothurians. Copepods and ciliates are the main predators of the auricularia, copepods also compete with juveniles for food and in some hatcheries have been controlled by the use of pesticides (Battaglene 1999). Routine filtering of seawater and the regular transfer of juveniles to clean tanks can prevent copepod infestation (Battaglene 1999). Fungal infections of the skin are also a problem in the cage culture of wild-collected *H. scabra* in Indonesia.

4 Bioactive Compounds from Echinoderms

Like many other marine organisms, echinoderms have been, and continue to be, examined as a source of biologically active compounds with biomedical applications. Sea cucumber has been valued in Chinese medicine for hundreds of years as a cure for a wide variety of ailments. More recently isolated compounds, mainly from sea cucumbers and starfish, are summarised below.

4.1 Triterpene Glycosides

Triterpenes are mainly synthesised by higher plants, but in animals, cholesterol is an example of a triterpene-like structure. Triterpenes and a group of plant steroids are broadly classified together as saponins (glycosidic surfactants) which affect the solubility of membrane proteins. A well-described toxic effect of plant saponins, when given in high doses, is the haemolysis of red blood cells by the disruption of their membranes.

Novel triterpene glycosides, both sulphated and non-sulphated, have been isolated from sea cucumbers from polar, temperate and tropical regions, some of which have been reported to have significant cytotoxicity against human tumour cell lines (Zou et al. 2003), virucidal activity (Maier et al. 2001), anti-tumour and antiviral activity (Rodriguez et al. 1991), antifungal activity (Murray et al. 2001; Chludil et al. 2002) and to cause haemolysis by membrane disruption (Kalinin et al. 1996).

4.2 Glycosaminoglycans: Chondroitin Sulphate

Glycosaminoglycans (GAGs) (mucopolysaccharides) are polymers of acidic disaccharides containing derivatives of the amino sugars glucosamine or galactosamine. Sulphated polysaccharides abound in vertebrate tissues, and some invertebrate species are a rich source of sulphated GAGs with novel structures. The anticoagulant and antithrombotic characteristics are among the most widely studied properties of the sulphated polysaccharides, for example the anticoagulant GAG heparin is an important therapeutic agent in the prevention and treatment of thrombosis. A replacement agent for heparin is sought as there are problems with both allergy to heparin and heparin resistance. Recently isolated sulphated polysaccharides from the body wall of sea cucumbers, fucosylated chondroitin sulphates (FucCS), have structures analogous to heparin and have been investigated for possible biological activity in mammalian systems. Tapon-Bretaudiere et al. (2002) found that FucCS from a sea cucumber promoted the proliferation of blood vessels and had a concomitant capacity to prevent venous and arterial thrombosis. Mourao et al. (1996), Mourao and Pereira (1999) and Li et al. (2000) have described novel FucCS from sea cucumbers that possess anticoagulant activity in vivo and as such are promising drugs for antithrombotic therapy.

Sea cucumbers have long been used in traditional Chinese medicine for prevention of disease and as a longevity tonic. Products containing sea cucumber chondroitin sulphate are now available through natural product outlets for the treatment of arthritic pain and to promote healthy joints and mobility (Natural Products website, 2004, http://www.psoriasis.com/seacucumber.html).

4.3 Neuritogenic Gangliosides

Sphingolipids are structural lipids where the parent structure is sphingosine (a long-chain amino alcohol) rather than glycerol. Glycosphingolipids (GSLs) contain at least one monosaccharide residue and they are found in the plasma membranes of all animal and some plant cells. Where sialic acid (*N*-acetyl neuraminic acid) is present, these compounds are termed gangliosides. Gangliosides are found in highest concentration in the nervous system where they can constitute 5 % of the lipid. Many new and biologically active gangliosides have been described from starfish. To cite but a few, by way of examples, Higuchi et al. (1991, 1993, 1995) describe compounds with neuritogenic and antitumour activity from *Asterina pectinifera*, *Asterias amurensis* and *Astropecten latespinosus*. New neuritogenic gangliosides have also been described from sea cucumbers, for example *Stichopus japonicus* and *Holothuria* species (Yamada et al. 2001; Kaneko et al. 2003).

4.4 Antimicrobial Activity

Results from some recent studies suggest that echinoderms are a potential source of novel antibiotics. Haug et al. (2002) found antibacterial activity in different body parts of the sea urchin *Strongylocentrotus droebachiensis*, the starfish *Asterias rubens* and the sea cucumber *Cucumaria frondosa*. Antibacterial and antifungal activity has been found in alcoholic extracts of a range of holothurian species from the Tamil Nadu coast, India. The bacteria *Aeromonas hydrophila*, *Escherichia coli*, *Enterococcus* sp., *Klebsiella pneumoniae*, *Pseudomonas aeruginosa*, *Salmonella typhi*, *Staphylococcus aureus* and *Vibrio harveyi* and the fish-borne mould *Aspergillus* sp. were inhibited at varying levels by extracts of the sea cucumbers *Actinopyga miliaris*, *Holothuria atra* and *H. scabra* (Abraham et al. 2002). Antibacterial activity has also been found in extracts of the body wall, coelomocytes and eggs in a variety of species (Stabili and Pagliara 1994; Stabili et al. 1996; Haug et al. 2002).

4.5 Other Types of Bioactive Compounds: Branched-Chain Fatty Acids, Lectins, Opsonins, Analgesics and Anti-ulcer Compounds

Branched-chain fatty acids or fatty alcohols have been reported to possess antitumour activity in various tumour models. Yang et al. (2003) reported that a branched-chain fatty acid, 12-methyltetradecanoic acid (12-MTA), isolated from a sea cucumber, inhibited proliferation of prostate cancer cell lines in culture via apoptosis. The authors suggest that this agent may be a novel adjunctive therapy for selected malignancies including prostate cancer.

Lectins are proteins that possess binding sites for specific mono- and oligosaccharides. A lectin has been isolated from sea cucumber which exhibited cytotoxicity against mouse cancer cells and human lung cancer cells (Gana and Merca 2002). Opsonins are proteins known to bind to the surface of many kinds of pathogens and tag them as targets for phagocytosis. A new opsonin-like molecule in the coelomic fluid of the sea cucumber *H. leucospilota* has been reported (Xing and Chia 2000). A crude water extract of a Malaysian sea cucumber has been shown to possess analgesic activity with a relatively long duration of action (Yaacob et al. 1994). Anti-ulcerative effects on rat stomach cells have been demonstrated in an aqueous/ethanolic extract of *Stichopus japonicus* body-wall muscle (Migas and Klemenchenko 1990).

4.6 Regeneration of Nerve Tissue and Arm Regrowth in Crinoids

Sea cucumbers, starfish and crinoids (feather stars or sea lilies) are well known for their striking regenerative potential. Crinoids can rapidly and completely regenerate arms lost following self-induced or traumatic amputation. Thus they provide a valuable experimental model for investigation of the regenerative process from the macroscopic to the molecular level (Candia Carnevali and Bonasoro 2001; see Candia Carnevali, this Vol.) and for the identification of the genes involved in the process of neural regeneration (Thorndyke et al. 2001). Echinoderm regeneration also provides a convenient model for examining the effects of persistent micropollutants on the developmental physiology (cell proliferation, morphogenesis, differentiation, tissue renewal) of marine animals. The regeneration response of the crinoid *Antedon mediterranea* is especially sensitive to endocrine disrupters such as polychlorinated biphenyls (PCBs), and exposure to these chemicals induces significant variations in the timing and mode of arm regeneration (Candia Carnevali et al. 2001).

5 Sustainable Development

5.1 The Research Requirement

Although there has been a recent increase in research effort into echinoculture, the technology developed for many species outside of Japan has thus far largely been at a research scale and it is likely to require some adaptation to allow commercial operations. Being able to guarantee a supply of seed (juveniles) for on-growing underpins any successful aquaculture operation, and seed supply may prove to be a bottleneck initially as the growing industry scales up. Not all the technologies developed so far (and in particular diet formulations) may be totally transferable between species, and further refinements will be needed.

The aspects still requiring further research for commercial cultivation of sea urchins can be summarised as the need to:

- complete the life cycle in culture;
- improve larval diets and shorten larval life;
- provide suitable settlement substrates that maximise survival at metamorphosis and of the post-larval stages;
- refine artificial diet formulations for juveniles and adults to maximise growth rates and survivorship and produce gonads of the desired taste, texture, flavour and colour;
- optimise grow-out facilities for juveniles and adults either at sea (in containers or 'ranched') or land based;
- attend to packing, food hygiene, transport and marketing requirements.

The first three points in the list are also relevant for developing sea cucumber cultivation where there is a similar bottleneck in seed supplies. Countries already marketing fished produce will have experience in transport and marketing; others will have to establish trade outlets.

There is potentially the need to culture a range of echinoderm species for the growing market in compounds for biomedical research, and culture could play a conservation role in meeting the needs of the home aquarium trade.

5.2 Environmental Considerations

While major reseeding/sea ranching programs, such as that for sea urchins in Japan and for sea cucumbers in China, may be an appropriate way to enhance stocks in areas formerly depleted by fishing, releasing large numbers of captive-bred animals into the wild will undoubtedly impact the genetic composition of those populations. The FAO recommends a minimum breeding population size of 50 for short-term conservation and 500 for longer-term conservation. Relatively small numbers of brood stock are used in hatcheries (Agatsuma 2004; Sakai et al. 2004), and the release of their juveniles will decrease the genetic diversity in local populations. Similarly, where cultured urchins are caged on the seafloor or in suspended culture, their gametes will still be shed to the surrounding seawater. Therefore, consideration should be given to both (1) the desire to genetically manipulate brood stock for better growth characteristics of their progeny and (2) the preservation of genetic diversity in sea urchin populations (Robinson 2004a).

Echinoculture is now poised to expand at a time when globally the aquaculture industry is receiving bad publicity, accused of a range of negative environmental impacts. Therefore, to succeed, echinoculture must develop effectively in a framework of increased legislation affecting businesses operating in the marine environment. This may provide the incentive for polyculture of sea urchins or sea cucumbers with other species that feed at different trophic levels. In such systems, the echinoderms, feeding on uneaten feeds, detritus or seaweeds grown on waste nutrients from the polyculture systems, may serve to reduce the environmental impact of the aquaculture activity (Hagen 1996; Kelly et al. 1998; Shpigel et al. 2004).

5.3 Economic Considerations

It is a general trend that aquaculture operations for marine species do not start until the (wild) fished stock has been diminished to a point where earnings and lifestyle of the people involved are affected (Robinson 2004b). When wild stocks decline, the demand created in the market place raises the price of the product and consequently culturing is more likely to become viable economically. As this review of culture methods has shown, there have been dramatic advances in the culture methods of sea urchins and sea cucumbers in the last 10–15 years, to the extent that one can conclude that currently the major obstacles to successful cultivation are indeed economic rather than biological. For example, it is the cost of producing seed, infrastructure for grow-out systems and artificial diets for growing juveniles to market size rather than the technical difficulty of these operations that will constrain the growth of the industry. At present, it is the reseeding operations that are cost-effective, the 'cost' of the grow-out period being borne by the environment. One of the few examples, outside of Japan, of farmed urchins reaching the market place comes from Southern Ireland, where hatchery-reared juveniles are transplanted to rock pools where they can be fed drift algae, until they reach market size (J. Chamberlain, Dunmanus Seafoods Ltd., pers. comm.).

Hence the future of the echinoculture industry is closely linked to that of the fisheries, whose fate will ultimately determine the market forces that will shape this growing industry.

Acknowledgements. Biochemical definitions were in part drawn from Lackie and Dow (2000).

References

Aas K (2004) Technology for sea-based farming of sea urchins. In: Lawrence JM, Guzman O (eds) Sea urchins: fisheries and aquaculture. DEStech Publications, Lancaster, Pennsylvania, pp 366–373

Abraham TJ, Nagarajan J, Shanmugam SA (2002) Antimicrobial substances of potential biomedical importance from holothurian species. Indian J Mar Sci 31:161–164

Agatsuma Y, Sakai Y, Andrew NL (2004) Enhancement of Japan's sea urchin fisheries. In: Lawrence JM, Guzman O (eds) Sea urchins: fisheries and aquaculture. DEStech Publications, Lancaster, Pennsylvania, pp 18–36

Andrew NL, Agatsuma Y, Ballesteros E, Bazhin AG, Creaser EP, Barnes DKA, Botsford LW, Bradbury A, Campbell A, Dixon J, Einarsson S, Gerring PK, Hebert K, Hunter M, Hur SB, Johnson CR, Juinio-Menez MA, Kalvass P, Miller RJ, Moreno CA, Palleiro JS, Rivas D, Robinson SML, Schroeter SC, Steneck RS, Vadas RL, Woodby DA, Xiaoqi Z (2002) Status and management of world sea urchin fisheries. Oceanogr Mar Biol Annu Rev 40:343–425

Barnes DKA, Crook AC (2001) Implications of temporal and spatial variability in *Paracentrotus lividus* populations to the associated commercial coastal fishery. Hydrobiologia 465(1–3):95–102

Barnes DKA, Verling E, Crook A, Davidson I, O'Mahoney M (2002) Local population disappearance follows (20 yr after) cycle collapse in a pivotal ecological species. Mar Ecol Prog Ser 226:311–313

Battaglene SC (1999) Culture of tropical sea cucumbers for stock enhancement. Naga, ICLARM Q 22(4):4–11

Battaglene SC, Seymour JE (1998) Detachment and grading of the tropical sea cucumber sandfish, *Holothuria scabra*, juveniles from settlement substrates. Aquaculture 159:263–274

Battaglene SC, Seymour JE, Ramofafia C (1999) Survival and growth of cultured juvenile sea cucumbers. Aquaculture 178:293–322

Bell MV, Dick JR, Kelly MS (2001). Biosynthesis of eicosapentaenoic acid in the sea urchin *Psammechinus miliaris*. Lipids 36(1):79–82

Borisovets EE, Zadorozhny PA, Kalinina MV, Lepskaya NV, Yakush EV (2002) Changes of major carotenoids in gonads of sea urchins (*Strongylocentrotus intermedius* and *S. nudus*) at maturation. Comp Biochem Physiol B 132(4):779–790

Bruckner AW, Johnson KA, Field JD (2003) Conservation strategies for sea cucumbers: can CITES Appendix II listing promote sustainable international trade? SPC Beche-de-Mer Info Bull 18:24–36

Byrne M (1993) Annual reproductive cycles of the commercial sea urchin *Paracentrotus lividus* from an exposed intertidal and a sheltered subtidal habitat on the west coast of Ireland. Mar Biol 104:275–289

Caers M, Couteau P, Lombeida P, Sorgeloos P (1998) The effect of lipid supplementation on growth and fatty acid composition of *Tapes philippinarum* spat. Aquaculture 162(3–4): 287–299

Candia Carnevali MD, Bonasoro F (2001) Microscopic overview of crinoid regeneration. Microscopy Res Tech 55:403–426

Candia Carnevali MD, Bonasoro F, Patruno M, Thorndyke MC, Galassi S (2001) PCB exposure and regeneration in crinoids (Echinodermata). Mar Ecol Prog Ser 215:155–167

Chludil HD, Muniain CC, Seldes AM, Maier MS (2002) Cytotoxic and antifungal triterpene glycosides from the Patagonian sea cucumber *Hemoiedema spectabilis*. J Nat Prod 65:860–865

Conand C, Byrne M (1993) A review of recent developments in the world of sea cucumber fisheries. Mar Fish Rev 55:1–13

Cook EJ (1999) *Psammechinus miliaris* (Gmelin) (Echinodermata: Echinoidea): factors effecting its somatic growth and gonadal growth and development. PhD Thesis, Napier University, Edinburgh

Cook EJ, Kelly MS, Mckenzie JD (1998) Gonad development and somatic growth in the green sea urchin, *Psammechinus miliaris* fed artificial and natural diets. J Shellfish Res 17(5):1549–1555

Cuthbert FM, Hooper RG, McKeever T (1995) Sea urchin aquaculture phase I: sea urchin feeding and ranching experiments. Report of Project AUI-503. Canadian Centre for Fisheries Innovation, Government of Newfoundland

Dalgaard P, Buch P, Silberg S (2002) Seafood spoilage predictor – development and distribution of a product specific application software. Int J Food Microbiol 73(2–3):343–349

Dalzell P, Adams TJH, Polunin NVC (1996) Coastal fisheries in the Pacific Islands. Oceanogr Mar Biol Annu Rev 34:395–531

De Jong-Westman M, Qian P-Y, March BE, Carefoot TH (1995) Artificial diets in sea urchin culture: effects of dietary protein level and other additives on egg quality, larval morphometrics and larval survival in the green sea urchin *Strongylocentrotus droebachiensis*. Can J Zool 73:2080–2090

Ebert TA (1998) An analysis of the importance of allee effects in management of the red sea urchin *Strongylocentrotus franciscanus*. In: Mooi R, Telford T (eds) Echinoderms: San Francisco. AA Balkema, Rotterdam, pp 619–627

Fenaux L, Cellario C, Rassoulzadegan F (1988) Sensitivity of different morphological stages of the larva of *Paracentrotus lividus* (Lamarck) to quantity and quality of food. In: Burke R, Mladenov P, Lambert P, Parsley R (eds) Echinoderm biology. AA Balkema, Rotterdam, pp 259–266

Fernandez C (1997) Effect of diet on the biochemical composition of *Paracentrotus lividus* (Echinodermata: Echinoidea) under natural and rearing conditions (effect of diet on biochemical composition of urchins). Comp Biochem Physiol A 118(4):1377–1384

Fernandez C, Boudouresque CF (2000) Nutrition of the sea urchin *Paracentrotus lividus* (Echinodermata: Echinoidea) fed different artificial food. Mar Ecol Prog Ser 204:131–141

Floreto EAT, Techima S, Ishikawa M (1996) The effects of seaweed diets on the growth, lipid and fatty acids of juveniles of the white sea urchin *Tripneustes gratilla*. Fish Sci 62(4):589–593

Gana AE, Merca FE (2002) Isolation and purification of a cytotoxic lectin from brown sea cucumber by affinity chromatography. Philipp Agric Sci 85:236–247

George SB, Lawrence JM, Lawrence AL, Smiley J, Plank L (2001) Carotenoids in the adult diet enhance egg and juvenile production in the sea urchin *Lytechinus variegatus*. Aquaculture 199(3–4):353–369

Goebel N, Barker MF (1998) Artificial diets supplemented with carotenoid pigments as feeds for sea urchins. In: Mooi R, Telford M (eds) Echinoderms: San Francisco. AA Balkema, Rotterdam, pp 667–672

Gonzalez LP, Castilla JC, Guisado C (1987) Effect of larval diet and rearing temperature on metamorphosis and juvenile survival of the edible sea urchin *Loxechinus albus* (Molina, 1782) (Echinoidea, Echinidae). J Shellfish Res 6(2):109–115

Griffiths M, Perrott P (1976) Seasonal changes in the carotenoids of the sea urchin *Strongylocentrotus droebachiensis*. Comp Biochem Physiol B 55:435–441

Grosjean P, Spirlet C, Gosselin P (1998) Land-based, closed-cycle echinoculture of *Paracentrotus lividus* (Lamarck) (Echinoidea: Echinodermata): a long-term experiment at a pilot scale. J Shellfish Res 17(5):1523–1531

Guillou M, Lumingas LJL (1999) Variation in the reproductive strategy of the sea urchin *Sphaerechinus granularis* (Echinodermata: Echinoidea) related to food availability. J Mar Biol Assoc UK 79:131–136

Hagen NT (1996) Echinoculture: from fishery enhancement to closed cycle cultivation. World Aquacult 27(4): 6–19

Hagen NT (1998) Effect of food availability and body size on out-of-season gonad yield in the green sea urchin, *Strongyocentrotus droebachiensis*. J Shellfish Res 17(5):1533–1539

Hagen NT (2000) Echinoderm culture. In: Stickney RR (ed) Encyclopedia of aquaculture. Wiley-Interscience, New York, pp 247–253

Hammer HS, Watts SA, Lawrence JM, Lawrence AL, McClintock JB (2000) The effect of dietary protein concentration on gonad composition and gametic condition in the sea urchin *Lyechinus variegatus*. Am Zool 40(6):1042–1043

Harris LG, Tyrell MC, Williams CT, Sisson CG, Chavanich S, Chester CM (2001) Declining sea urchin recruitment in the Gulf of Maine: is overharvesting to blame? In: Barker M (ed) Echinoderms 2000. AA Balkema, Rotterdam, pp 439–444

Haug T, Kjuul AK, Styrvold OB, Sandsdalen E, Olsen OM, Stensvag K (2002) Antibacterial activity in *Strongylocentrotus droebachiensis* (Echinoidea), *Cucumaria frondosa* (Holothuroidea), and *Asterias rubens* (Asteroidea) J Invertebr Pathol 81(2):94–102

Havardsson B, Imsland AK (1999) The effect of astaxanthin in feed and environment temperature on carotenoid concentration in the gonads of the green sea urchins *Strongylocentrotus droebachiensis* Müller. J World Aqua Soc 30(2):208–218

Higuchi R, Inagaki K, Natori T, Komori T, Kawajiri S (1991) Biologically-active glycosides from Asteroidea 25. Glycosphingolipids from the starfish *Asterina pectinifera*. 2. Structure of three ganglioside molecular-species and a homogeneous ganglioside, and biological-activity of the ganglioside. Liebigs Ann 1:1–10

Higuchi R, Inukai K, Jhou JX, Honda M, Komori T, Tsuji S, Nagai Y (1993) Biologically-active glycosides from Asteroidea 31. Glycosphingolipids from the starfish *Asterias amurensis versicolor* Sladen. 2. Structure and biological-activity of ganglioside molecular-species Liebigs Ann 4:359–366

Higuchi R, Matsumoto S, Fujita M, Komori T, Sasaki T (1995) Biologically-active glycosides from Asteroidea 32. Glycosphingolipids from the starfish *Astropecten latespinosus*. 2. Structure of two new ganglioside molecular-species and biological-activity of the ganglioside. Liebigs Ann 3:545–550

Himmelman JH (1977) Reproductive cycle of the green sea urchin *Strongylocentrotus droebachiensis*. Can J Zool 56:1828–1836

Ito S, Kitamura H (1997) Induction of larval metamorphosis in the sea cucumber *Stichopus japonicus* by periphytic diatoms. Hydrobiologia 358:281–284

Jensen M (1969) Breeding and growth of *Psammechinus miliaris*. Ophelia 7:65–78

Jimmy RA, Kelly MS, Beaumont AR (2003) The effect of diet type and quantity on the development of common sea urchin larvae *Echinus esculentus*. Aquaculture 220(1–4):261–275

Kalinin VI, Prokofieva NG, Likhatskaya GN, Schentsova EB, Agafonova IG, Avilov SA, Drozdova OA (1996) Hemolytic activities of triterpene glycosides from the holothurian order dendrochirotida: some trends in the evolution of this group of toxins. Toxicon 34(4): 475–483

Kaneko M, Kisa F, Yamada K, Miyamoto T, Higuchi R (2003) Structure of a new neuritogenic-active ganglioside from the sea cucumber *Stichopus japonicus*. Eur J Org Chem 6:1004–1008

Kawakami T, Tsushima M, Katabami Y, Mine M, Ishida A, Matsuno T (1998) Effect of beta, beta-carotene, beta-echinenone, astaxanthin, fucoxanthin, vitamin A and vitamin E on the biological defence of the sea urchin *Pseudocentrotus depressus*. J Exp Mar Biol Ecol 226(2):165–174

Keesing JK, Hall KC (1998) Review of the status of world sea urchin fisheries points to opportunities for aquaculture. J Shellfish Res (5):1597–1604

Kelly MS (2000) The reproductive cycle of the sea urchin *Psammechinus miliaris* (Gmelin) (Echinodermata: Echinoidea) in a Scottish sea loch. J Mar Biol Assoc UK 80:909–919

Kelly MS (2001) Environmental parameters controlling gametogenesis in the echinoid *Psammechinus miliaris*. J Exp Mar Biol Ecol 266(1):67–80

Kelly MS (2002) Survivorship and growth rates of hatchery-reared sea urchins. Aquacult Int 10:309–316

Kelly MS (2004) *Echinus esculentus*, a fully farmed approach for a designer roe? In: Lawrence JM, Guzman O (eds) Sea urchins: fisheries and aquaculture. DEStech Publications, Lancaster, Pennsylvania, pp 256–263

Kelly MS, Brodie CC, McKenzie JD (1998) Somatic and gonadal growth of the sea urchin *Psammechinus miliaris* (Gmelin) maintained in polyculture with the Atlantic salmon. J Shellfish Res 17(5):1557–1562

Kelly MS, Hunter AJ, Scholfield C, McKenzie JD (2000) Morphology and survivorship of larval *Psammechinus miliaris* (Gmelin) (Echinoidea: Echinodermata) in response to varying food quantity and quality. Aquaculture 183:223–240

Kelly MS, Owen PV, Pantazis P (2001) The commercial potential of the common sea urchin *Echinus esculentus* from the west coast of Scotland. Hydrobiologia 465(1–3):85–94

Kennedy E (2002) Effect of proteins, lipids, minerals, and pigment in prepared diets on the somatic growth of juvenile green sea urchins, *Strongylocentrotus droebachiensis*. MSc thesis, Memorial University of Newfoundland

Klinger TS, Lawrence JM, Lawrence AL (1994) Digestive characteristics of the sea urchin *Lytechinus variegatus* (Lamarck) (Echinodermata: Echinoidea) fed prepared feeds. J World Aqua Soc 25(4):489–496

Lackie JM, Dow JAT (2000) The dictionary of cell and molecular biology, 3rd edn. Academic Press, London

Lawrence JM (2001) Edible sea urchins: biology and ecology. Developments in aquaculture and fisheries science 32. Elsevier, Amsterdam, 419 pp

Lawrence JM, Olave S, Otaiza R, Lawrence AL, Bustos E (1997) Enhancement of gonad production in the sea urchin *Loxechinus albus* in Chile fed extruded feeds. J World Aqua Soc 28(1):91–96

Leighton P (1995) Contributions towards the development of echinoculture in North Atlantic waters with particular reference to *Paracentrotus lividus* (Lamarck). PhD thesis, National University of Ireland, Galway

Lessios HA, Cubit JD, Robertson DR, Shulman MJ, Parker MR, Garrity SD, Levings SC (1984) Mass mortality of *Diadema antillarum* on the Caribbean coast of Panama. Coral Reefs 3(4):173–182

Li ZG, Wang HL, Li JZ, Zhang GS, Gao CJ (2000) Basic and clinical study on the antithrombotic mechanism of glycosaminoglycan extracted from sea cucumber Chinese Med J Peking 113(8):706–711

Liyana-Pathirana CM, Shahidi F (2003) Effect of an artificial diet on lipid, free amino acid, and carotenoid composition of green sea urchin gonads. Off-flavors in aquaculture. ACS Symp Ser 848:83–93

Liyana-Pathirana CM, Shahidi F, Whittick A (2002) Effect of an artificial diet on the biochemical composition of the gonads of the sea urchin *Strongylocentrotus droebachiensis*. Food Chem 79(4):461–472

MacBride EW (1903) The development of *Echinus esculentus* together with some points on the development of *E. miliaris* and *E. acutus*. Philos Trans R Soc B 195:285–330

Maier MS, Roccatagliata AJ, Kuriss A, Chludil H, Seldes AM, Pujol CA, Damonte EB (2001) Two new cytotoxic and virucidal trisulfated triterpene glycosides from the Antarctic sea cucumber *Staurocucumis liouvillei*. J Nat Prod 64(6):732–736

Matsuno T, Tsushima M (2004) Carotenoids in sea urchins. In: Lawrence JM (ed) (2001) Edible sea urchins: biology and ecology. Developments in aquaculture and fisheries science 32. Elsevier, Amsterdam, pp 115–138

McBride SC, Lawrence JM, Lawrence AL, Mulligan TJ (1998) The effect of protein concentration in prepared diets on growth, feeding rate, total organic absorption, and gross assimilation efficiency of the sea urchin *Strongylocentrotus franciscanus*. J Shellfish Res 17(5):1562–1570

McEdward LR (1984) Morphometric and metabolic analysis of the growth and development of an echinopluteus. J Exp Mar Biol Ecol 82:259–287

McLaughlin G, Kelly MS (2001) Effect of artificial diets containing carotenoid-rich microalgae on gonad growth and colour in the sea urchin *Psammechinus miliaris* (Gmelin). J Shellfish Res 20(1):377–382

Mercier A, Battaglene SC, Hamel J-F (2000) Settlement preferences and early migration of the tropical sea cucumber *Holothuria scabra*. J Exp Biol Mar Ecol 249:89–110

Migas EA, Klemenchenko EG (1990) Pharmacological effects of the sea-cucumber *Stichopus japonicus* extract. Biol Morya-Mar Biol 5:72–76

Mourao PAS, Pereira MS (1999) Searching for alternatives to heparin – sulfated fucans from marine invertebrates. Trends Cardiovas Med 9(8):225–232

Mourao PAS, Pereira MS, Pavao MSG, Mulloy B, Tollefsen DM, Mowinckel MC, Abildgaard U (1996) Structure and anticoagulant activity of a fucosylated chondroitin sulfate from echinoderm – sulfated fucose branches on the polysaccharide account for its high anticoagulant action. J Biol Chem 271(39):23973–23984

Moylan E (1997) Gonad conditioning and wild stock enhancement of the purple sea urchin *Paracentrotus lividus* on the west coasts of Ireland. Bull Aquacult Assoc Can 97(1):38–45

Murata Y, Yokoyama M, Unuma T, Sata NU, Kuwahara R, Kaneniwa M (2002) Seasonal changes of bitterness and pulcherrimine content in gonads of green sea urchin *Hemicentrotus pulcherrimus* at Iwaka in Fukushima prefecture. Fish Sci 68(1):184–189

Murray AP, Muniain C, Seldes AM, Maier MS (2001) Patagonicoside A: a novel antifungal disulfated triterpene glycoside from the sea cucumber *Psolus patagonicus*. Tetrahedron 57(47):9563–9568

Olave S, Bustos E, Lawrence JM, Carcamo P (2001) The effect of size and diet on gonad production by the Chilean sea urchin *Loxechinus albus*. J World Aqua Soc 32(2):210–214

Otero-Villanueva M, Kelly MS, Burnell G (2004) How diet influences energy partitioning in the regular echinoid *Psammechinus miliaris*; constructing an energy budget. J Exp Mar Biol Ecol 304(2):159–181

Pantazis P, Kelly MS, Connolly JG, Black KD (2000) Effect of artificial diet on growth, lipid utilisation, and gonad biochemistry in the adult sea urchin *Psammechinus miliaris*. J Shellfish Res 19(2):995–1001

Pearce CM, Daggett TL, Robinson SMC (2002a) Effect of protein source ratio and protein concentration in prepared diets on gonad yield and quality of the green sea urchin, *Strongylocentrotus droebachiensis*. Aquaculture 214(1–4):307–332

Pearce CM, Daggett TL, Robinson SMC (2002b) Effect of binder type and concentration on prepared feed stability and gonad yield and quality of the green sea urchin *Strongylocentrotus droebachiensis*. Aquaculture 205:301–323

Plank LR, Lawrence JM, Lawrence AL, Olvera RM (2002) The effect of dietary carotenoids on gonad production and carotenoid profiles in the sea urchin *Lytechinus variegatus*. J World Aqua Soc 33(2):127–137

Ramofafia C, Foyle TP, Bell JD (1996) Growth of juvenile *Actinopyga mauritania* (Holothuroidea) in captivity. Aquaculture 152:119–128

Ramofafia C, Battaglene SC, Bell JD, Byrne M (2000) Reproductive biology of the commercial sea cucumber *Holothuria fuscogilva* in the Solomon Islands. Mar Biol 136:1045–1056

Reichenbach N, Nishar Y, Saeed A (1996) Species and size-related trends in asexual propagation of commercially important species of tropical sea cucumbers (Holothuria). J World Aqua Soc 27(4):475–482

Robinson SM (2004a) A roundtable discussion of the future of aquaculture of sea urchins. In: Lawrence JM, Guzman O (eds) Sea urchins: fisheries and aquaculture. DEStech Publications, Lancaster, Pennsylvania, pp 387–392

Robinson SM (2004b) The evolving role of aquaculture in the global production of sea urchins. In: Lawrence JM, Guzman O (eds) Sea urchins: fisheries and aquaculture. DEStech Publications, Pennsylvania, pp 343–357

Robinson SMC, Castell JD, Kennedy EJ (2002) Developing suitable colour in the gonads of cultured green sea urchins (*Strongylocentrotus droebachiensis*). Aquaculture 206:289–303

Rodriguez J, Castro R, Riguera R (1991) Holothurinosides – new antitumor nonsulfated triterpenoid glycosides from the sea-cucumber *Holothuria forskalii*. Tetrahedron 47(26):4753–4762

Sakai Y, Tajima KI, Agatsuma Y (2004) Mass production of seed of the Japanese edible sea urchins *Strongylocentrotus droebachiensis* and *Strongylocentrotus nudus*. In: Lawrence JM, Guzman O (eds) Sea urchins: fisheries and aquaculture. DEStech Publications, Lancaster, Pennsylvania, pp 287–298

Scheibling RE, Hennigar AW (1997) Recurrent outbreaks of disease in sea urchins *Strongylocentrotus droebachiensis* in Nova Scotia: evidence for a link with large-scale meteorologic and oceanographic events. Mar Ecol Prog Ser 152(1–3):155–165

Sivertsen K (1996) Incidence, occurrence and distribution of the nematode *Echinodermata matsi* in its echinoid host, *Strongylocentrotus droebachiensis*, in northern Norway. Mar Biol 126:703–714

Sivertsen K (2004) Harvestable sea urchin *Strongylocentrotus droebachiensis* resources along the Norwegian coast. In: Lawrence JM, Guzman O (eds) Sea urchins: fisheries and aquaculture. DEStech Publications, Lancaster, Pennsylvania, 85 pp

Shpigel M, McBride SC, Marciano S, Lupatsch I (2004) Propagation of the European sea urchin *Paracentrotus lividus* in Israel. In: Lawrence J, Guzman O (eds) Sea urchins: fisheries and aquaculture. DEStech, Publications, Lancaster, Pennsylvania, 85 pp

Sloan NA (1984) Echinoderm fisheries of the world: a review. In Keegan BF, O'Connor BDS (eds) Proc 5th Int Echinoderm Conf, Galway. AA Balkema, Rotterdam, pp 109–124

Spirlet C, Grosjean P, Jangoux M (2001) Cultivation of *Paracentrotus lividus* (Echinodermata: Echinoidea) on extruded feeds: digestive efficiency, somatic and gonadal growth. Aquacult Nutr 7(2):91–99

Stabili L, Pagliara P (1994) Antibacterial protection in *Marthasterias glacialis* eggs – characterization of lysozyme-like activity. Comp Biochem Physiol B 109(4):709–713

Stabili L, Pagliara P, Roch P (1996) Antibacterial activity in the coelomocytes of the sea urchin *Paracentrotus lividus*. Comp Biochem Physiol B 113:639–644

Strathmann RR, Fenaux L, Strathmann MF (1992) Heterochronic developmental plasticity in larval sea urchins and its implications for evolution of non-feeding larvae. Evolution 46:972–986

Tajima K, Lawrence M (2001) Disease in edible urchins. In: Lawrence JM (ed) Edible sea urchins: biology and ecology. Developments in aquaculture and fisheries science 32. Elsevier, Amsterdam, pp 139–148

Tapon-Bretaudiere J, Chabut D, Zierer M, Matou, S, Helley D, Bros A, Mourao PAS, Fischer AM (2002) A fucosylated chondroitin sulfate from echinoderm modulates in vitro fibroblast growth factor 2-dependent angiogenesis. Mol Cancer Res 1(2):96–102

Thorndyke MC, Chen WC, Beesley PW, Patruno M (2001) Molecular approach to echinoderm regeneration. Microsc Res Tech 55(6):474–485

Watts SA, Boettger SA, McClintock JB, Lawrence JM (1998) Gonad production in the sea urchin *Lytechinus variegatus* (Lamark) fed prepared diets. J Shellfish Res 17(5):1591–1595

Xing J, Chia FS (2000) Opsonin-like molecule found in coelomic fluid of a sea cucumber, *Holothuria leucospilota*. Mar Biol 136(6):979–986

Yaacob HB, Kim KH, Mustapha MS (1994) Antinociceptive effect of the water extract of Malaysian sea-cucumber. Asia Pac J Pharmacol 9(1):23–28

Yamada K, Matsubara R, Kaneko M, Miyamoto T, Higuchi R (2001) Constituents of Holothuroidea. 10. Isolation and structure of a biologically active ganglioside molecular species from the sea cucumber *Holothuria leucospilota*. Chem Pharm Bull 49(4):447–452

Yang PY, Collin P, Madden T, Chan D, Sweeney-Gotsch B, McConkey D, Newman RA (2003) Inhibition of proliferation of PC3 cells by the branched-chain fatty acid, 12-methyltetradecanoic acid, is associated with inhibition of 5-lipoxygenase. Prostate 55(4):281–291

Young A, Phillips D, Waring C, Kelly MS (2004) Carotenoid profiles of the edible echinoid *Paracentrotus lividus*. In: Lawrence J, Guzman O (eds) Sea urchins: fisheries and aquaculture. DEStech Publications, Lancaster, Pennsylvania, 172 pp

Zou ZR, Yi YH, Wu HM, Wu JH, Liaw CC, Lee KH (2003) Intercedensides A–C, three new cytotoxic triterpene glycosides from the sea cucumber *Mensamaria intercedens* Lampert. J Nat Prod 66(8):1055–1060

Regenerative Response and Endocrine Disrupters in Crinoid Echinoderms: An Old Experimental Model, a New Ecotoxicological Test

M.D. Candia Carnevali

Abstract The regenerative phenomena that reproduce developmental processes in adult organisms and are regulated by endocrine and neurohumoral mechanisms can provide new sensitive tests for monitoring the effects of exposure to anthropogenic chemicals such as *endocrine disrupter* (ED) contaminants. These pollutants in fact can be bioaccumulated by the organisms, causing dysfunctions in steroid hormone production/metabolism and activities and inducing dramatic effects on reproductive competence, development and growth in many animals, man included. Current research is exploring the effects of exposure to different classes of compounds well known for their ED activity, such as polychlorinated biphenyls (PCBs), nonylphenols and organotins, on regenerative potential of echinoderms, a relatively unexplored and promising applied approach which offers the unique chance to study physiological developmental processes in adult animals. The selected test species is the crinoid *Antedon mediterranea*, which represents a valuable experimental model for investigation into the regenerative process from the macroscopic to the molecular level. The present study employs an integrated approach which combines exposure experiments, chemical analysis and biological analysis utilizing classical methods of light (LM) and electron (TEM and SEM) microscopy and immunocytochemistry. The experiments were carried out on experimentally induced arm regenerations in controlled conditions with exposure concentrations comparable to those of moderately polluted coastal zones in order to reproduce common conditions of exposure to environmental contaminants. The results of the exposure tests were analysed in terms of effects at the whole organism, at the tissue and cellular level, and possible sites of action of EDs. Our results show that prolonged

M.D. Candia Carnevali (e-mail: daniela.candia@unimi.it)
Dipartimento di Biologia, Universita' degli Studi di Milano, Via Celoria 26, 20133 Milano, Italy

Progress in Molecular and Subcellular Biology
Subseries Marine Molecular Biotechnology
V. Matranga (Ed.), Echinodermata

exposure to these compounds significantly affects the regenerative mechanisms by inducing appreciable anomalies in terms of regeneration times, overall growth, general morphology and histological and cellular pattern. A concentration/effect relationship could be found for all the substances. Interestingly, contrasting results in terms of inhibition or acceleration of regeneration phenomenon were obtained for the different chemicals.

1 Regeneration and Its Biological Implications

Development in animals does not necessarily mean embryogenesis and does not always start from an egg, fertilized or not. Regeneration is in fact a distinct type of developmental process typically occurring in adult animals: it can involve limited processes of cell turnover and tissue repair, replacement of lost parts or organs, and even complete regrowth of whole individuals from small body fragments. Due to its obvious close relation with fission phenomena and cloning processes, regeneration can be considered the specific developmental strategy complementary to asexual reproduction, in the same way as embryogenesis is complementary to sexual gametic reproduction. Therefore, although regeneration unavoidably involves analogous problems of cellular identity and positioning and often superficially resembles embryogenesis in an accelerated form, this basic difference in its intrinsic asexual starting-point makes regeneration a significantly different biological process. In embryogenesis, the structure is totally created ex novo, whereas in regeneration an already existing structure is reformed after its loss or severe injury and the new cells develop in an established context of mature tissues and cells. Therefore, in all animals, the regenerative processes related to different organs and structures should be regarded as fundamentally distinct developmental processes rather than as an accelerated recapitulation of ontogenetic processes (Candia Carnevali and Bonasoro 2001a).

Although a response to injury is evoked in all animals, the degree of morphological and functional recovery can show a remarkable variability not only between unrelated groups, but also between closely related species. In contrast to the old traditional views which regard the regenerative potential as a prerogative of the simplest and most primitive animals, regeneration is actually a widespread phenomenon through phylogeny and its quite heterogeneous distribution from the lowest to the highest phyla appears to be independent of their organization and complexity level (Ferretti and Géraudie 1998; Thouveny and Tassava 1998; Candia Carnevali and Bonasoro 2001a). In fact, the regenerative capabilities appear to depend upon the individual potential for histogenetic and morphogenetic plasticity expressed in terms of recruitment of stem cells and/or dedifferentiated cells, cell proliferation and migration, supply of specific regulatory/trophic factors, and finally expression or re-expression of the developmental programme in adult animals.

1.1 Regeneration in Echinoderms

In spite of the wide choice of potential models for studying regeneration, this phenomenon has been extensively explored only in a few animals, which traditionally monopolized the attention of the developmental biologists. In contrast, with regard to many animal groups well known for their regenerative capabilities, there are surprising gaps in knowledge in terms of not only cellular and molecular aspects of regeneration, but also basic mechanisms (Candia Carnevali and Bonasoro 2001a).

Regarding general mechanisms, regeneration in all animals is traditionally considered to involve one or other of two basic processes, epimorphosis and morphallaxis. In epimorphosis, new tissues arise from undifferentiated cells (stem cells or dedifferentiated cells) which form a typical blastema. This is a discrete centre of proliferative activity providing a pool of new cells which can give rise to all the regenerated structures. In morphallaxis, extensive phenomena of rearrangement/recycling from existing tissues take place: no blastema is involved and there is only limited and localized proliferation of cells derived from existing tissues by dedifferentiation, transdifferentiation and/or migration. In spite of this apparently clear and well-established difference between epimorphosis and morphallaxis, recent results obtained in the same model in different experimental conditions (see crinoid arm regeneration, Candia Carnevali and Bonasoro 2001b) suggest that the traditional classification of these two processes is too reductive and that the mechanisms at the tissue/cellular level can be largely interchangeable and interpretable in a more plastic and dynamic light.

Echinoderms, which thanks to their spectacular regenerative capabilities were the favourite models of the pioneer *regenerationists* of the 19th and early 20th centuries, after a long period of unexplainable neglect, were reproposed to our attention by a series of recent papers (for review see Candia Carnevali and Bonasoro 2001b; Candia Carnevali et al. 2001 c; Thorndyke and Candia Carnevali 2001) exploring the basic mechanisms of the regenerative phenomenon and its cellular and molecular aspects. Regenerative potential finds in echinoderms its maximum expression (Hyman 1955). It is a common phenomenon in all the classes, extensively employed to reconstruct external parts (arms or other appendages) and internal organs (gonads, gut, visceral mass) often subjected to amputation, self-induced or traumatic, rapidly followed by complete successful regrowth of the lost parts. Regeneration in echinoderms is largely a predicted phenomenon and in most cases follows autotomy, which can be considered the most important proximate cause of structural loss and depends on the presence and properties of "mutable collagenous tissues" (MCTs) (for a review see Wilkie 2001; Wilkie, this Vol.) at the level of the autotomy plane. Under physiological conditions regeneration is prompted by autotomy and proceeds from the retained side of a fractured autotomy plane.

Reconstitutive regeneration of arms is particularly frequent in crinoids and ophiuroids which have fragile arms often involved in self-induced or traumatic mutilations. Interestingly, in many cases, the detached body fragments can survive in good health for a long time and undergo phenomena of partial or total regeneration independently of the *donor* animal (Candia Carnevali et al. 1998). These phenomena, which are also quite common in asteroids, provide evidence of the wide exploitation and implications of regenerative potential in echinoderms. In particular in asteroids, besides the extensive application in common repair mechanisms, arm regeneration offers in fact the most complete example of cloning strategies. As well known, in a few starfish species, individual autotomized arms can regenerate to produce new complete adults. This extreme case clearly shows that in echinoderms regeneration is an indispensable complement of the programme of asexual reproduction which leads to the development of new individuals through fission mechanisms (Emson and Wilkie 1980; Mladenov and Burke 1994). Besides asteroids, also many ophiuroids and holothuroids undergo asexual propagation involving the splitting of adults into two or three pieces, with subsequent regenerative development of complete individuals from each isolated portion. This extensive and strategic employment of regenerative phenomena throughout the phylum indicates that in echinoderms regeneration actually represents an essential component of the life cycle, and this has a wide range of biological implications. In fact, if a close correlation between the regenerative potential of the individual and its possibility of survival can be inferred easily, self-repair abilities not only appear to be an undoubted advantage for the individual, but also give a fundamental contribution to the adaptive capacities of the species and its fitness, since they increase the individual's chances of reproducing, sexually or asexually, even when it is dramatically compromised in its body integrity.

Regeneration by both epimorphosis and morphallaxis is found in echinoderms (Candia Carnevali and Bonasoro 2001a). Epimorphosis with blastema formation seems to typically occur whenever regeneration is a widely predictable, rapid and effective phenomenon, which takes place following autotomy (for instance in crinoids and ophiuroids; Candia Carnevali and Bonasoro 2001b; Thorndyke et al. 2001). These epimorphic mechanisms appear to reproduce rather closely what usually happens in embryonic developmental processes. In contrast, morphallaxis seems to be a more complicated and slower regenerative process which tends to follow traumatic mutilations, for instance in arm tip regeneration of asteroids (Mladenov et al. 1989; Moss et al. 1998). In this case, amputation is not a predictable event and the morphogenetic mechanisms imply phenomena of substantial rearrangement of the old structures, which appear to be unique to regeneration and not shared by embryonic development. As stated above, investigation of crinoid regeneration under different experimental conditions (see Candia Carnevali and Bonasoro 2001b) suggests that this distinction is quite artificial, and that the borders between these two processes are not so defined, especially in terms of mechanisms at the tissue/cellular level.

Although occurring in all echinoderm classes, regeneration has been studied most thoroughly in crinoids, ophiuroids and asteroids (for review see Candia Carnevali and Bonasoro 2001b; Candia Carnevali et al. 2001 c; Thorndyke and Candia Carnevali 2001). In holothuroids it is also ubiquitous but has been explored in terms of mechanisms only in a few cases (Dolmatov and Ginanova 2001; Garcia-Arraras and Greenberg 2001). Regeneration also occurs in echinoids, but is limited in terms of extent and degree of capabilities and only a few examples have been investigated (Dubois and Ameye 2001; Bonasoro et al. 2004). In the species examined so far most results throw light on aspects related to wound healing, growth, morphogenesis and differentiation, but in most cases many crucial questions remain largely unanswered, especially those related to specific cellular and molecular aspects, and much work is still required. In fact, also in echinoderm regeneration, a complete interpretation of the real biological significance and evolutionary implications of the phenomenon can be achieved only by a broad-spectrum integrated understanding which emerges from diverse perspectives and from different experimental approaches applied to the most representative models.

1.2 The Regenerative Potential of Crinoids

Crinoids are well known for their spectacular regenerative capabilities extensively and successfully employed to reconstruct both external parts, namely arms, pinnules and cirri, and internal organs, such as digestive apparatus, gonads and even complete visceral mass, which can be frequently lost following traumatic injury, predation or spontaneous autotomy. Specimens collected in nature always show regenerating arms at different stages of growth (Fig. 1a). These regenerative phenomena can be easily reproduced in the laboratory by mimicking the autotomy conditions and amputating the arms at the level of the autotomy plane (sutures) (Fig. 1b–d). Arm regeneration in crinoids represents the most thoroughly explored model in echinoderm regeneration studies (Candia Carnevali and Bonasoro 2001b). Recently, we have carried out a comprehensive study of the overall process of arm regeneration in the comatulid *Antedon mediterranea*, a valuable and flexible experimental model previously successfully employed by old classical studies of developmental biology (Perrier 1873; Minckert 1905; Reichensperger 1912), which was re-explored in all its aspects from the macroscopic to the molecular level.

This phenomenon can be described on the whole as a typical blastemal regeneration in which new structures develop from migratory pluripotential, actively proliferating cells in the presence of presumptive regulatory factors. The overall process can be subdivided into three main phases (Fig. 2a): a repair phase, an early regenerative phase and an advanced regenerative phase,

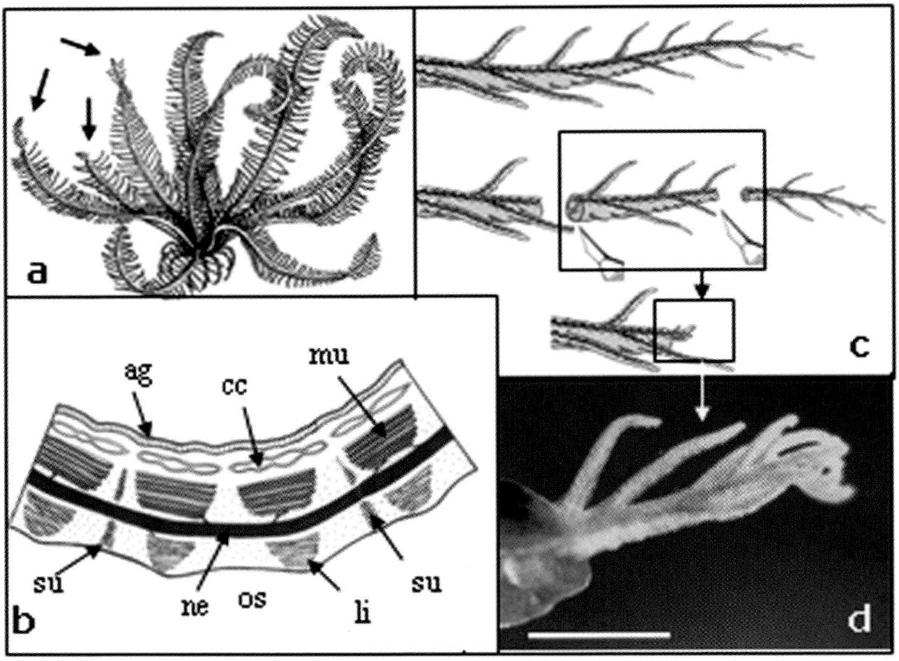

Fig. 1. Schematic presentation of the experimental model. **a** Specimen of *Antedon mediterranea* with three regenerating arms at different stages (*arrows*). **b** Main anatomical features of the arm in sagittal section. *ag* Ambulacral groove; *cc* coelomic canal; *li* ligament; *mu* muscle; *os* ossicle; *su* suture. **c** Normal arm, experimental amputations and regenerating arm. **d** Stereomicroscopic view of regenerating arm (2 weeks post-amputation). *Bar* 1 mm

whose crucial aspects are related to common fundamental mechanisms such as (1) intervention of stem cells and/or employment of dedifferentiated cells (Fig. 2 b–e), (2) cell migration and proliferation, (3) contribution of putative growth factors, particularly in terms of specific neurally derived factors and (4) mechanisms of pattern formation. The data obtained so far are derived from an integrated approach which utilizes different methods (first of all classical methods of microscopy – LM, confocal, TEM, SEM – and specific methods of immunocytochemistry, but also basic methods of biochemistry and molecular biology) on experimentally induced arm regenerations (standard or abnormal) obtained under significantly different experimental conditions, including extreme mutilations (*explants*) or exposure to specific types of environmental contaminants. In particular, the normal mechanisms and pattern of the regenerative processes under standard conditions have been established in serial experiments of regeneration at different stages following pseudo-autotomic amputations (Candia Carnevali et al. 1993, 1995, 1997). A parallel analysis has been carried out on the regenerative processes of both the normal regenerating arms and the respective amputated arm segments

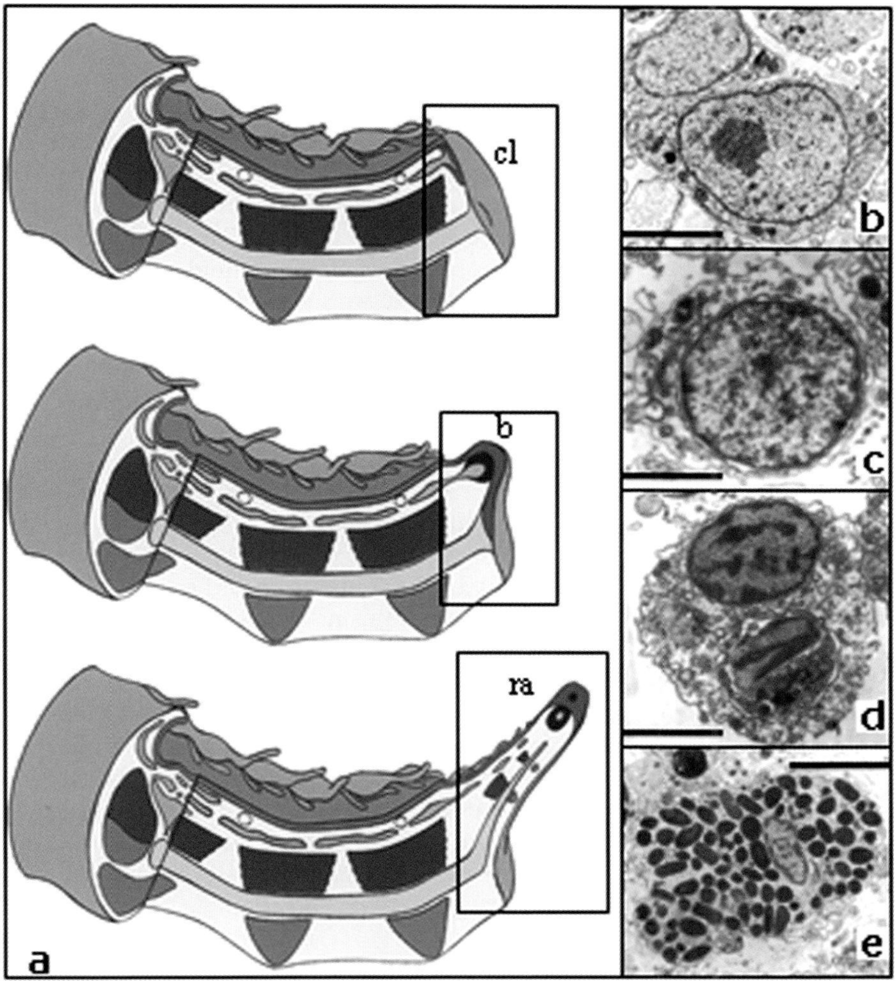

Fig. 2. a Schematic reconstruction of the main phases of arm regeneration. *Top downwards* Repair phase [0–24 h post-amputation (pa)], early regenerative phase (24–72 h pa), advanced regenerative phase (72 h to 3 weeks pa). *b* Regenerative blastema; *cl* cicatricial layer; *ra*: regrowing arm. **b–e** TEM: migratory cells involved in regenerative processes. Presumptive stem cells: amoebocyte (**b**), coelomocyte (**c**), phagocyte (**d**), granulocyte (**e**). *Bars* 4 μm

(*explants*, Fig. 1 c; Candia Carnevali et al. 1998), which can be maintained under good living conditions for about 3 weeks and represent excellent models for testing the arm regenerative potential in terms of autonomy of resources and control and for comparing regenerative mechanisms in the same individual. Different types of isolated explants have been successfully employed: during the culture period they are able to undergo extensive repair and regenerative processes in parallel with their donor arms. Comparison between the regenerative processes of arm explants and normal regenerating arms of corresponding stages highlights that beside general similarities in the basic regenerative processes there are some meaningful differences in terms of mechanisms employed and cellular/tissue elements involved. The regenerative potential, mechanisms and pattern have also been explored and compared under other experimental conditions, particularly with regard to aberrant regenerations resulting from arms deliberately subjected to traumatic mutilations which do not reproduce autotomy (Candia Carnevali and Bonasoro 2001b). The bulk of the results obtained so far in crinoids not only throw light on the most relevant aspects related to wound healing, morphogenesis, differentiation and growth in echinoderm regeneration, but also strongly suggest employing this fascinating and promising experimental model for a successful applied approach.

2
Endocrine Disrupters and Echinoderms

Some persistent and widely diffused contaminants of anthropogenic origin can be easily bioaccumulated by the organisms and exert their effects as "*endocrine disrupters*" (EDs). According to a recent definition (Holmes et al. 1997; Stahl et al. 1999), an ED is "an exogenous substance that causes adverse health effects in an intact organism, or its progeny, secondary to changes in endocrine function". This means that these compounds can affect significantly the exposed organisms, by modifying the homeostatic status of hormones, particularly steroids, mimicking the action of natural hormones, interfering synergistically or antagonistically with their synthesis, activity and metabolism, interacting with their nuclear receptors and, finally, inducing dramatic effects on gene expression (Colborn et al. 1993; Soto et al. 1995; Fairley et al. 1996; Gray et al. 1996; Cooper and Kavlock 1997; Cadbury 1998).

There is a long list of compounds (including pesticides, fungicides, insecticides, commercial and industrial chemicals, medical drugs, contraceptives) in some cases well known for their adverse effects on organisms, which exert their action by modifying or modulating natural hormonal activity in humans and wildlife populations: this list includes compounds with established estrogenic/anti-estrogenic activity, as well as chemicals with androgenic/anti-androgenic activity. Detailed knowledge of routes of uptake, and pattern of bioactivation, biotransformation and excretion is far from being available for most of these compounds. Another unresolved problem is the equivocal relationship between hormonal disturbances and toxicity. Endocrine dysfunctions, in fact, can be in some cases a secondary response related to toxicity rather than an independent primary response to the chemical (Depledge and Billinghurst 1999). Several wildlife species have demonstrated a significant susceptibility to endocrine-disrupting compounds. In some extensively explored animal models, the exposure to EDs results in appreciable disruption of steroid hormone production accompanied by related effects on reproductive and developmental mechanisms which can vary a great deal according to the species. Since there are many categories of pollutants displaying these potential actions, it is particularly difficult to find a direct correlation between the exposure levels of the pollutant and its observable effects on organisms, and we are still far from a broad detailed understanding of their potential effects in terms of endocrine-regulated processes. This is particularly true with invertebrates. In fact, in spite of the increasing number of standardized laboratory tests with invertebrate species, there is still a great demand for bioassay systems and/or biomarkers specific for evaluating the potential threats of EDs to invertebrates, which actually represent more than 95 % of the extant animal species in natural ecosystems. It is important to underline that, among invertebrates, there are relevant differences in terms of physiology and biochemistry and the same process can be regulated by different mechanisms of endocrine controls. So, the use of the same biomarker in one taxon has limited value in predicting similar effects in other taxa. Even when a common mechanism is involved in regulating a physiological process among taxa, sufficient differences among organisms with respect to the mechanism itself may significantly alter the level of susceptibility to EDs. Another crucial point in exploring the effects of EDs in invertebrates is related to their mode of action which changes with the seasonal cycle and/or the life stage of the target animal. In invertebrates, in general, ecotoxicological testing has been mainly addressed to measuring adverse effects on early developmental stages, with particular reference to larval stages (Schweitzer et al. 1997; Novelli et al. 2002), which are more sensitive than adults, or to identifying specific biomarkers of a number of relevant endocrine systems in adult organisms.

In recent years, interest has grown in using aquatic invertebrates as tools for monitoring environmental hazards. Although acute and chronic toxicity tests with benthic freshwater organisms are being developed for many compounds, the list of representative organisms for the marine environment is still very incomplete. In spite of the number and variety of potentially useful invertebrate models, and the structural and functional key role of such animals in the marine ecosystem, the attention tends always to focus on vertebrates, even though several excellent examples of ED effects on marine invertebrates are actually available in the literature (for a review see Depledge and Billinghurst 1999). It is enough to quote the phenomena of *imposex* and *intersex* described in gastropod molluscs, which can be regarded as some of the most dramatic and well-documented examples of adverse effects of ED contaminants in the marine field (Bryan et al. 1986; Gibbs et al. 1988; Horiguchi et al. 1995). However, with regard to important marine invertebrates such as echinoderms, the available information is still rather limited, particularly as far as adults are concerned (Kobahashi 1984; den Besten et al. 1989; 1990; 1991a,b; Anderson et al. 1994; den Besten 1998; Coteur et al. 2001; Békri and Pelletier 2004). In fact, the teratogenic effects of various chemical agents and drugs on the development of embryos and larvae have been slightly more extensively studied, namely in sea urchins (Schweitzer et al. 1997; Novelli et al. 2002). For a review on the effects of physical and chemical pollutants on fertilization and embryogenesis see also Angelini et al., and Matranga et al., this book.

Echinoderms are prime candidates for being selected as marine target macroinvertebrates and utilized as test animals because of their ubiquitous distribution, tractability and sensitivity. They offer a wide range of models for studying the effect of exposure to ED contaminants. A series of important aspects make echinoderms particularly relevant and amenable for this ecotoxicological approach. First, echinoderms are benthic animals and are particularly susceptible to the presence of micropollutants stored in marine sediments. Primary uptake across external epithelia (respiratory surfaces, epidermis, etc.) or secondary uptake from food represent important routes of entry for many dissolved aquatic pollutants which can be rapidly bioaccumulated by these organisms.

Second, regulatory factors and hormones, including peptides and steroids, similar to those of vertebrates have been recently detected and characterized in echinoderms. In particular, considerable published evidence indicates that vertebrate-type steroids can be synthesized (both androgens and estrogens) and used as terminal hormones along the neuro-endocrine cascades regulating reproductive, growth and developmental processes (Schoenmakers 1979, 1980; Schoenmakers and Voogt 1980; Voogt et al. 1984; 1990; 1991; Shirai and Walker 1988; den Besten et al. 1989; Aminin et al. 1995; Shubina et al. 1997; LeBlanc et al. 1999), and they have been demonstrated to control reproductive activities and growth at cellular and tissue level. Current research is actually focusing on echinoderm endocrinology and the improved knowledge in terms of comparative physiology and biochemistry and specific mechanisms involved, including steroid metabolism and possible modulations by ED compounds, is providing a rather good background of necessary information (Janer et al. 2004; Lutz et al. 2004). It is relevant to point out that although individual components of endocrine systems have undergone significant evolutionary divergence in response to specific adaptations, endocrine control strategies and basic hormonal regulatory mechanisms have been, on the whole, rather conserved among closely related animal groups. In terms of evolutionary relationships, echinoderms are deuterostome invertebrates and are phylogenetically more related to chordates than to other invertebrate groups. A short analysis of the phylogenetic distance among genes is given in the chapter by Zito et al., in this book. This means that there is less divergence between echinoderms and vertebrates than between echinoderms and the major protostome invertebrates (annelids, arthropods, molluscs): it is not surprising, therefore, that echinoderms possess control mechanisms of physiological processes rather similar, and maybe nearly homologous, to those of vertebrates, in terms of molecules and actions. In this light, echinoderms may therefore share with vertebrates similar targets in terms of EDs and be susceptible to the same chemicals known for causing reproductive dysfunctions in vertebrates. It is worth noting that in some echinoderm classes there is limited but significant published evidence of the disruptive effects of contaminants on steroid metabolism and steroid levels and on the effects on the mono-oxygenase (MO) system (den Besten et al. 1989, 1990, 1991a,b; Schweitzer et al. 1997; den Besten 1998). For this reason, echinoderms can be considered key organisms in both basic and applied research in this field and can be usefully employed for developing new successful experimental approaches and strategies.

A third noteworthy point is that many echinoderm species are ecologically relevant and occupy important positions in the food chain of vertebrate wildlife: their loss could have dramatic consequences on the marine ecosystem. In addition, some echinoderm species are edible and are highly valued by humans, whereas other species are relevant for commercial and recreational purposes. So it is important to understand the factors that might influence long-term population viability of these valued resources and to develop a testing programme, in the field and in the laboratory, in order to protect ecologically and economically important species from the potential effects of this dangerous class of environmental contaminants.

A last important characteristic is that echinoderms, as already seen, besides the normal processes of sexual reproduction, have spectacular and unique capacity for regeneration and offer a wide range of models for studying this phenomenon. Regenerating echinoderms can be regarded as particularly valuable experimental models to test the effects of exposure to different types of EDs. In fact, the regenerative phenomena which, as explained above, have the peculiarity of representing developmental processes in the adult organism, are characterized by enhanced and active phenomena of cell proliferation, morphogenesis, differentiation and tissue renewal. They are typically modulated by endocrine and neurohumoral mechanisms comparable to, if not the same as, those usually involved in reproductive and developmental processes. Vertebrate-type regulatory factors, including peptides and steroids, which have been demonstrated to play a role in these processes, are very likely to modulate cellular differentiation and tissue growth during regeneration processes (Candia Carnevali et al. 2001 c; Thorndyke and Candia Carnevali 2001). For this reason, and on the basis of what has been suggested by a few previous data (Walsh et al. 1986; Fingerman 1997), it was assumed that regenerative processes can be susceptible to EDs present in the environment. It is very probable, in fact, that exposure to pseudo-hormonal contaminants can induce variations, in terms of time, mechanisms and actions, in the physiology of regenerative development which is amongst the most sensitive phenomena with respect to environmental stress and consists of stages that can be directly and conveniently monitored for cellular damage. This idea is in agreement with the general trend of ecotoxicological testing with EDs, which has been frequently addressed to detect early effects in developmental stages, embryonic or larval, more sensitive than adults from this point of view. Toxicity tests on the sea urchin *Paracentrotus lividus*, for instance, highlighted that

ED compounds such as organotins can cause critical and consistent damage in early life stages (Schweitzer et al. 1997; Novelli et al. 2002). On the other hand, previous data obtained from the ophiuroid *Ophioderma brevispina* (Walsh et al. 1986) showed clearly that exposure to organotin compounds significantly affected arm regeneration processes, demonstrating successfully how useful the study of regenerative development in adult organisms can be. In the light of what has been seen above, regenerating echinoderms appear to represent ideal bioindicators of ED-induced stress at the whole-organism, cellular and molecular level.

It is relevant to recall that an important goal in studying EDs is not only determining the specific disruptive activity of each compound, but also establishing the most sensitive test species and the most specific forms of response (endpoints) at which the hormonal dysfunction is unequivocally expressed. For this reason, unique endocrine-regulated processes, such as echinoderm regeneration, can provide an important target for toxic action and an original and easily quantifiable endpoint that makes the regenerating animal a very sensitive bioindicator: in fact, it can provide at the same time precious indications of the specific effects of these persistent pollutants on biological mechanisms at the tissue and cellular level and an early assessment of possible degenerative modifications of the marine ecosystem.

3 Crinoid Regeneration and Endocrine Disrupters

On the basis of their spectacular regenerative capabilities, crinoid echinoderms are prime candidates for this ecotoxicological approach. In particular, as previously explained, in the common feather star, *A. mediterranea*, arm regeneration is a common phenomenon which can be easily studied under laboratory conditions thanks to the favourable possibility of inducing experimentally arm autotomy and subsequent regeneration (Fig. 1 c,d), and following the complete regrowth of the lost arm in a relatively short period of time (about 3 weeks). *A. mediterranea* is a widespread species in the Mediterranean Sea, representative of the benthic fauna and easily found. It is a microsuspension-feeding organism on which persistent sediment-bound pollutants have an immediate impact: it can, in fact, concentrate persistent and hydrophobic pollutants directly by primary uptake from water and from suspended matter and sediment, and secondarily from food, thus giving rise to significant biomagnification phenomena. These animals can be maintained in the laboratory for many months and, due to their remarkable capacity for survival, are particularly amenable to the experimental approach. For all these reasons, *A. mediterranea* should be considered as a suitable ecotoxicological model to test endocrine-disrupting activity of selected compounds on a typical marine species. This idea has been recently confirmed by a series of preliminary results obtained with specimens exposed to specific contaminants belonging to the ED class, which have clearly shown a wide range of interesting potential effects of the exposure to these pollutants on the regenerative processes at all levels, macroscopic and microscopic (Candia Carnevali et al. 2001a,b, 2003).

In this chapter, we analyse and compare the effects of exposure to different classes of ED compounds on the process of arm regeneration in *A. mediterranea*. The selected types of test compounds are: polychlorinated biphenyls (PCBs), nonylphenols (4-NP) and tripheniltin (TPT-Cl) (Fig. 3). PCBs are commercial products widely used as heat transfer fluids, hydraulic fluids, flame retardants and dielectric fluids because of their unique properties, including resistance to biological and chemical oxidation and nonflammability. Nonylphenols are products of the slow and incomplete biodegradation of nonylphenol polyethoxylates which are widely employed for domestic and industrial uses. TPT-Cl is a typical organotin compound, extensively used in agriculture and in antifouling paints, which tends to accumulate in the soil and sediments.

The widespread use of these different compounds coupled with improper disposal has led to significant global environmental contamination. Although acute and chronic toxicity tests with benthic freshwater animals have being developed for these contaminants, they have to be still definitely established for marine organisms. These three classes of chemicals have widely documented endocrine-disrupting activities which can be expressed by contrasting effects on the exposed animal models. The first two com-

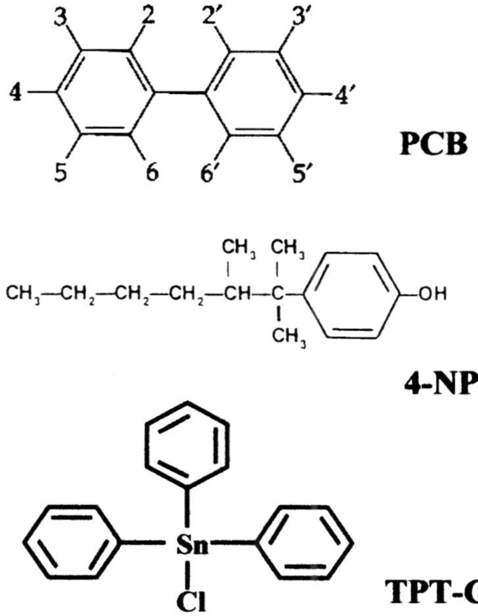

Fig. 3. The tested molecules: typical *endocrine disrupter* compounds

pounds (PCBs, 4-NP) are included in the category of contaminant with estrogenic effects (anti-estrogenic or estrogenic, respectively) (Granmo 1991; Arnold et al. 1996; Dickey 1997; Depledge and Billinghurst 1999;); the third (TPT-Cl) is a typical compound with androgenic effect (Fait et al. 1994; Matthiessen and Gibbs 1998). The aim of this study is (1) to offer a comparative account of the possible insidious effects of exposure to ED contaminants on developmental physiology of typical marine animals, with particular reference to mechanism(s) of action of EDs on growth, differentiation, and repair/regeneration processes at the whole organism, tissue and cellular level in the model species, and to possible homeostatic mechanisms through which the organism can interact and adapt to adverse environmental conditions; and (2) to assess the validity of the regenerative response for bioassay. A relevant point is that in our experiments, exposure to EDs was performed in seawater at concentration ranges reproducing those of moderately polluted coastal areas. The idea was to simulate realistic exposure conditions for wildlife marine fauna.

3.1 Experimental Approach

Our experimental approach employed laboratory tests with individual compounds based on exposure experiments with regenerating crinoid samples. During the overall exposure period, lasting 2 weeks, the concentrations of the selected compounds were measured in the exposure water and in the animal tissues in order to follow bioconcentration kinetics.

3.1.1 Exposure Experiments

Specimens of *A. mediterranea*, collected from along the Tyrrhenian coast of Italy (Giglio Island), were maintained in closed glass aquaria containing 50 l of artificial seawater with an internal circulation system, at 14 °C, and fed with InverteMin (Tetra Marin). Exposure to the selected pollutants was performed under controlled conditions, in terms of environmental parameters and contamination levels, under static or semistatic conditions (20 % water renewal in 24 h) as explained below. The exposure concentration was selected on the basis of available data on real contamination levels in the coastal areas. Groups of 30 specimens were employed in each aquarium (exposure, control and solvent-control aquaria). In each exposed or control specimen, experimental regeneration was induced in two to three arms, mimicking the autotomy conditions. Immediately after the amputation, the experimental animals were put in the test aquaria and exposed to different concentrations of the selected compound for prefixed periods (72 h, 1 week and 2 weeks): in this way, the exposure period corresponded to well-defined and established regenerative stages. At each stage, chemical analyses of water and echinoderm tissues were performed in order to check the variability of exposure concentration and the bioaccumulation.

3.1.2
PCB Exposure

The exposure to PCBs tests were performed under static conditions by employing a commercial mixture of Aroclor 1260. Some replicates of the experiments were performed at different exposure concentrations. The exposure medium was obtained by adding concentrated water solutions of PCBs (mother solutions) to the exposure aquaria (Candia Carnevali et al. 2001a,b), reaching initial concentrations of 23, 77 and 81 ng/l respectively.

3.1.3
4-NP Exposure

The exposure to 4-NP tests were performed under static conditions. Some replicates of the experiments were performed at different exposure concentrations (2, 5 and 10 μg/l), which were obtained by adding suitable small aliquots (less than 2 ml) of 4-NP acetone solutions in the exposure aquarium until the pre-established final concentration was reached (Candia Carnevali et al. 2003).

3.1.4
TPT-Cl Exposure

The exposure tests to TPT-Cl (Merck) were performed under semistatic conditions (20 % water renewal in 24 h). The experimental animals were exposed to different concentrations of TPT-Cl (50, 100 and 225 ng/l). The exposure medium was obtained by adding to the aquaria 1.25 ml ethanol-TPT-Cl solution at the start of the experiment and 0.250 ml ethanol-TPT-Cl solutions day by day for the 20 % renewal (Barbaglio et al. 2004). The final ethanol concentration in exposure aquaria was 0.025 ml/l (lower than that officially allowed in long-term ecotoxicity tests with aquatic invertebrates). As far as the selected TPT-Cl exposure concentrations are concerned, the maximum concentration was close to LC_{50} experimental values quoted in the literature (Rippen 1990), whereas the minimum concentration was close to NOEC (no observed effect concentration) experimental values known for echinoderms (Walsh et al. 1986).

3.2
Biological Analysis

At whole-organism level the analysis focused on the effects of ED pollutants on crinoid regenerative processes by taking into account abnormal morphological aspects related to growth and development. This analysis was suitably associated with an accurate histopathological investigation focusing on specific anomalies at tissue and cellular levels. This integrated analysis provided extensive information on the possible alterations, resulting in a highly sensitive condition index related to ecologically relevant, individual-level responses. The histological examination, in particular, still provides one of the most rapid and sensitive methods of detecting adverse acute or chronic effects of exposure which represent intermediate levels of response between those detectable at the whole-organism level and those at the molecular level, and can reflect prior toxicant-induced molecular and biochemical aspects of cell physiological alteration (Hinton 1997). Thus the biological aspects of the regenerative processes were investigated in regenerating crinoids at whole-organism, tissue and cellular levels by employing classical morphological methods, both macroscopic and microscopic (stereo, light, electron – TEM, SEM – and confocal microscopy) as described in detail in previous papers (Candia Carnevali et al. 1993; Candia Carnevali and Bonasoro 2001b), associated with specific immunocytochemical protocols and/or statistical analyses whenever appropriate.

Exposed and control regenerating arms were prefixed with 2 % glutaraldehyde in 0.1 M cacodylate buffer for 4–5 h, and then, after overnight washing in the same buffer, post-fixed with 1 % osmium tetroxide in the same buffer. After standard dehydration in an ethanol series, the samples were embedded in Epon-Araldite 812. The semithin and thin sections, cut with a Reichert Ultracut E (diamond knife), were stained by conventional methods (crystal violet-basic fuchsin for LM; uranyl acetate and lead citrate for TEM) and then observed in a Jenaval light microscope and Jeol 100 SX electron microscope respectively. A range of specific immunocytochemistry (ICC) techniques utilizing commercial monoclonal antibodies was employed to monitor cell proliferation (BrdU method; see Candia Carnevali et al. 1995, 1997) or pattern distribution of common cellular biomarkers (cytochrome P450; see Candia Carnevali et al. 2001a,b 2003). The results of the exposure tests were compared with those obtained by a parallel analysis of normal regenerating samples under standard conditions.

3.3
Chemical Analysis: Summary of Analytical Procedures and Results

The chemical analysis performed in parallel with the biological analysis is an important component of the integrated ecotoxicological approach presented here and the results add great value to those derived from the biological analysis. Nevertheless, the details of the analytical work are rather far from the aims of the present review and, for completeness, it seems to be sufficient to give only a schematic overview of the main points and to briefly summarize the most relevant results obtained so far. Analytical techniques used in quantification include high performance liquid chromatography (HPLC), high resolution capillary gas chromatography (GC) with flame ionization detector (FID) and selective detection, and mass spectrometric (MS) detection. Aquarium water samples were extracted by liquid–liquid extraction or solid-phase extraction according to the polarity of the investigated pollutants.

3.3.1 *PCBs*

Three exposure concentrations were tested: 23, 77 and 81 ng/l. The concentration in the exposure aquaria was controlled daily during 2 weeks of experiments by extracting the PCB solution in seawater with n-hexane (pesticide grade). The PCB congener concentration decreases with the exposure time: in terms of total PCBs, at the end of the experiments, the analytical values of exposure water were about 20 % of the initial nominal concentrations. For PCB determination in the animal tissues, three animals were collected from each exposure and control aquarium at 72 h, 1 week and 2 weeks. The concentration of chemicals in water and animal tissue extracts was measured by GC-ECD for PCB (Candia Carnevali et al. 2001a,b) after clean-up on a Florisil column (4×0.7 cm i.d.). As far as bioconcentration is concerned, in the tests with a nominal exposure concentration of 77 ng/l the 14-day total PCB concentration was 2,257 ng/g (on a lipid basis), not far from those measured in other filter feeders collected along the Mediterranean coasts.

3.3.2 *4-NP*

Three exposure concentrations have been tested: 2, 5 and 10 μg/l. The water concentration was controlled at the end of the exposure period by extracting the seawater 4-NP solution with an SPE (solid phase extraction) procedure. A LiChrolut EN cartridge was used as solid-phase column and 500 ml of sample was passed through the column to extract 4-NP which was then recovered with 5 ml of methanol. The 4-NP concentration decreases in the exposure medium with the exposure time: at the end of the experiments the analytical values of exposure water were about 50 % of the initial nominal concentrations. For 4-NP determination in the animal tissues three animals were collected from each exposure and control aquarium at 72 h, 1 week and 2 weeks. HPLC-fluorescence was used to analyse both water and tissues extracts (Candia Carnevali et al. 2003). As far as bioconcentration is concerned, this was quite rapid, the tissue concentration reaching high values (0.4–0.6 μg/g) after a short exposure period (72 h) with the lowest exposure concentration (2 μg/l).

3.3.3
TPT-Cl

Three exposure concentrations were tested: 50, 100 and 225 ng/l. TPT-Cl analyses were performed by gas-chromatographic separation and mass-spectrometry detection after derivatization of the original compound (TPT-Cl) in the extraction medium. In terms of chemical parameters, a detailed chemical analysis of water and tissue samples from our exposure experiments is still in progress (Dagnac et al., unpubl.). Nevertheless, preliminary analytical results clearly indicate that (1) actual concentrations measured in exposure medium are much lower than the nominal ones (i.e. 1.3 and 9.3 ng/l for nominal exposure concentrations of 50 and 225 ng/l respectively; Tremolada et al. 2004); and (2) TPT-Cl concentration in the animal tissues appears to be significantly high (53 ng/g fresh wt.) with the lowest exposure concentration (50 ng/l), in spite of the low final concentrations in the exposure water.

4
Exposure Effects of EDs and Biological Implications on Regeneration

Arm regeneration in *Antedon* consists of a typical epimorphic blastemal process which has been reconstructed in its main phases – repair phase, early regenerative phase and advanced phase – particularly with respect to its cellular and molecular aspects (Candia Carnevali et al. 2001b, 2003). These phases are schematically shown in Fig. 2a in relation to the basic anatomy of the arm, whose main components are detailed in Fig. 1b: a segmental series of brachial ossicles, connected by muscles and ligaments, a central brachial nerve, a multiple system of coelomic canals, and an ambulacral epithelium. Our recent studies have also shown that arm regeneration is a typical nerve-dependent phenomenon in which the nervous system acts as a primary source of regulatory factors involved not only in the regenerative processes of the neural tissue itself, but also to a large extent in development and regrowth of all other structures. In addition, the nervous system and the coelomic canals act as important sources/vehicles for the different types of migratory cells, which are responsible for the regenerative processes, including presumptive stem elements (undifferentiated amoebocytes and coelomocytes), phagocytes and granulocytes (Fig. 2b–e). It should be noted here that coelomic fluids of echinoderms contain different types of cells, generically called coelomocytes, which correspond to structurally and functionally distinct elements. Due to their capability to respond to injuries, host invasion and cytotoxic agents, coelomocytes are regarded as the immune effectors of echinoderms (Chia and Xing 1996; Gross et al. 1999). For a review on coelomocytes see Matranga et al., in this book. Coelomocytes are not yet well characterized functionally in crinoids, but are morphologically distinguishable as undifferentiated (presumptive stem cells, Fig. 2b,c) and differentiated (phagocytes and granulocytes, Fig. 2d,e) cells. These latter cells are apparently typical of crinoids and are extensively employed in repair processes (Fig. 2e). They are characterized by a massive content of dense granules and represent well-differentiated migratory elements, present in the coelomic fluids or scattered in the connective tissues, particularly around the brachial nerve.

In our experiments, the effects of exposure to ED compounds on the regeneration process have been monitored at three regenerative stages, 72 h, 1 week and 2 weeks post-amputation, which are considered significant stages in terms of developmental processes (growth, morphogenesis and differentiation). Here, we give an account of our recent results on the effects of the exposure to different contaminants representative of distinct classes of ED compounds on crinoid regeneration, with particular reference to the following points:

- effects at whole-organism level (mortality, growth, malformations);
- effects at tissue and cellular level (histological pattern, morphogenesis, differentiation);
- possible sites of action, particularly in terms of steroid dysfunction.

4.1 Mortality

No toxic effects were observed in PCB-exposed animals in the sets of experiments: all the experimental animals survived in good condition until the end of the prefixed exposure period (72 h, 1 week, 2 weeks). In contrast, some mortality was observed for 4-NP at the highest exposure concentration (5 and 10 μg/l), particularly for the longest exposure periods (1 week and 2 weeks). In particular, a mean mortality of 30 % was recorded for 5 μg/l exposure concentration and 60 % for the 10 μg/l concentration. As far as TPT-Cl is concerned, according to the data obtained so far (Barbaglio et al. 2004), no toxic effects were observed with the concentration range from 50–225 ng/l. In all these experiments, the mortality was less than 20 % in both exposed and control samples, a value that is considered acceptable for the validity of the ecotoxicological results. The TPT-Cl concentration of 1,000 ng/l appeared to be close to acute toxicity threshold for *A. mediterranea*; 50 % mortality was evident in specimens exposed to this concentration within the first 72 h (LC_{50}), and, in the surviving samples, growth did not progress further and regenerative development appeared to stop at a very early stage (24–72 h post-amputation).

4.2 Growth

The overall growth of the regenerating arms of *A. mediterranea* was significantly affected by PCBs, 4-NP and TPT-Cl exposure in comparison with control animals (Candia Carnevali et al. 2001a,b; 2003; Barbaglio et al. 2004). However, contrasting results were obtained for the three classes of chemicals: in fact, some compounds accelerated the regenerative growth, whereas others appreciably inhibited the regrowth process. As far as PCB exposure is concerned, if during the early regenerative phase (72 h post-amputation) the exposed samples did not show significant anomalies in terms of growth and development, at the advanced regenerative phase (1–2 weeks post-amputation) an unusual accelerated growth was clearly evident in all the exposed arms in comparison with the growth of the standard regenerating arms at the same stages (Fig. 4a,b,e,f). This involved the overall size of the regenerating arm, which appeared much more developed, and the differentiation of its anatomical structures, external or internal, such as the lateral pinnules and the brachial skeletal components (ossicles). In contrast, 4-NP appeared to cause an appreciable delay in the overall growth of the regenerating blastema observable with all the exposure concentration tested (Fig. 4a,d,e,g). With regard to TPT-Cl-exposed samples, a clear effect in terms of enhanced growth, even though less marked than that of PCB samples, could be also detected (Fig. 4a,c,e,h). A quantitative analysis carried out in parallel with the qualitative morphological analysis on the measured lengths of all the experimental regenerates, in both exposed and control samples, showed that in terms of average overall sizes of the regenerating arms, a clear correlation between growth and exposure could be observed for all the test compounds: the differences in the overall growth were particularly evident at 2-week post-amputation; at this stage, which corresponded to a long-term exposure period, the exposed samples also displayed a remarkable variability in terms of overall growth if compared with the uniform standard size of the controls. Quantitative data showed that in PCB-exposed samples the growth increase was significant at both 1 week and 2 weeks post-amputation (Candia Carnevali et al. 2001a,b), whereas in TPT-Cl-exposed samples a significant effect in terms of enhanced growth could be detected only at the advanced stage of 2 weeks and with the higher concentration employed (225 ng/l) (Barbaglio et al. 2004).

Fig. 4a–h. Whole mount stereomicroscopic views of control and exposed regenerating arms of *Antedon mediterranea* at different stages. Exposure concentrations: PCB: 14 ng/l; 4-NP: 2 μg/l; TPT-Cl: 100 or 225 ng/l. **a–d** 1 week pa (post-amputation): **a** control sample; **b** PCB-exposed sample; **c** TPT-Cl-exposed sample; **d** 4-NP-exposed sample. In the PCB-

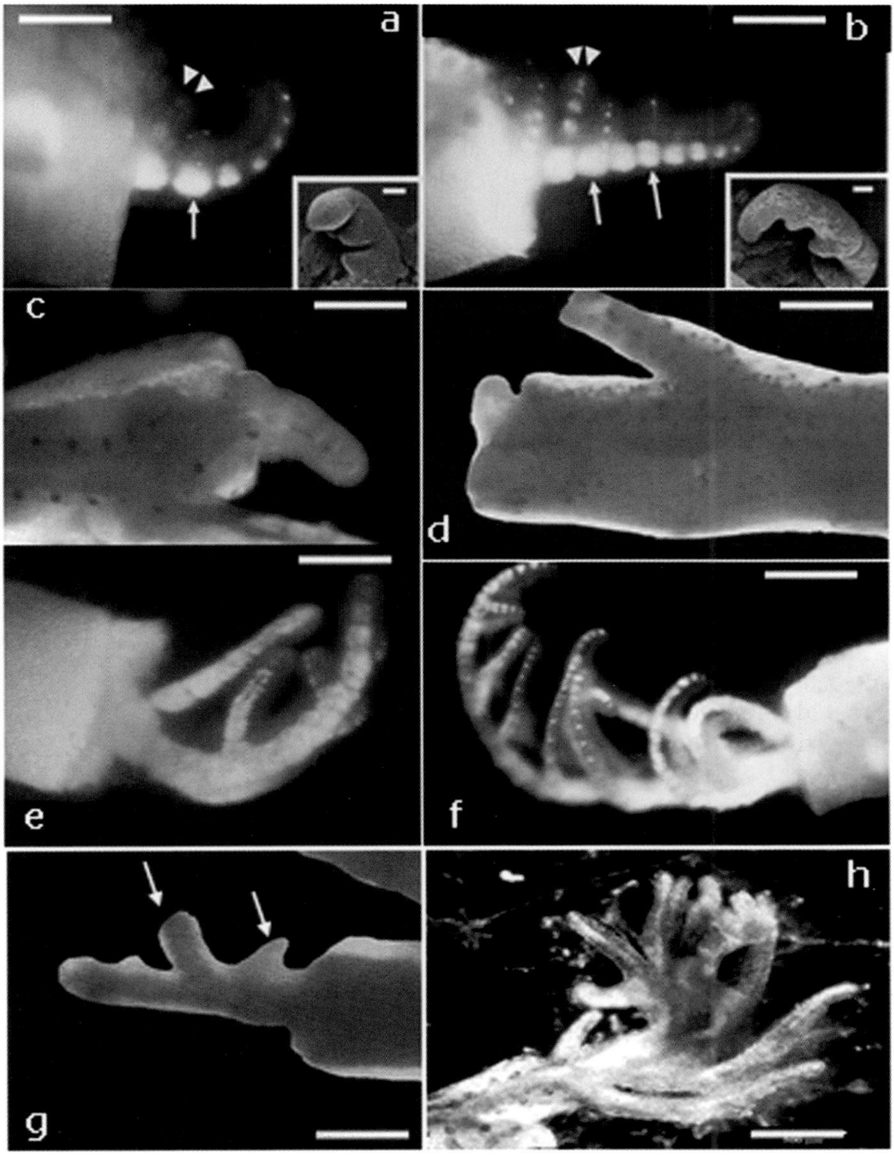

exposed sample, abnormal growth of the regenerate and advanced development of its anatomical features are evident. In the TPT-Cl- and 4-NP-exposed samples, delayed and anomalous regrowth is evident. *Arrow* Brachial ossicles; *double arrowheads* pinnules. *Insets* show details of the respective regenerating arms from SEM. **a** and **b**, *bars* 240 µm; *inset bars* 100 µm. **e–h** Regenerating arms at 2 weeks pa: **e** control sample; **f** PCB-exposed sample; **g** 4-NP-exposed sample; **h** TPT-Cl-exposed sample. Growth of the PCB-exposed sample is more pronounced than in the controls, whereas that of the 4-NP and TPT-Cl samples is markedly delayed. In addition, 4-NP and TPT-Cl samples show evident malformations. This can be appreciated in terms of both overall size and shape of the regenerate and differentiation of anatomical structures. *Bars* 240 µm (**e**, **g** and **h**); 250 µm (**f**)

4.3 Malformations

The morphological analysis under the stereomicroscope pointed out specific malformations in the exposed samples which could be attributed to individual compounds (Candia Carnevali et al. 2001a,b, 2003; Barbaglio et al. 2004). Results obtained from the different series of tests performed with PCBs, 4-NP and TPT-Cl, respectively, were in good agreement from a qualitative point of view. Appreciable malformations in terms of general morphology and external anatomy were observed in specimens exposed to 4-NP and TPT in comparison with controls (Fig. 4c,d,g,h). In particular, during the advanced regenerative phase (1–2 weeks), the evident delayed development of the regenerate was expressed not only by the smaller overall size of the regenerating arm but also by a partial lack or abnormal development of its lateral pinnules (Fig. 4g,h). The same range of anomalies was observed with all the exposure concentrations tested. As far as TPT-Cl exposure experiments are concerned, except for the 50 ng/l samples, all the exposed samples showed evident anomalies in terms of both external malformations in their overall shape and anatomical features. Beside a generalized reduced growth of the regenerative blastema, remarkable atypical features could be seen in the regenerating arms of the exposed specimens which appeared to be twisted and coiled or unusually tuft-shaped (Fig. 4c,h). This aberrant anatomy of the regenerate was consistent with the presence of relevant anomalies in the corresponding histological sections (see below), which were characterized by an unusual pattern of specific tissues, with particular reference to a pronounced abnormal development of the skeletal components.

4.4 Histological Pattern

The effects observed at the level of general morphology can be correlated with relevant atypical features in terms of microscopic anatomy of both the stump and the regenerating arm. These alterations are the most sensitive index of ecologically relevant, individual-level responses such as growth and development (Hinton 1997). With all the test compounds employed so far, the histological sections, in spite of a general good preservation of tissue integrity in all the exposed samples, showed clear signs of anomalies, which were detectable since the early regenerative phase and became more and more evident at the more advanced regenerative stages (1 and 2 weeks post-amputation). It is significant that these histological alterations always involved the same target structures, but showed rather characteristic aspects that could be considered specific effects of the exposure to the individual pollutant (Candia Carnevali et al. 2001a,b, 2003; Barbaglio et al. 2004). Apart from a number of other minor atypical features, the following anomalies appeared to be particularly relevant:

1. Development of an atypical blastema (pseudo-blastema), flattened and/or ectopic, often including foreign non-blastemal elements (myocytes and/or skeletal spicules) (in 4-NP- and TPT-exposed samples; Fig. 5a,b).
2. Unusual hypertrophy and swelling of the coelomic canals of both the stump and the regenerate (particularly in PCB- and 4-NP-exposed samples; Fig. 5d,e,f,i).
3. Pronounced abnormal development of the skeletal components in the regenerate (particularly in TPT-exposed samples; Fig. 5 c).
4. Enhanced and prolonged phenomena of cell migration (coelomocytes, amoebocytes, phagocytes and granulocytes) evident at the level of both coelomic canals and tissues (in all the exposed samples; Fig. 5b,i,k). In particular, this intense migratory activity tended to involve a large number of granulocytes: their diffuse presence in the tissues, even at the advanced regenerative stages, and the massive unusual occurrence of degranulation phenomena in the stump tissues, particularly close to the coelomic canals, were distinctive features of all the exposed samples. Under standard conditions, in fact, the granule cells are extensively but limitedly employed during the repair processes, their degranulation processes occurring specifically at the level of the amputation surface.
5. Massive recruitment of dedifferentiated or semi-dedifferentiated cells, particularly myocytes, that were frequently involved in migration phenomena, particularly in the coelomic canals close to the amputation area (in all the exposed samples, but mostly in 4-NP and TPT samples; Fig. 5b,i,k). The muscles of the stump contributed extensively to this cell recruitment/migration through phenomena of tissue rearrangement and dedifferentiation during which the muscle fibres were massively replaced by other elements (coelomocytes and phagocytes; Fig. 5g,h,i,j). At TEM (Fig. 6a) many individual myocytes appeared to progressively dedifferentiate to acquire the features of undifferentiated coelomocytes actively involved in cell division (see below), whereas other myocytes were obviously involved in apoptosis (Fig. 6b). These processes of muscular dedifferentiation and turnover were always accompanied by related phenomena of massive cell migration/proliferation in the adjacent coelomic canals (Fig. 6 c). This is strongly in contrast to what was observed under standard conditions, where regeneration is typically accomplished by the contribution of undifferentiated stem cells which are responsible for the blastema formation.
6. Enhanced or inhibited cell proliferation. The first effect was particularly evident in PCB-exposed samples, the second in 4-NP- and in TPT-exposed samples. As indicated by our BrdU incorporation experiments, although cell proliferation in both exposed and control samples was always localized in the usual specific cytopoietic sites, mainly at the level of the apical blastema and the coelomic epithelium of both the regenerate and the stump, there was a substantial difference in labelling intensity and distribution among the different samples: the labelling in fact was much stronger and more widely distributed in PCB samples (Fig. 6f,h), whereas it was

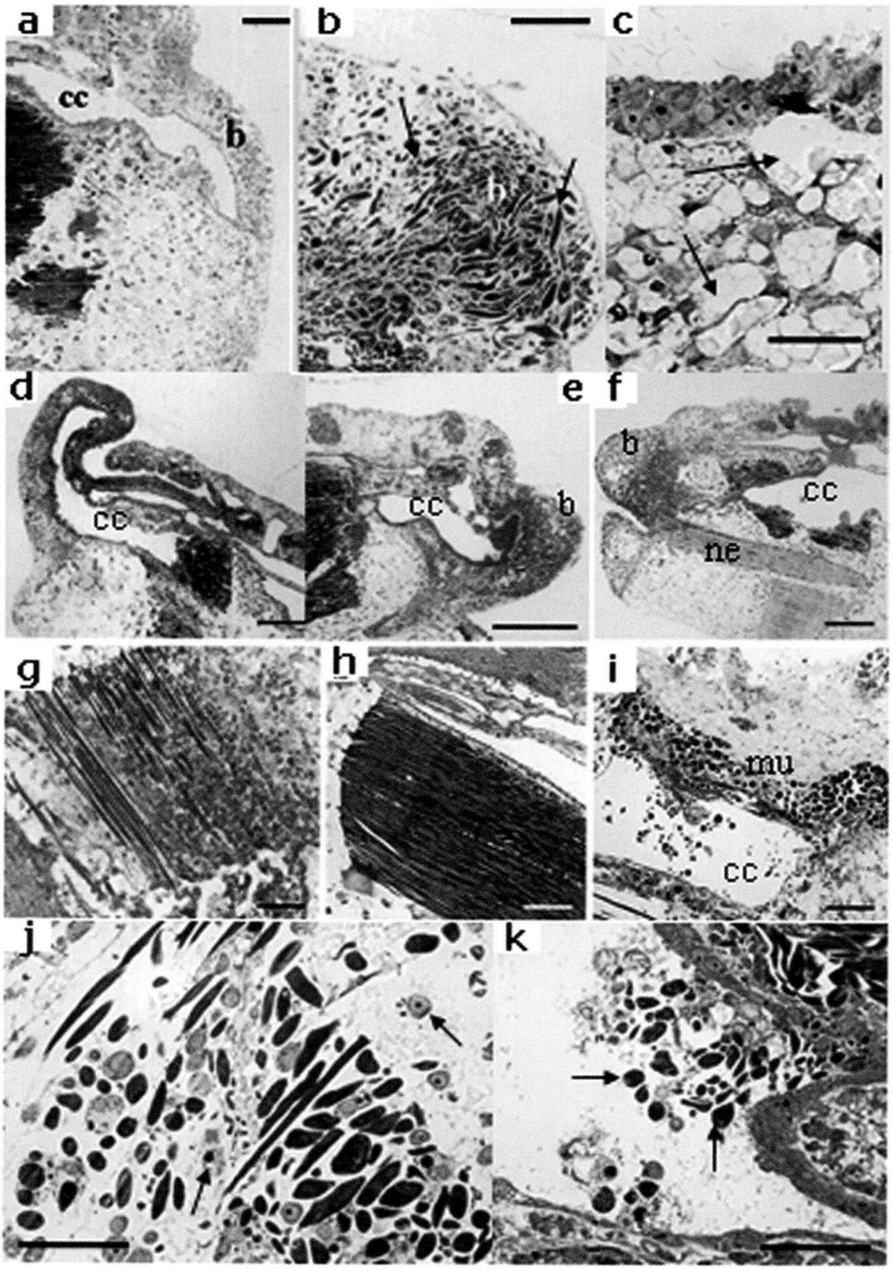

Fig. 5a–k. LM sagittal sections of control and exposed regenerating arms of *Antedon mediterranea* at different stages. Exposure concentrations: PCB: 14 ng/l; 4-NP: 2 μg/l; TPT-Cl: 100 or 225 ng/l. **a–c** Details of the blastema regions at 72 h pa (post-amputation): control (**a**) and TPT-Cl samples (**b**, **c**). In **b**, an atypical pseudo-blastema which mainly consists of ectopic migratory myocytes (*arrows*) is evident; **c** shows the unusual presence of well-

rather modest and less diffuse in 4-NP and TPT samples (Fig. 6g) with respect to the controls. Unexpectedly, in PCB-exposed specimens, the labelling involved to a relevant extent also the muscles of the stump (Fig. 6h). This seems to indicate that, at least in the PCB-exposed samples, the muscles can directly provide a significant contribution to regeneration also in terms of cell proliferation.

7. Recycling and turnover of the stump tissues, namely the endoskeleton and the connective tissue (mostly in PCB- and 4-NP-exposed samples; Fig. 6d). These phenomena were significantly different from those related to the muscles and involved extensive degenerative phenomena which led to vacuolization/vesiculation of both the extracellular matrix and the cells (fibroblasts or scleroblasts respectively). The massive presence of phagocytes at advanced regenerative stages of 2 weeks indicates that these tissues are employed as a secondary indirect source of reserve materials for new synthesis.
8. Atypical ultrastructural features of specific cell types, including the blastemal cells (mostly in PCB- and 4-NP-exposed samples; Fig. 6e). In particular, a marked development of endoplasmic reticulum (both RER and SER) and Golgi complexes and an unusual abundance of lipid granules and empty vacuoles were typical features of the exposed samples and could be correlated with a cytological pattern of steroid dysfunction. Interestingly, in the exposed samples, besides these histological and cytological features, the integrity/preservation of other tissues of both the stump and regenerate (epithelia, nerve tissue, etc.) was rather good, even in the long-tem exposed samples of 2 weeks post-amputation, without any significant morphological variation with respect to the standard conditions seen in normal regeneration;
9. Possible induction of protective biochemical responses. Immunocytochemical results showed, in fact, that in PCB- and 4-NP-exposed samples there was an appreciable increase in the expression pattern of specific

---

developed skeletal spicules in the blastema (*arrows*). **a** and **b**, *bar* 100 μm; **c** 20 μm. **d–f** Exposed regenerating samples of 1 week pa: **d** PCB; **e** TPT-Cl; **f** 4-NP. Abnormal growth of the regenerate is evident in PCB samples. In contrast, the regenerative blastema is heterogeneous and poorly developed in TPT-Cl and 4-NP specimens. In all the exposed samples the coelomic canals (*cc*) are hypertrophic. *b* Blastema; *ne* nerve. *Bars* 200 μm. **g–i** Exposed and control samples at 1 week pa: **g** PCB; **h** control; **i** 4-NP. Details of the muscle bundles of the stump. In contrast to the compact structure of the control muscle, the PCB- and 4-NP-exposed samples show extensive muscle rearrangement/dedifferentiation. *cc* Coelomic canal; *mu* muscle. *Bars* 30 μm. **j, k** Details of 4-NP-exposed samples of 1 week pa. **j** The muscle bundle consists of semi-dedifferentiated myocytes and undifferentiated coelomocytes (*arrows*); **k** the coelomic canals of the stump are involved in extensive cell proliferation/ migration (*arrows*). A number of myocytes are released through the coelomic wall into the lumen. *Bars* 20 μm

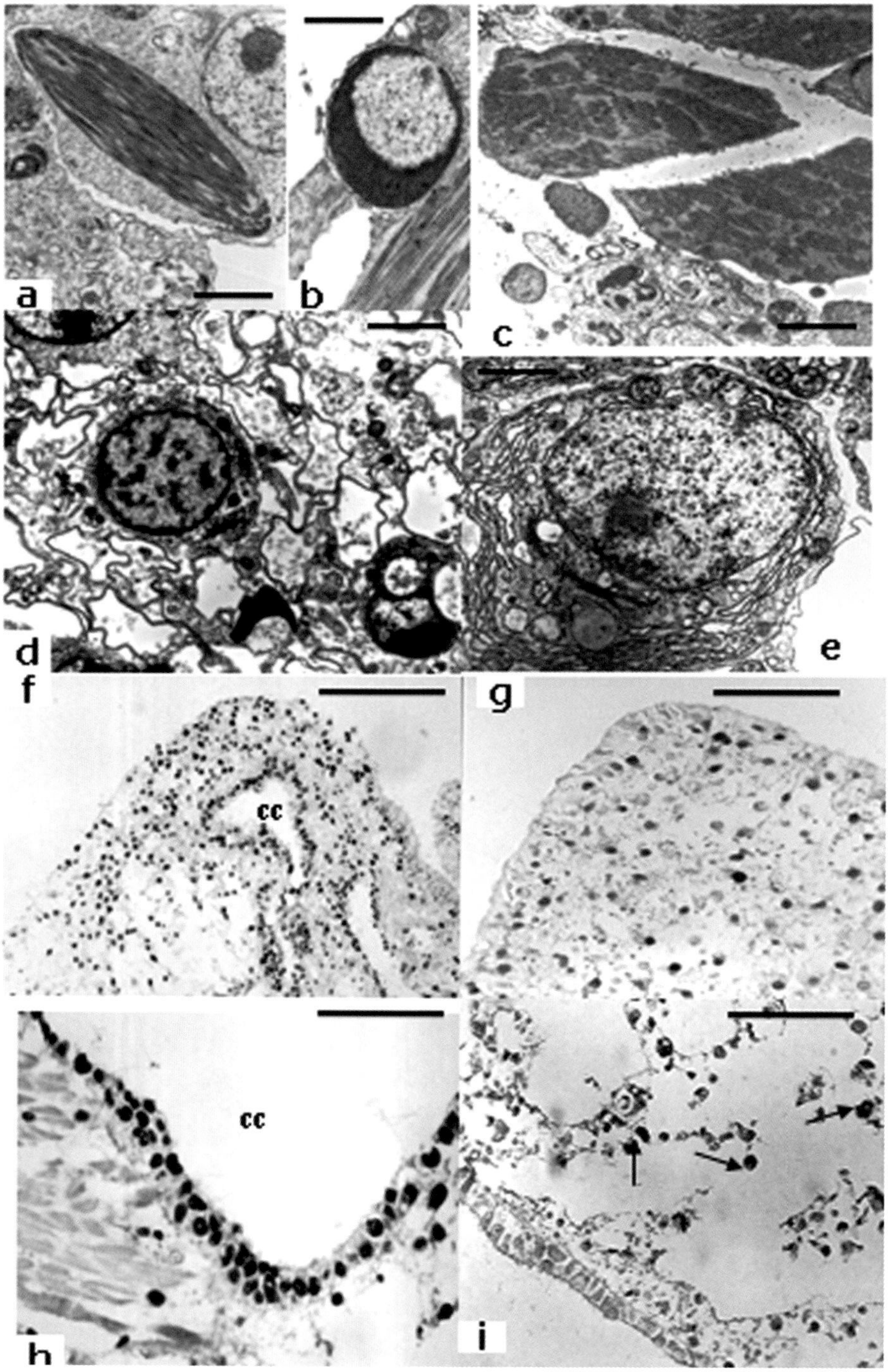
a
b
c
d
e
f
cc
g
cc
h
i

Fig. 6. TEM and ICC sections of control and exposed regenerating arms of *Antedon mediterranea* at different stages. Exposure concentrations: PCB: 14 ng/l; 4-NP: 2 μg/l; TPT-Cl: 100 or 225 ng/l. **a–d** TEM details of 1 and 2 weeks exposed samples. *Bars* 2 μm. PCB (**a**, **e**, **f**), 4-NP (**b**, **c**). **a** Detail of the muscle bundle of the stump. **b** Myocyte showing a typical apoptotic nucleus. **c** Migratory semi-dedifferentiated myocytes inside the coelomic canal. **d** Connective tissue of the stump showing extensive processes of cellular rearrangement/vacuolization. **e** Blastemal cell showing unusual development of endoplasmic reticulum and Golgi complexes. *Bar* 1 μm. **f–h** ICC for BrdU. The number of BrdU-labelled cells is massive in the blastema of PCB samples (**f**) and rather limited in that of 4-NP samples (**g**). A strong and extensive labelling is also found in the coelomic epithelium (*cc*) of PCB samples (**h**). **i** ICC for cytochrome P450. The details show an intense immunoreaction in many cells (*arrows*) at the level of the tissues of the stump, particularly in connective and skeletal tissue. *Bars*: **f** 25 μm; **g** 30 μm; **h** 10 μm; **i** 30 μm

enzymes such as the microsomal cytochrome P-450 mono-oxygenase system, i.e. the main enzymes responsible for biotransformation and metabolism of the majority of lipophilic xenobiotics (den Besten 1998). Interestingly, the reaction was intense and diffuse in the tissues of the stump more involved in turnover activity (Fig. 6i).

5 Conclusions and Future Prospects

In the light of the present results, the regenerative processes of crinoids appear to provide valuable and flexible experimental models for studies exploring the effects of exposure to exogenous substances on the regenerative growth with particular reference to endocrine disrupter contaminants (EDs), which induce significant dysfunctions in terms of processes of growth and development from the whole organism level up to the tissue, cellular and molecular level.

In particular the experiments of prolonged exposure to typical ED compounds, such as PCBs, 4-NP and TPT, employed at low concentrations reproducing the contamination levels of moderately polluted coastal zones (Geyer et al. 1994), indicate significant variations in timing and intrinsic mechanisms of regeneration in terms of:

- accelerated or delayed growth
- enhanced or inhibited cell proliferation
- massive and prolonged cell migration
- extensive cell/tissue recycling and rearrangement
- aberrant histogenesis
- cytological disorder
- induction of protective biochemical responses

These effects are associated to a rapid and persistent bioaccumulation of all the test compounds in the animal tissues (Candia Carnevali et al., 2001a,b; Dagnac et al. unpublished; Tremolada et al. in press). Although these aspects often indicate specific and contrasting effects caused by the individual contaminant (Barbaglio et al. 2004; Candia Carnevali et al. 2001a,b; 2003), they point out that all these compounds 1) affect growth and development by interfering with the same basic cellular mechanisms of regeneration, such as cell proliferation, migration and differentiation/dedifferentiation, which are possibly controlled and stimulated, directly or indirectly, by steroid hormones (Marsh and Walker 1995); 2) can induce a number of significant modifications in timing, modalities and pattern of arm regeneration which, in terms of histopathology, can be interpreted in the light of a significant activation of cell mechanisms related to steroid synthesis/metabolism and/or to specific detoxification processes (den Besten 1998; Schoenmakers 1980). The possible effects of exposure tests with suitable mixture of ED compounds, such as PCBs, 4-NP, organotins and others, could be an intriguing point to investigate in a further step of the research work in order to progress in exploring in realistic terms the potential ecological significance of endocrine disruption in marine invertebrates.

Taking into account the increasing importance for our future prospects of life to determine the impacts of chemicals in the environment, the bulk of our results not only presents a new valuable model for specific ecotoxicological applications, but also clearly reveal how appropriate can be to employ an integrated approach centered on individual responses at more than one level of biological organization, from cell to whole animals, for testing the effects of exposure to insidious exogenous substances on growth and development. Useless to say, our multilevel approach, can be obviously helpful to identify, assess and validate suitable biomarkers at tissue and cellular level. In addition this approach can also be naturally expanded in the area of *in vitro* aquatic toxicology if suitably associated to applications of in vitro techniques, a field which needs urgently to be developed in aquatic invertebrates for conducting controlled laboratory experiments on well-characterized cell lines also derived from adult regenerating animals. The potential for *in vitro* techniques in this area is clearly shown by Matranga et al., this book. Last but not least, our approach can also help to throw light on basic problems such as that of the regeneration-competent cells, in terms of origin (stem cells or dedifferentiated cells), activities (proliferation and/or migration) and fate (derived cell lineages), or that of the functional implication of regulatory mitogenic or morphogenic factors. An integrated understanding of the cellular and molecular basis of these processes emerging from basic research associated to complementary experimental approaches, such as a modern ecotoxicological approach, in vivo and in vitro, appears in fact to be quite timely and appropriate. The crucial point of the "progenitor" elements involved in regenerative processes, in terms of cell recruitment, sources and fate, and totipotentiality, pluripotentiality or unipotentiality, is a central problem, particularly relevant for its topical interest, which deserves to be explored appropriately also in experimental models less commonly used but potentially very amenable for this approach. The identification of the cellular events induced by wounding and repair in regenerating animals in normal and extreme conditions of environmental stress can significantly help to understand the conditions and factors that allow developmental fields to be established de novo in terminally differentiated tissues. The stimulation of regeneration of new tissues from old tissues in vivo, constitutes, in fact, an emerging and exciting interdisciplinary field called "regenerative biology" (Stocum 1998) whose wide potential of expansion is strictly depending on the knowledge of the fundamental mechanisms at the cellular and molecular level. The identification of the distribution and characteristics of the reserve of progenitor cells and of the signals, stimulatory or inhibitory, to regeneration will be major avenues of research in the new field of regenerative medicine (Pearson 2001) and, in our attempt to regenerate human tissues in vivo, we can learn from crinoids how to induce the formation of progenitor cells by cell dedifferentiation.

On the whole our ecotoxicological approach throw new light into the wide field of environmental regulation of development by confirming that also regenerative development is "critically keyed to the environment" (Gilbert 1997). The bulk of results obtained so far provide significant indications on sub-lethal effects of exposure to ED compounds and mechanisms of toxicity related to developmental physiology, in terms of both regulatory mechanisms of growth, morphogenesis, differentiation and morphological anomalies at whole organism, tissue and cellular level. In conclusion, the regenerative response of crinoids can be considered a new, quite original and sensitive test for studying ecotoxicological effects of persistent ED pollutants and assessing their disrupting activity, which not only perfectly responds to the 5-*R* requisites indicated by Hopkin (1993): *R*elevant, *R*obust, *R*eliable, *R*esponsive, *R*eproducible, but even extends the definition of the ideal ecotoxicological model by adding a possible sixth *R* : *R*egenerating.

Acknowledgements. The present work was carried out with the financial support provided by the EU (COMPRENDO Project no. EVK1-CT-2002-00129) and the MIUR COFIN-2003 Research Project. In addition, the author is particularly grateful to the following colleagues for their help and profitable collaboration in the research work: Alice Barbaglio, Francesco Bonasoro, Silvana Galassi, Daniela Mozzi, Michela Sugni, Paolo Tremolada.

References

Aminin DL, Agafonova IG, Federov SN (1995) Biological activity of disulfated polyhydroxysteroids from the Pacific brittle star *Ophiopholis aculeata*. Com Biochem Physiol 112C:201–204

Anderson S, Hose JE, Knezovic JP (1994) Genotoxic and developmental effects in sea urchin are sensitive indicators of effects of genotoxic chemicals. Environ Toxicol Chem 13:1033–1041

Arnold SF, Klotz DM, Collin BM, Vonier PM, Gillette L Jr, McLachlan JA (1996) Synergistic activation of estrogen receptor with combination of environmental chemicals. Science 272:1489–1492

Barbaglio A, Sugni M, Mozzi D, Invernizzi A, Doria A, Pacchetti G, Tremolada P, Bonasoro F, Candia Carnevali MD (2004) Exposure effects of organotin compounds (TPT-Cl) on regenerative potential of crinoids. In: Proc 11th Int Echinoderm Conf, Munich 2003. Swets and Zeitlinger, Lisse

Békri K, Pelletier E (2004) Trophic transfer and in vivo immunotoxicological effects of tributyltin (TBT) in polar sea star *Leptasterias polaris*. Aquat Toxicol 66:39–53

Bonasoro F, Ferro P, Di Benedetto C, Sugni M, Mozzi D, Candia Carnevali MD (2004) Regenerative potential of echinoid test. In: Heinzeller T (eds) Proc 11th Int Echinoderm Conf, Munich 2003. Swets and Zeitlinger, Lisse

Bryan GW, Gibbs PE, Hummerstone LG, Burt GR (1986) The decline of the gastropod *Nucella papillus* around south-west England: evidence for the effect of tributyltin from antifouling paints. J Mar Biol Assoc UK 66:611–640

Cadbury D (1998) The feminisation of nature. Our future at risk. Penguin, London

Candia Carnevali MD, Bonasoro F (2001a) Introduction to the biology of regeneration in echinoderms. Microsc Res Tech 55:365–368

Candia Carnevali MD, Bonasoro F (2001b) Microscopic overview of crinoid regeneration. Microsc Res Tech 55:403–426

Candia Carnevali MD, Lucca E, Bonasoro F (1993) Mechanism of arm regeneration in the feather star *Antedon mediterranea*: healing of wound and early stages of development. J Exp Zool 267:299–317

Candia Carnevali MD, Bonasoro F, Lucca E, Thorndyke MC (1995) Pattern of cell proliferation in the feather star *Antedon mediterranea*. J Exp Zool 272:464–474

Candia Carnevali MD, Bonasoro F, Biale A (1997) Pattern of bromodeoxyuridine incorporation in the advanced stages of arm regeneration in the feather star *Antedon mediterranea*. Cell Tissue Res 289:363–374

Candia Carnevali MD, Bonasoro F, Patruno M, Thorndyke MC (1998) Cellular and molecular mechanisms of arm regeneration in crinoid echinoderms: the potential of arm explant. Dev Gen Evol 208:421–430

Candia Carnevali MD, Bonasoro F, Patruno M, Thorndyke MC, Galassi S (2001a) PCB exposure and regeneration in crinoids (Echinodermata). Mar Ecol Prog Ser 215:155–167

Candia Carnevali MD, Galassi S, Bonasoro F, Patruno M, Thorndyke MC (2001b) Regenerative response and endocrine disrupters in crinoid echinoderms: arm regeneration in *Antedon mediterranea* after experimental exposure to polychlorinated biphenyls. J Exp Biol 204:835–842

Candia Carnevali MD, Bonasoro F, Patruno M, Thorndyke MC (2001 c) Role of the nervous system in echinoderm regeneration. In: Barker M (ed) Proc 10th Int Echinoderm Conf, Dunedin 2000. AA Balkema, Rotterdam, pp 5–20

Candia Carnevali MD, Bonasoro F, Ferreri P, Galassi S (2003) Regenerative potential and effects of exposure to pseudo-estrogenic contaminants (4-nonylphenol) in the crinoid *Antedon mediterranea*. In: Feral JP, David B (eds) Echinoderm research 2001. Swets and Zeitlinger, Lisse, pp 201–207

Chia FS, Xing J (1996) Echinoderm coelomocytes. Zool Stud 35:231–254

Colborn T, von Saal FS, Soto AM (1993) Developmental effects of endocrine disrupting chemicals in wildlife and humans. Environ Health Perspect 101:378–384

Cooper RL, Kavlock RJ (1997) Endocrine disruptors and reproductive development: a weight-of-evidence overview. J Endocr 152:159–166

Coteur G, Danis B, Fowler SW, Teyssié J-L, Dubois Ph, Warnau M (2001) Effects of PCBs on reactive oxygen species (ROS) production by the immune cells of *Paracentrotus lividus* (Echinodermata). Mar Pollut Bull 42:667–672

den Besten PJ (1998) Cytochrome P450 monooxigenase system in echinoderms. Comp Biochem Physiol 121C:139–146

den Besten PJ, Herwig HJ, Zandee DI, Voogt PA (1989) Effects of cadmium and PCBs on reproduction of the sea star *Asterias rubens*: aberrations in the early development. Ecotox Environ Safety 18:173–180

den Besten PJ, Herwig HJ, Small AC, Zandee DI, Voogt PA (1990) Interference of polychlorinated biphenyls with gametogenesis in the sea star *Asterias rubens* L. Aquat Toxicol 28:231–246

den Besten PJ, Elanbaas JML, Maas EJR, Dieleman SJ, Hervig HJ, Voogt PA (1991a) Effects of cadmium and polychlorinated biphenyls on steroid metabolism and cytochrome P-450 monooxigenase system in the sea star *Asterias rubens* L. Aquat Toxicol 20:95–110

den Besten PJ, Maas EJR, Livingstone DR, Zandee BI, Voogt PA (1991b) Interference of benzo[α]pirene with cytochrome P-450 mediated steroid metabolism in pyloric caeca microsome of the sea star *Asterias rubens* L. Comp Biochem Physiol 100C:165–168

Depledge MH, Billinghurst Z (1999) Ecological significance of endocrine disruption in marine invertebrates. Mar Pollut Bull 39:32–38

Dickey P (1997) Troubling bubbles. The case for replacing alkylphenol ethoxylate surfactants. Washington Toxics Coalition, Seattle, pp 1–59

Dolmatov IY, Ginanova TT (2001) Muscle regeneration in holothurians. Micro Res Tech 55:452–463

Dubois P, Ameye L (2001) regeneration of spines and pedicellariae in echinoderms: a review. Microsc Res Tech 55:427–437

Emson RH, Wilkie IC (1980) Fission and autotomy in echinoderms. Oceanogr Mar Biol Annu Rev 18:155–250

Fairley P, Roberts M, Stringer J (1996). Endocrine disruptor. Chem Week 8 May:29–36

Fait A, Ferioli A, Barbieri F (1994) Organotin compounds. Toxicology 91:77–82

Ferretti P, Géraudie J (1998) Preface. In: Ferretti P, Geraudie J (eds) Cellular and molecular basis of regeneration: from invertebrates to humans. Wiley, New York, pp xix–xx

Fingerman M (1997) Crustacean endocrinology: a retrospective, prospective and introspective analysis. Physiol Zool 70:257–269

Garcia-Arraras JE, Greenberg MJ (2001) Visceral regeneration in holothurians. Microsc Res Tech 55:438–451

Gibbs PE, Pascoe PL, Burt GR (1988) Sex changes in the female dog-whelk, *Nucella lapillus*, induced by tributyltin from antifouling paints. J Mar Biol Assoc UK 68:715–731

Gilbert SF (1997) Developmental biology, 5th edn. Sinauer Associates, Sunderland, Massachusetts

Granmo Å (1991) Toxicity of 4-nonylphenol to aquatic organisms and potential for bioaccumulation. In: Proc Seminar on Nonylphenolethoxylates (NPE) and Nonylphenol (NP), Saltsjöbaden, Sweden. Ingvar Bingham, Stockholm, pp 53–75

Gray LE Jr, Monosson E, Kelce WR (1996) Emerging issues: the effects of endocrine disrupters on reproductive development. In: Di Giulio RT, Monosson E (eds) Interconnection between human and ecosystem health. Chapman and Hall, London, pp 47–81

Gross PS, Al-Sharif WZ, Cow LA, Smith LC (1999) Echinoderm immunity and the evolution of the complement system. Dev Comp Immunol 23:429–442

Hinton DE (1997) Histological responses signifying exposure or effect and their application to aquatic ecotoxicology. In: Zelikoff JT (ed) Proc OECD Worksh on Ecotoxicology: Responses, Biomarkers and Risk Assessment. SOS Publications, Fairhaven, New Jersey

Holmes P, Harrison P, Bergman A, Brandt I, Brouwer B, Keiding N, Randall G, Sharpe R, Skakkebaek N, Ashby J, Barlow S, Dickerson R, Humfrey C, Smith LM (1997) Proc European Workshop on The Impact of Endocrine Disrupters on Human Health and Wildlife, 2–4 Dec 1996. Rep EUR 17549. MRC Institute for Environmental and Health, Weybridge

Hopkin SP (1993) In situ biological monitoring of pollution in terrestrial and aquatic ecosystems. In: Calow P (ed) Handbook of ecotoxicology, vol I. Blackwell, Oxford, pp 397–427

Horiguchi T, Shiraishi H, Shimizu M, Morita M (1995) Imposex in Japanese gastropods (Neogastropoda and Mesogastropoda): effects of tributyltin and triphenyltin from antifouling paints. Mar Pollut Bull 31:402–405

Hyman HL (1955) The invertebrates: Echinodermata, vol 4. McGraw-Hill, New York

Janer G, LeBlanc GA, Porte C (2004) A comparative study on the metabolism of androgens in invertebrates and its modulation by xenoandrogens. In: Proc CREDO Cluster Worksh on Ecological Relevance of Chemically Induced Endocrine Disruption in Wildlife, Exeter, p 23

Kobahashi N (1984) Marine ecotoxicological testing with echinoderms. In: Persoone G, Jaspers F, Claus C (eds) Ecotoxicological testing for the marine environment, vol 1. State University of Ghent and Institute of Marine Scientific Research, Bredene, Belgium, pp 341–405

LeBlanc GA, Campbell PM, den Besten P, Brown RP, Chang ES, Coats JR, deFur PL, Dhadialla T, Edwards J, Riddiford Lm, Simpson MG, Snell TW, Thorndyke MC, Matsumura F (1999) The endocrinology of invertebrates. In: deFur P, Crane M, Ingersoll C, Tattersfield L (eds) Proc Worksh on Endocrine Disruption in Invertebrates: Endocrinology, Testing and Assessment, 12–15 Dec 1998. SETAC, Noordwijkerhout, The Netherlands, pp 23–106

Lutz I, Sugni M, Candia Carnevali MD, Lavado R, Porte C, Schulte-Oehlmann U, Kloas W (2004) First evidence for specific [3H]-testosterone and [3H]-estradiol binding sites in echinoderms. In: Proc CREDO Cluster Workshop on Ecological Relevance of Chemically Induced Endocrine Disruption in Wildlife, Exeter, p 43

Marsh A, Walker C (1995) Effect of estradiol and progesterone on *c-myc* expression in the sea star testis and the seasonal regulation of spermatogenesis. Mol Reprod Dev 40:62–68

Matthiessen P, Gibbs PE (1998) Critical appraisal of the evidence for tributyltin-mediated endocrine disruption in molluscs. Environ Toxicol Chem 17(1):37–43

Minckert W (1905) Über Regeneration bei Comatuliden nebst Ausführungen über die Auffassung und Bedeutung der Syzygien. Arch Naturg Jahrg 71(1):163–244

Mladenov PV, Burke RD (1994) Echinodermata: asexual propagation. In: Adiyodi KG, Adyodi RG (eds) Reproductive biology of invertebrates, vol 6, part B. Asexual propagation and reproductive strategies. Oxford and Hill, New Delhi, pp 339–383

Mladenov PV, Bisgrove B, Asotra S, Burke RD (1989) Mechanisms of arm-tip regeneration in the sea-star *Leptasterias hexactis*. Roux's Arch Dev Biol 189:19–28

Moss C, Hunter J, Thorndyke MC (1998) Pattern of bromodeoxyuridine incorporation and neuropeptide immunoreactivity during arm regeneration in the starfish *Asterias rubens*. Philos Trans R Soc Lond B 353:421–436

Novelli AA, Argese E, Tagliapietra D, Bettiol C, Volpi Ghirardini A (2002) Toxicity of tributyltin and triphenyltin to early life-stages of *Paracentrotus lividus* (Echinodermata: Echinoidea). Environ Toxicol Chem 21:859–864

Pearson H (2001) The regeneration gap. Nature 414: 388–390

Perrier E (1873) L'anatomie et la régenération des bras de la comatula. Arch Zool Exp Gen 2:29–86

Reichensperger A (1912) Beiträge zur Histologie und zum Verlauf der Regeneration bei Crinoiden. Z Wiss Zool 101:1–69

Rippen G (1990) Handbuch Umweltechemikalien. Stoffdaten, Pfüfverfahren, Vorschirften. 3. Auflage, 5. Ergänzungslieferung 2/90. Ecomed, Landsberg am Lech

Schoenmakers HJN (1979) In vitro biosynthesis of steroids from cholesterol by the ovaries and pyloric caeca of the starfish *Asterias rubens*. Comp Biochem Physiol 63(B):179–184

Schoenmakers HJN (1980) The variation of 3ß-hydroxysteroid dehydrogenase activity of the ovaries and pyloric caeca of the starfish *Asterias rubens* during the annual reproductive cycle. J Comp Physiol 138:27–30

Schoenmakers HJN, Voogt PA (1980) In vitro biosynthesis of steroids from progesterone by the ovaries and pyloric caeca of the starfish *Asterias rubens*. Gen Comp Endocr 41:408–416

Schweitzer LE, Hose JL, Suffet IH, Bay SM (1997) Differential toxicity of three polychlorinated biphenyl congeners in developing sea urchin embryos. Environ Toxicol Chem 16:1510–1514

Shirai H, Walker CW (1988) Chemical control of asexual and sexual reproduction in echinoderms. Alan R Liss, New York

Shubina LK, Fedorov SN, Levina EV, Andriyaschenko PV, Kalinovsky AI, Stonik VA, Smirnov IS (1997) Comparative study on polyhydroxylated steroids from echinoderms. Comp Biochem Physiol 119(B):505–511

Soto AM, Sonnenschein C, Chung KL (1995) The E-Screen as a tool to identify estrogens: an update on estrogenic environmental pollutants. Environ Health Perspect 103(Suppl): 113–122

Stahl RG, Tattersfield LJ, Campbell PM, Horiguchi T, deFur PL, Dick Vethaak A (1999) Introduction to the workshop on endocrine disruption in invertebrates: endocrinology, testing and assessment. In: deFur P, Crane M, Ingersoll C, Tattersfield L (eds) Proc Worksh on Endocrine Disruption in Invertebrates: Endocrinology, Testing and Assessment, 12–15 Dec 1998. SETAC, Noordwijkerhout, The Netherlands, pp 7–21

Stocum DL (1998) Bridging the gap: restoration of structure and function in humans. In: Ferretti. and Geraudieu J (eds) Cellular and Molecular Basis of Regeneration: from Invertebrates to Humans. New York: John Wiley & Sons. pp. 411–450

Thorndyke MC, Candia Carnevali MD (2001) Regeneration neurohormones and growth factors in echinoderms. Can J Zool 79:1171–1208

Thorndyke MC, Patruno M, Moss C, Beesley PW (2001) Cellular and molecular bases of arm regeneration in brittlestars. In: Barker M (ed) Proc 10th Int Echinoderm Conf, Dunedin, 2000. AA Balkema, Rotterdam, pp 323–326

Thouveny Y, Tassava RA (1998) Regeneration through phylogenesis. In: Ferretti P, Geraudie J (eds) Cellular and molecular basis of regeneration: from invertebrates to humans. Wiley, New York, pp 9–43

Tremolada P, Mozzi D, Sugni M, Barbaglio A, Dagnac T, Candia Carnevali MD (2004) A simple model for evaluating compound loss processes in aquatic ecotoxicological tests: the case of TPT-Cl. In: Albanis TA (ed) Proc 3rd European Conf on Pesticides and Related Organic Micropollutants in the Environment, Halkidiki, Greece, pp 407-410

Voogt PA, Oudejans RCHM, Broertjes JJS (1984) Steroids and reproduction in starfish. In: Engels W, Clark WH, Fischer A, Olive PJM, Went DF (eds) Advances in invertebrate reproduction. Elsevier North-Holland, New York, pp 151–161

Voogt PA, den Besten PJ, Jansen M (1990) The Δ^5-pathway in steroid metabolism in the sea star *Asterias rubens* L. Comp Biochem Physiol 97(B):555–562

Voogt PA, den Besten PJ, Jansen M (1991) Steroid metabolism in relation to the reproductive cycle in *Asterias rubens* L. Comp Biochem Physiol 99(B):77–82

Walsh GE, McLaughlin LL, Louie MK, Deans CH, Lores EM (1986) Inhibition of arm regeneration by *Ophioderma brevispina* (Echinodermata, Ophiuroidea) by tributyltin oxide and triphenyltin oxide. Ecotoxicol Environ Saf 12:95–100

Wilkie IC (2001) Autotomy as a prelude to regeneration in echinoderms. Microsc Res Tech 55:369–396

Echinoderm Adhesive Secretions: From Experimental Characterization to Biotechnological Applications

P. Flammang, R. Santos, D. Haesaerts

Abstract. Adhesion is a way of life in echinoderms. Indeed, all the species belonging to this phylum use adhesive secretions extensively for various vital functions. According to the species or to the developmental stage considered, different adhesive systems may be recognized. (1) The tube feet or podia are organs involved in attachment to the substratum, locomotion, feeding or burrowing. Their temporary adhesion relies on a duo-gland adhesive system resorting to both adhesive and de-adhesive secretions. (2) The larval adhesive organs allow temporary attachment of larvae during settlement and strong fixation during metamorphosis. (3) The Cuvierian tubules are sticky defence organs occurring in some holothuroid species. Their efficacy is based on the instantaneous release of a quick-setting adhesive. All these systems rely on different types of adhesion and therefore differ in the way they operate, in their structure and in the composition of their adhesive. In addition to fundamental interests in echinoderm bioadhesives, a substantial impetus behind understanding these adhesives are the potential technological applications that can be derived from their knowledge. These applications cover two broad fields of applied research: design of water-resistant adhesives and development of new antifouling strategies. In this context, echinoderm adhesives could offer novel features or performance characteristics for biotechnological applications. For example, the rapidly attaching adhesive of Cuvierian tubules, the releasable adhesive of tube feet or the powerful adhesive of asteroid larvae could each be useful to address particular bioadhesion problems.

P. Flammang (e-mail: patrick.flammang@umh.ac.be), R. Santos
Laboratoire de Biologie Marine, Université de Mons-Hainaut, Pentagone 2B//6, Avenue du Champ de Mars, 7000 Mons, Belgium
D. Haesaerts
Laboratoire de Biologie Marine, Université Libre de Bruxelles, Brussels, Belgium

Progress in Molecular and Subcellular Biology
Subseries Marine Molecular Biotechnology
V. Matranga (Ed.), Echinodermata

1 Introduction

Among mechanisms allowing marine organisms to attach to or manipulate a substratum, one can distinguish mechanical attachments (e.g. hooks or suckers) from chemical attachments (with adhesive substances) (Nachtigall 1974). The way the former operates is generally obvious, whereas the functioning of the latter remains enigmatic. Yet, adhesion (attachment with adhesive substances) is a way of life in the sea. Indeed, representatives of bacteria, protoctists (including macroalgae) and all animal phyla living in the sea attach to natural or artificial surfaces. Adhesion is particularly developed and diversified in invertebrates, who use it during their larval life as well as during their adult life (Walker 1987). It is involved in various functions such as attachment to the substratum, handling of food or building tubes or burrows (Walker 1987; Tyler 1988; Flammang 1996; Whittington and Cribb 2001).

Seawater, being a dense medium, denies gravity the power to hold organisms to the bottom; thus, if they want to withstand the hydrodynamism, marine organisms must have adhesive mechanisms. Attachment to the substratum is therefore the most important use of adhesion by marine invertebrates. Adhesion to the substratum may be permanent, transitory or temporary (Tyler 1988; Flammang 1996; Whittington and Cribb 2001). Permanent adhesion involves the secretion of a cement and is characteristic of sessile organisms staying at the same place throughout their adult life (such organisms have representatives among sponges, hydrozoan cnidarians, cirripede crustaceans, bivalve molluscs, tubicolous polychaetes, bryozoans and tunicates) (Walker 1987). Transitory adhesion allows simultaneous adhesion and locomotion: the animals attach by a viscous film they lay down between their body and the substratum and they creep on this film, which they leave behind as they move. This type of adhesion is characteristic of invertebrates – mostly small soft-bodied invertebrates such as turbellarians, nemertines, gastrotrichs and polychaetes – moving along the substratum by ciliary gliding (Tyler 1988; Whittington and Cribb 2001). Larger animals such as sea anemones and gastropod molluscs also use transitory adhesion; they move by means of waves of muscular contractions running along their foot (Walker 1987). Temporary adhesion allows organisms to attach firmly but momentarily to a substratum. This type of adhesion is very frequently found in small invertebrates inhabiting the interstitial environment, e.g. in turbellarians, gastrotrichs, nematodes and polychaetes (Tyler 1988). A few macro-invertebrates, such as some cnidarians and most echinoderms, can also attach and detach repeatedly (Flammang 1996).

Recently, Whittington and Cribb (2001) have introduced the term "tissue adhesion" to describe the attachment of symbiotic organisms to the living tissues of their hosts. Although they suggested that tissue adhesion is a fourth type of adhesion, we rather propose that this is a subcategory of either permanent, transitory or temporary adhesion, which should be opposed to adhesion to abiotic substrata. Indeed, examples of tissue adhesion include permanent attachment of parasitic barnacles on whale skin (Ridgway et al. 1997), transitory attachment of parasitic gastropods on echinoderm epidermis (Vaitilingon et al. 2004) and temporary attachment of parasitic monogeneans on the gills or skin of fishes (Whittington and Cribb 2001). There are, however, invertebrate adhesive systems that do not fit into the three types of adhesion described above. These adhesive systems rely on single-use organs or cells and are used in functions other than attachment to the substratum requiring a very fast formation of adhesive bonds. Prey capture by collocyte-bearing tentacles of ctenophorans (Franc 1978; Eeckhaut et al. 1997) is a typical example of this type of adhesion for which we propose the term "instantaneous adhesion".

The phylum Echinodermata is quite exceptional in the sense that most species belonging to this group use adhesion extensively. Moreover, according to the species or to the developmental stage considered, different adhesive systems may be recognized. These include: (1) tube feet or podia, organs involved in attachment to the substratum, locomotion, feeding or burrowing; (2) larval adhesive organs allowing attachment of larvae during settlement and metamorphosis; and (3) Cuvierian tubules, sticky defence organs occurring in some holothuroid species. All these systems rely on different types of adhesion and therefore differ in the way they operate, in their structure and in the composition of their adhesive.

2 Tube Feet

Being exclusively benthic animals, echinoderms have activities and adaptations that are correlated with a benthic existence. Most of these activities, such as attachment to the substratum, locomotion, handling of food and burrow-building, rely on adhesive secretions allowing the animal to stick to or to manipulate a substratum. In post-metamorphic echinoderms, these adhesive secretions are always produced by specialized organs, the podia or tube feet. These are the external appendages of the ambulacral system and are also probably the most advanced hydraulic organs in the animal kingdom. Tube foot attachment is typically temporary adhesion. Indeed, although tube feet can adhere very strongly to the substratum, they are also able to detach easily and voluntarily from the substratum before reinitiating another attachment–detachment cycle (Thomas and Hermans 1985; Flammang 1996)

Tube feet have diversified into a wide variety of morphotypes, which were classified by Flammang (1996) into disc-ending, penicillate, knob-ending, lamellate, ramified and digitate. In terms of adhesion, however, for practical considerations only disc-ending tube feet involved in attachment to the substratum and locomotion have been studied in detail. These tube feet consist of a basal extensible cylinder, the stem, and an enlarged and flattened apical extremity, the disc (Fig. 1A). Tube foot adhesive strength was evaluated by measuring their tenacity, which is the adhesion force per unit area and is expressed in Pascals (Pa). Tenacity of a single tube foot has been quantified in several species of asteroids and echinoids. The mean normal tenacities measured on a glass substratum were 170 kPa in *Asterias vulgaris* (Paine 1926), 198 kPa in *Asterias rubens* (Flammang and Walker 1997) and 59, 120 and 290 kPa in *Arbacia lixula*, *Sphaerechinus granularis* and *Paracentrotus lividus*, respectively (Santos 2003). Tenacities of whole individuals were also measured in the same echinoid species and respectively average 190, 260 and 330 kPa (Santos 2003). All these values are in the same range as those observed in other marine invertebrates known to adhere very strongly to the substratum (e.g. 230 kPa in limpets, 520 kPa in barnacles, 750 kPa in mussels; see Walker 1987 for review). Tube foot adhesive secretions therefore appear to be well tailored to provide an efficient attachment to the substratum, allowing echinoderms to resist hydrodynamically generated forces.

The histological structure of the tube feet is remarkably constant for all echinoderm species. Their tissue stratification consists of four layers: an inner myomesothelium surrounding the ambulacral lumen, a connective tissue layer, a nerve plexus and an outer epidermis covered externally by a cuticle (Flammang 1996). At the level of the tube foot tip, these tissue layers are specialized in adhesion and sensory perception: the connective tissue layer and the nerve plexus are thickened, and the epidermis is differentiated into a well-developed sensory-secretory epithelium. The latter comprises two types of secretory cells: non-ciliated secretory cells (NCS cells) enclosing large heterogeneous granules and ciliated secretory cells (CS cells) enclosing small homogeneous electron-dense granules (see Flammang 1996 for review). In some species, two types of NCS cells co-occur in the sensory-secretory epidermis. The study of the ultrastructure of these different types of secretory cells during a complete cycle of attachment–detachment of the tube foot in *A. rubens* (Fig. 2) demonstrated that they function as a duo-gland adhesive system as originally proposed by Hermans (1983), and in which NCS cells release an adhesive secretion and CS cells a de-adhesive secretion (Flammang et al. 1994; Flammang 1996). The adhesive is present as a thin film between the tube foot cuticle and the substratum and, when detachment occurs, it takes place at the level of the outermost layer of the cuticle, the fuzzy coat, leaving the adhesive material strongly attached to the substratum as a footprint (Fig. 1B,C) (Flammang 1996). In *A. rubens*, polyclonal antibodies have been raised against footprint material and were used to locate the origin of footprint constituents in

Fig. 1A–C. Tube foot adhesion and adhesive in the echinoid *Paracentrotus lividus* (originals). SEM photograph of a disc-ending tube foot (**A**), and 3-D and 2-D topographical views of adhesive footprints deposited on a glass substratum by this tube foot (**B** and **C**, respectively). *D* Disc; *S* stem; *SSE* sensory-secretory epidermis

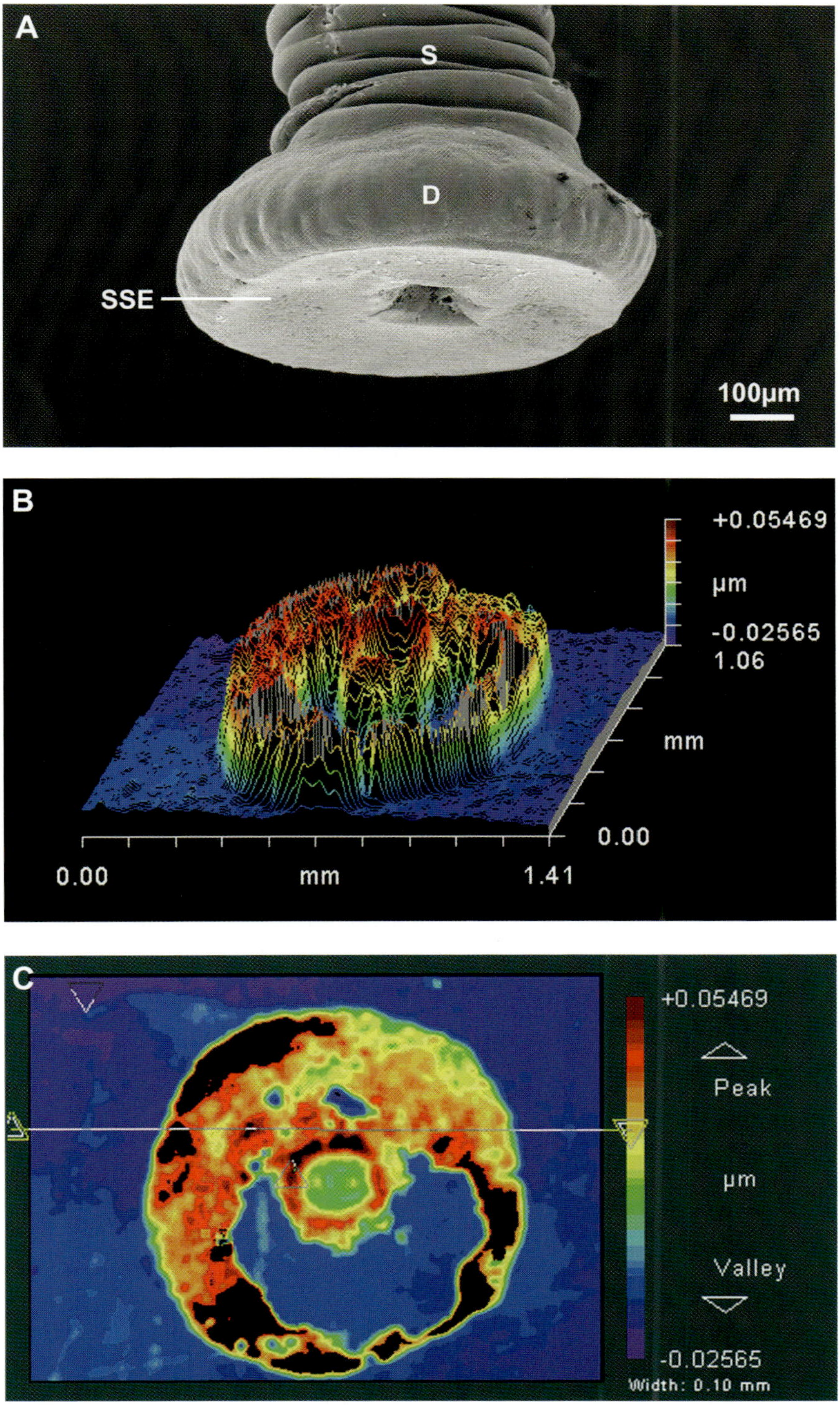
A
S
D
SSE
100μm
B
+0.05469
μm
-0.02565
1.06
mm
0.00
0.00
mm
1.41
C
+0.05469
Peak
μm
Valley
-0.02565
Width: 0.10 mm

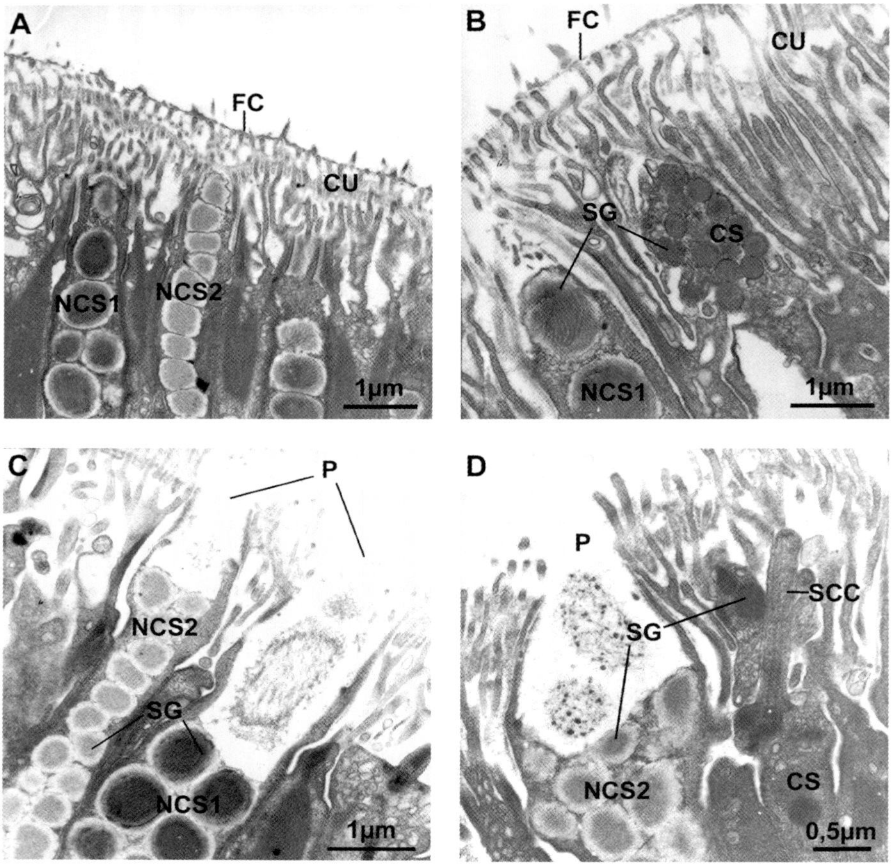

Fig. 2A–D. Ultrastructure of the tube foot sensory-secretory epidermis in the asteroid *Asterias rubens* during an attachment–detachment cycle (originals). TEM photographs showing non-ciliated secretory cells (**A**) and ciliated secretory cells (**B**) before attachment. During attachment to the substratum, non-ciliated secretory cells release some of their granules (**C**), while ciliated secretory cells remain unchanged (not illustrated). After voluntary detachment from the substratum, ciliated secretory cells have released their most apical granules (**D**). *CS* Ciliated secretory cell; *CU* cuticle; *FC* fuzzy coat; *NCS1* type 1 non-ciliated secretory cell; *NCS2* type 2 non-ciliated secretory cell; *P* pore; *SCC* subcuticular cilium; *SG* secretory granule

the tube feet (Flammang et al. 1998a). Extensive immunoreactivity was detected in the secretory granules of both NCS1 and NCS2 cells, suggesting that their secretions make up together the bulk of the adhesive material. No immunoreactivity was detected in the secretory granules of CS cells and the only other structure strongly labelled was the fuzzy coat. This pattern of immunoreactivity suggests that secretions of CS cells are not incorporated into the footprints, but instead might function enzymatically to jettison the fuzzy coat, thereby allowing the tube foot to detach (Flammang 1996; Flammang et al. 1998a).

Footprints in echinoderms consist of a sponge-like material deposited as a thin layer on the substratum (Thomas and Hermans 1985; Flammang 1996; Flammang et al. 1998a). Although their diameter is easily measured after staining of the adhesive material (Flammang 1996), footprint thickness is difficult to estimate. Using an interference-optical profilometer, which generated three-dimensional images of the footprint surface, the mean maximum footprint thickness was found to be 100 nm in the echinoid *P. lividus* and 230 nm in the asteroid *A. rubens* (Figs. 1B,C; Santos, Gorb and Flammang, unpubl. data). The chemical composition of the footprint material was analysed in *A. rubens*. Leaving inorganic residue apart, this material is made up mainly of proteins and carbohydrates (Flammang et al. 1998a). The protein moiety contains significant amounts of both charged (especially acidic) and uncharged polar residues as well as cysteine. The carbohydrate moiety is also acidic, comprising both uronic acids and sulphate groups. Adhesive interactions with the substratum could be through ionic bonds, presumably involving the acidic residues of both carbohydrate and protein moieties (Waite 1987), whereas cohesive strength could be achieved by intermolecular disulphide bonds. So far, *A. rubens* is the only species in which the tube foot adhesive has been studied biochemically and nothing is known about other echinoderm species. Regarding the asteroids, however, a comparative immunohistochemical study of the tube feet from 14 species representing five orders and ten families revealed that the adhesives of all these species are closely related, and this independently of the taxon considered, of the species habitat and of the tube foot morphotype or function (Santos et al. 2005).

Comparison of the composition of the temporary adhesive of *A. rubens* with that of other marine invertebrates shows that it is closer to the transitory adhesive of limpets, also composed of an association of proteins and acidic glycans, than to the permanent adhesives of mussels and barnacles, made almost exclusively of proteins (Flammang et al. 1998a; Flammang 2003). A similar relationship between non-permanent adhesives is also observed when one compares the amino acid compositions of the temporary adhesive from asteroids with the temporary adhesive from monogenean flatworms and the transitory adhesive from limpets (see Flammang 2003). This relationship indicates convergence in composition because of common function (i.e. non-permanent attachment to the substratum) and selective pressures.

3
Larval Adhesive Organs

For most echinoderms, metamorphosis transforms a bilaterally symmetrical and pelagic larva into a radially symmetrical and benthic postmetamorphic individual. Settlement always takes place during the so-called perimetamorphic period (Gosselin and Jangoux 1998; Haesaerts et al. 2003), but either before or after the metamorphic stage according to the class considered (Strathmann 1978). In both cases, adhesive organs attach either the competent larva or the postlarva to the substratum during settlement. In three of the five extant echinoderm classes, these organs are the tube feet, viz. the five primary tube feet of competent echinoplutei in echinoids, the five primary tentacles (and, for some species, two posterior tube feet) of pentactulae in holothuroids, the five primary tube feet and the five first pairs of tube feet of ophiuroid postlarvae (Strathmann 1978). These tube feet are similar in structure and function to tube feet of adults (Cameron and Fankboner 1984; Flammang et al. 1998b). Larval adhesive organs of crinoids and asteroids are, on the other hand, unique and have no equivalent in the postmetamorphic stage (Strathmann 1978).

The perimetamorphic period of crinoids comprises three stages: the doliolaria (free-swimming larval stage), the cystidean (attached metamorphic stage) and the pentacrinoid stages (attached postlarval stage) (Mladenov and Chia 1983; Lahaye and Jangoux 1987; Nakano et al. 2003). Competent doliolariae are small barrel-shaped larvae. They possess an attachment complex at their anterior end which consists of a ciliary cap surrounding an apical tuft of elongated cilia and a ventrally located and slightly depressed adhesive pit (Fig. 3A,B). The ultrastructure of this attachment complex has been studied in comatulids (Chia et al. 1986; Jangoux and Lahaye 1990). It is strictly epidermal and made up of elongated ciliated cells associated with a thick basiepidermal nerve plexus. The four cell types forming the complex are sensory cells, covering cells and two types of secretory cells. Sensory cells and secretory cells of the first type occur exclusively in the ciliary cap. The former bear a long vibratile cilium whereas the latter are filled with secretory granules, which contain a flocculent mucopolysaccharidic material. Secretory cells of the second type are restricted to the adhesive pit where they are the most abundant cell type. These cells are filled with secretory granules with an electron-dense fibrillar proteinaceous content. At the beginning of the settlement phase, the doliolaria becomes demersal and brushes the substratum with its apical tuft (sensory structure) (Mladenov and Chia 1983; Lahaye and Jangoux 1988). This implies the occurrence of a mechanism allowing the larva to combine loose adhesion to the substratum with movement. This transitory adhesion is achieved by the combined action of the secretory cells of the ciliary cap which produce a thin mucous film retaining the larva at the water–substratum interface, and of the covering cells whose cilia beat in this mucus (Jangoux and Lahaye 1990; Flammang 1996). When reaching a suitable site, the larva stops moving and turns itself round to have its body directed obliquely (the adhesive pit facing the substratum). It then becomes permanently fixed and transforms into a cystidean larva (Mladenov and Chia 1983; Lahaye and Jangoux 1988). Permanent adhesion starts with the release of the proteinaceous cement by the secretory cells of the adhesive pit and continues during both cystidean and pentacrinoid stages (Chia et al. 1986; Jangoux and Lahaye

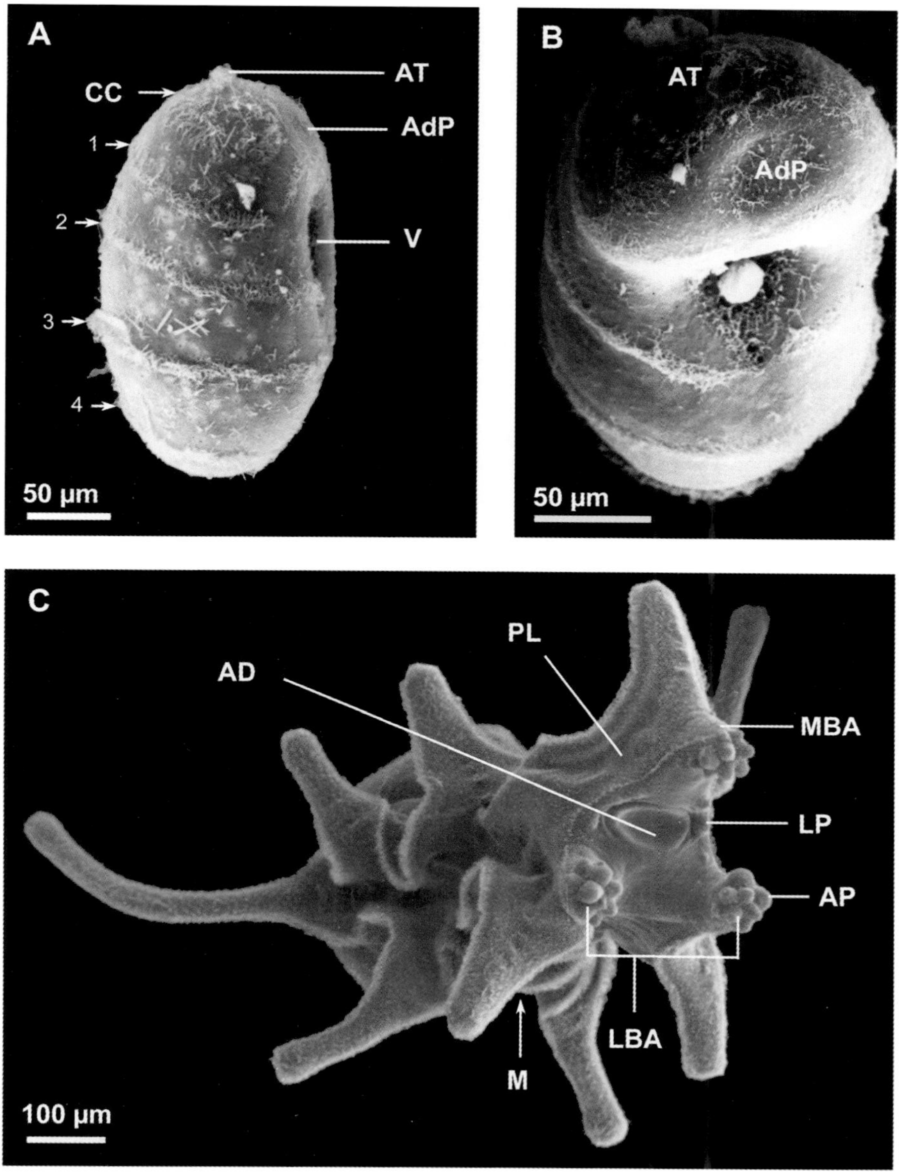

Fig. 3A–C. Larval adhesive organs of crinoids and asteroids. SEM photographs of the doliolaria larva of *Antedon bifida* (**A**) and of its anterior adhesive pit (**B**) (from Lahaye 1987), and of the brachiolaria larva of *Asterias rubens* (**C**) (original). *AD* Adhesive disc; *AdP* adhesive pit; *AP* apical papilla; *AT* apical tuft; *CC* ciliary cap; *LBA* lateral brachiolar arm; *LP* lateral papilla; *M* mouth; *MBA* median brachiolar arm; *PL* preoral lobe; *V* vestibule; *1–4* ciliary bands

1990). After development of the cirri during this last stage, the juvenile detaches from its cemented stalk (Lahaye and Jangoux 1987).

Competent larvae in asteroids are called brachiolariae because they possess a specialized attachment complex on their anterior part comprising three brachiolar arms and an adhesive disc (Fig. 3C; Barker 1978; Haesaerts et al. 2003). Brachiolar arms are hollow tubular structures occupied by an extension of the larval anterior coelom. Their histological organization comprises four tissue layers: an inner myomesothelium, a connective tissue layer, a subepidermal nerve plexus and an outer epidermis. Each brachiolar arm is tipped by several sensory-secretory areas named papillae, where both the epidermis and the nerve plexus are greatly thickened. The papillary epidermis encompasses three types of secretory cells (cell types A, B and C), sensory cells and support cells (Barker 1978; Haesaerts et al. 2005). Type A and B secretory cells are numerous and occupy most of the volume of the papilla, while type C secretory cells are scarce and occur only at the base of the papilla. Type A secretory cells bear an apical cilium and contain large ovoid granules that enclose an electron-dense heterogeneous material staining histochemically as neutral mucopolysaccharides. Type B secretory cells bear a subcuticular cilium and are filled with small granules containing a homogeneous electron-dense material. The adhesive disc is a round, concave structure lying between the brachiolar arms. It is an epidermal structure composed of two main cell types: ciliated secretory cells and support cells (Barker 1978; Haesaerts et al. 2005). The former are full of large secretory granules enclosing a fibrous proteinaceous content of woven aspect. When exploring the substratum, the competent larva orients itself ventral side down and successively attaches and detaches its brachiolar arms (Barker 1978; Strathmann 1978; Haesaerts et al. 2003). Papillae, when making contact with the substratum, are responsible for sensory testing and temporary adhesion. Like adult tube feet, they function as a duo-glandular system, with type A and B secretory cells acting as adhesive and de-adhesive cells, respectively (Hermans 1983; Flammang 1996; Haesaerts et al. 2005). In addition, the contents of type A secretory cells cross-react with antibodies raised against tube foot adhesive of *A. rubens*, indicating that adhesives from both brachiolar arms and podia are related to each other and probably share identical molecules, or, at least, identical epitopes on their constituents (Haesaerts et al. 2005). Once the larva has found a suitable site for metamorphosis, brachiolar arms are gradually splayed out, enabling the disc to release its cement (Barker 1978; Haesaerts et al. 2003). This attaches the larva permanently to the substratum and marks the onset of the metamorphic stage. During this stage, tube feet become functional and ultimately help the newly formed postlarva to detach from the cemented disc (Haesaerts et al. 2003).

Among marine invertebrates, crinoids and asteroids are unique in using non-permanent adhesion during settlement (transitory adhesion for the doliolariae and temporary adhesion for the brachiolariae), permanent adhesion for fixation during metamorphosis, and then reversing to non-permanent adhesion for their whole postmetamorphic life (mechanical attachment for the comatulids and temporary adhesion for the asteroids). Indeed, in general, invertebrates that remain mobile as adults use a single type of adhesion throughout their perimetamorphic period. For example, during settlement, pediveligers of gastropod molluscs adhere to the substratum through a viscous film of mucus produced by their foot on which they creep (transitory adhesion; Koehl and Hadfield 2004). This type of adhesion is then conserved up until the adult form (Walker 1987). Sessile invertebrates, on the other hand, cannot rely on a single type of adhesion during their perimetamorphic period. These organisms, which as adults use permanent adhesion and live cemented to the substratum, need a non-permanent type of adhesion during settlement to enable them to move around while exploring the substratum (Crisp 1984). For premetamorphic attachment, they can therefore use either transitory adhesion like bryozoan larvae or temporary adhesion like barnacle cyprids. From metamorphic attachment onwards, all these organisms then rely on permanent adhesion to remain cemented to the substratum (Crisp 1984).

Adhesion strength of marine invertebrate larvae is difficult to measure due to the small size of these organisms. It can be estimated, however, from the water current required to wash larvae off a glass substratum. Using this technique, the nominal wall shear stress needed to dislodge temporarily attached individuals of the asteroid *Asterina gibbosa* was about 1 Pa for brachiolariae attached by the arms and 6 Pa for postmetamorphic individuals attached by tube feet (Haesaerts, Callow and Flammang, unpubl. data). These values are comparable to those required to dislodge newly settled barnacle cyprids (0.2–8.7 Pa) and nudibranch larvae (4.26 Pa) (Koehl and Hadfield 2004). On the other hand, a nominal wall shear stress of about 40 Pa was needed to detach metamorphic larvae of *A. gibbosa* permanently attached by the disc (Haesaerts, Callow and Flammang, unpubl. data), showing the very high adhesive strength of this permanent adhesive.

4 Cuvierian Tubules

Cuvierian tubules are peculiar organs found in several species of holothuroids (sea cucumbers), all belonging exclusively to the family Holothuriidae. Tubules (Fig. 4A) occurring in holothuroids of the genera *Bohadschia*, *Holothuria* and *Pearsonothuria* are expelled as sticky white threads that function as a defence mechanism against predators (Hamel and Mercier 2000; Flammang et al. 2002). Cuvierian tubule adhesion is a typical example of instantaneous adhesion, adhesion being achieved in a matter of seconds (less than 10 s; Zahn et al. 1973).

Cuvierian tubules occur in great numbers (between 200 and 600 in *H. forskali*; VandenSpiegel and Jangoux 1987) in the posterior part of the body cavity of the holothuroid. Proximally they are attached to the basal part of the

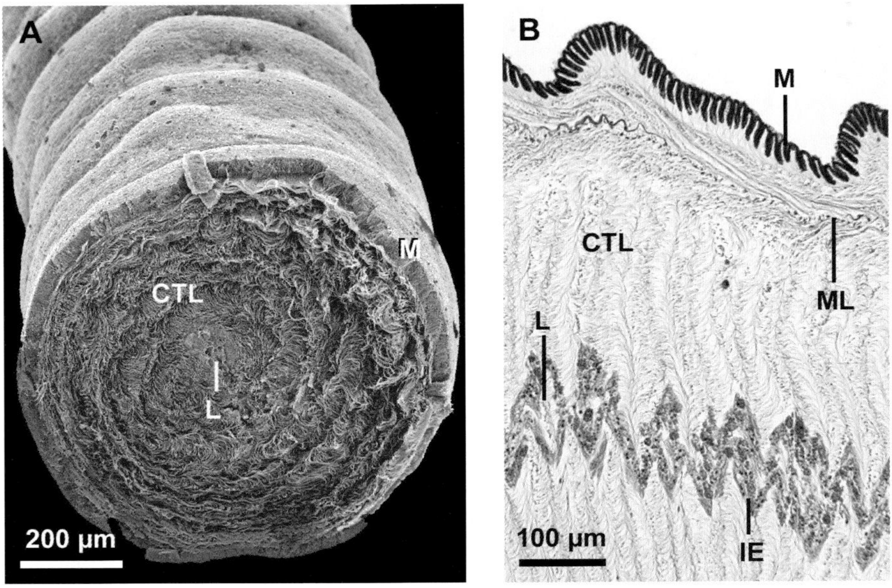

Fig. 4A, B. Morphology of Cuvierian tubules of *Holothuria impatiens* (originals). SEM photograph of a transversely sectioned tubule (**A**), and longitudinal histological section showing arrangement of tissue layers (**B**). *CTL* Connective tissue layer; *IE* inner epithelium; *L* lumen; *ML* muscle layer; *M* mesothelium

left respiratory tree and their distal, blind ends float freely in the coelomic fluid. When disturbed, the sea cucumber directs its aboral end toward the stimulating source and undergoes a general body contraction. The anus opens, the wall of the cloaca tears, and the free ends of a few tubules (usually 10 to 20 in *H. forskali*; VandenSpiegel and Jangoux 1987), together with coelomic fluid, are expelled through the tear and the anus. As water from the respiratory tree is forcefully injected into their lumen, the emitted tubules elongate up to 20 times their original length (VandenSpiegel and Jangoux 1987). Upon contact with any surface, the elongated tubules become instantly sticky. The adhesiveness of Cuvierian tubules combined with their tensile strength make them very efficient for entangling and immobilizing most potential predators (VandenSpiegel and Jangoux 1987; Hamel and Mercier 2000). Finally, the expelled tubules autotomize at their attachment point on the left respiratory tree and are left behind as the holothuroid crawls away (VandenSpiegel and Jangoux 1987). After expulsion and autotomy, Cuvierian tubules are readily regenerated. Cuvierian tubules thus constitute an efficient defensive mechanism. Their large number, sparing use and regeneration dynamics make them a formidable line of defence (Hamel and Mercier 2000; VandenSpiegel et al. 2000).

Cuvierian tubule adhesive strength on glass has been measured in seven species of sea cucumbers belonging to the genera *Bohadschia*, *Holothuria* and *Pearsonothuria* (Flammang et al. 2002). The mean normal tenacities observed varied from about 30 to 135 kPa. These tenacities fall within the range of adhesive strengths described for marine organisms. They are, however, among the lowest observed values (Flammang 2003).

Cuvierian tubules consist, from the inside to the outside, of an epithelium surrounding the narrow lumen, a thick connective tissue layer and a mesothelium lining the surface of the tubule that is exposed to the coelomic cavity (Fig. 4). The mesothelium is responsible for adhesion. In quiescent tubules, it is a pseudostratified epithelium made up of two superposed cell layers, an outer layer of peritoneocytes and an inner layer of granular cells which is highly folded along the long axis of the tubule (Fig. 4B). Granular cells are filled with densely packed membrane-bound granules enclosing a proteinaceous material (Endean 1957; VandenSpiegel and Jangoux 1987). During elongation, the structure of the mesothelium is modified: the protective outer layer of peritoneocytes disintegrates and the granular cell layer, now unfolded, thus becomes outermost on the tubule. Granular cells empty the contents of their granules when the elongated tubule comes into contact with a surface, resulting in adhesion (VandenSpiegel and Jangoux 1987; De Moor et al. 2003).

In *H. forskali*, tubule print material, i.e. the adhesive left on the substratum after mechanical detachment of the tubule, is composed of 60 % protein and 40 % neutral carbohydrate, a composition that is unique among the adhesive secretions of marine invertebrates (De Moor et al. 2003). The proteinaceous nature of the adhesive material is confirmed by the observation that proteolytic enzymes reduce the adhesive strength of Cuvierian tubules in *H. forskali* (Zahn et al. 1973). The amino acid compositions of the protein fraction in *H. forskali*, *H. leucospilota*, *B. subrubra* and *P. graeffei* indicate that their adhesives are closely related. All are rich in small side-chain amino acids, especially glycine, and in charged and polar amino acids (Table 1). Their compositions differ, however, from those of every other marine bioadhesive (Flammang 2003). Only a small fraction of the Cuvierian tubule adhesive can be extracted using denaturing buffers. This soluble fraction contains several proteins with different molecular masses but with closely related amino acid compositions, resembling the one of the whole adhesive (De Moor et al. 2003). As for the tube foot adhesive, charged and polar amino acids are probably involved in adhesive interactions with the substratum through hydrogen and ionic bonding (Waite 1987). Small side-chain amino acids, on the other hand, are often found in large quantities in elastomeric proteins (Tatham and Shewry 2000). These proteins are able to withstand significant deformations without rupture before returning to their original state when stress is removed (Smith et al. 1999). The composition of Cuvierian tubule adhesive has therefore all the characteristics of a strong and resistant underwater adhesive.

Table 1. Amino acid compositions of adhesive prints from the Cuvierian tubules of several species of holothuroids (values in residues per thousand)

Amino acid	*Holothuria forskali*[a]	*Holothuria leucospilota*[b]	*Bohadschia subrubra*[b]	*Pearsonothuria graeffei*[b]
HYP	0	24	8	8
ASX	78	74	64	62
THR	87	69	65	80
SER	60	42	58	58
GLX	91	122	106	124
PRO	55	74	69	63
GLY	266	267	298	254
ALA	88	115	91	85
CYS/2	14	3	9	4
VAL	38	29	35	37
MET	10	9	1	9
ILE	28	24	25	32
LEU	37	31	37	38
TYR	20	14	17	17
PHE	20	16	20	20
HIS	26	13	8	20
HLYS	0	5	12	3
LYS	31	12	29	22
ARG	50	57	46	63

[a] De Moor et al. (2003)
[b] Flammang and Waite (unpubl. data)

5 Applications of Marine Bioadhesives

In addition to fundamental interests in marine bioadhesives, a substantial impetus behind understanding these adhesives is the potential technological applications that can be derived from more detailed knowledge. These applications cover two broad fields of applied research: design of water-resistant adhesives and development of new antifouling strategies.

Sessile organisms such as mussels, barnacles and tube-dwelling worms are important macro-foulers and their adhesives are secreted as a fluid which then gradually solidifies to form a cement possessing high adhesive and cohesive strength (Walker 1987). Most studies of invertebrate adhesive systems have therefore focused on the characterization of the permanent adhesives from these organisms (see, e.g., Taylor and Waite 1997; Kamino et al. 2000). The best-characterized permanent adhesive is that from the blue mussel, *Mytilus edulis*. In this species, several proteins have been identified and characterized that co-occur as a complex blend in the byssal adhesion plaques

(Waite 2002). So far, however, only one of these proteins (*Mytilus edulis* foot protein 1; Mefp-1) has been used in biotechnological applications: in the form of crude or recombinant preparations of the protein and in the form of chemically synthesized, derived peptides (Burzio et al. 1997; Deming 1999; Taylor and Weir 2000).

5.1 Design of Water-Resistant Adhesives

The fact that marine invertebrates produce adhesives that act in the presence of water has aroused increasing scientific and technological attention because water, including moist air, is usually the most common subverter of man-made adhesive joints (Kinloch 1987; Waite 1987). Biomimetic materials inspired by marine adhesives are therefore sorely needed for applications in wet environments. Such materials could be used in underwater construction, of course, but the most important applications are certainly to be found in the biomedical field (Strausberg and Link 1990; Peppas and Langer 1994; Taylor and Waite 1997). Indeed, the environment of the sea is similar in many ways to the internal environment of mammalian organisms. Tissues are bathed in fluids with pH and ionic composition similar to salt water. Theoretically, the attachment mechanisms that some marine invertebrates have evolved to survive can be useful as biological adhesives for in vitro as well as in vivo uses.

For in vitro techniques, a crude preparation of Mefp-1 has been developed as a cell and tissue adhesive (Cell-Tak, BD Biosciences) for immobilization of biologically active moieties to inert substrata (Benedict and Picciano 1989; Taylor and Waite 1997). Cell-Tak is used to attach cells or tissue sections to many types of surfaces, including plastic, glass, metal and biological materials. It can simplify the manipulation of biological samples for a number of techniques, including in situ hybridization, immunoassays, microinjection, immunohistochemistry and establishing primary cells in culture (Benedict and Picciano 1989; Burzio et al. 1997; Taylor and Waite 1997). Cell-Tak has been used successfully with a variety of healthy cell types, tumour cells, permeabilized cells and subcellular components (Taylor and Waite 1997). This broad range of applications is explained by the fact that Mefp-1, in contrast to cell adhesion molecules, acts as a non-specific attachment factor that fastens onto a variety of functional groups present and accessible on the surface of all cells and tissues (Taylor and Waite 1997).

Surgical or topical reconnection of severed tissues is essential for restoration of their structure and function. The most widely used methods for joining tissues focus on mechanical fasteners such as sutures and staples. Surgical adhesives, however, provide attractive alternatives to mechanical fastening (Strausberg and Link 1990; Albala 2003; Ninan et al. 2003; Singer and Thode 2004). In addition to their rapid application, they are particularly useful in tissues that are difficult to reach, too difficult to cauterize or too delicate to withstand suturing. Moreover, they are relatively painless to apply and may not require the use of painful local anesthetics. They are also biodegradable, eliminating the need for suture removal. Currently, two types of surgical adhesives are commercially available – fibrin-based adhesives (see, e.g., Albala 2003) and cyanoacrylate-based adhesives (see, e.g., Singer and Thode 2004) – which have been used successfully in a growing number of surgical procedures. There are still several applications, however, for which these adhesives cannot be used, e.g. in areas continuously bathed in body fluids (mucous membranes, bladder, etc) (Albala 2003; Ninan et al. 2003; Singer and Thode 2004). Marine bioadhesives could be ideal candidates for such applications because they function in an aqueous environment, they possess the appropriate adhesive and cohesive properties and they are ultimately biodegradable (Strausberg and Link 1990; Ninan et al. 2003). However, any adhesive material targeted for medical applications should also be biocompatible (non-toxic, low immunogenicity) and should not interfere with the natural healing process (Strausberg and Link 1990). Mefp-1 seems to be biocompatible and non-toxic in vitro (Benedict and Picciano 1989). However, much has still to be learned before clinical trials are performed for human applications (Burzio et al. 1997).

Dentistry is another field in which there is current demand for non-toxic bioadhesives able to form durable adhesive bonds in the aqueous environment of the mouth (Peppas and Langer 1994; Burzio et al. 1997).

5.2 Development of New Antifouling Strategies

Biofouling is one of the most important problems currently facing marine technology. In the marine environment, any solid surface will become fouled. Materials submerged in seawater experience a series of discrete physical, chemical and biological events, which result in the formation of a complex layer of attached organisms known as biofouling (Abarzua and Jakubowski 1995; Callow and Callow 2002). Until now, antifouling methods involved the use of toxic self-polishing paints releasing tributyltin (TBT) or cuprous oxide. However, the impact of these compounds on the environment has led to a ban on the use of TBT-containing paints and to close monitoring of copper discharge from antifouling paints (Callow and Callow 2002). Fouling control is therefore increasingly a problem of managing adhesion, and molecular understanding of how marine bioadhesives work should open up novel technologies intended to specifically intervene in organism attachment (Taylor and Waite 1997; Callow and Callow 2002). One of these technologies is the use of enzyme-leaching paints. Indeed, it has been shown that enzymes can be active in a paint coating and hydrolyze the attachment glue which is secreted by the fouling organisms (Abarzua and Jakubowski 1995). Moreover, enzymes are completely degraded in seawater within several days and they are non-toxic for marine life. Another technology is based on the development of fouling-release coatings such as silicone coatings (Callow and Callow 2002). These coatings, which exhibit low critical surface tensions, function by minimizing the adhesion strength of attached organisms. Fouling organisms can settle on the coating but are then removed by either mechanical or hydrodynamic cleaning. Paradoxically, a recently developed fouling-release coating was based on a marine bioadhesive possessing strong fouling characteristics. Dalsin et al. (2003) synthesized hybrid molecules by combining a decapeptide derived from Mefp-1 and poly(ethylene glycol), and used these molecules to modify surfaces. The strategy exploits the adhesive characteristics of the decapeptide to anchor poly(ethylene glycol) onto surfaces, rendering the surfaces resistant to cell attachment in vitro. In the future, it may also be employed in preventing mussels, barnacles and other fouling organisms from attaching to ship hulls, piers and other man-made structures (Dalsin et al. 2003).

5.3 The Potential of Echinoderm Adhesives

In view of the examples mentioned above, it appears that Mefp-1 has been used successfully in several applications. However, there are some limitations, such as requirement of post-translational modifications to certain amino acids and the need for a separate enzyme for curing (Strausberg and Link 1990). New adhesive molecules that could overcome these problems are therefore sought. In this context, echinoderm adhesives could offer novel features or performance characteristics for biotechnological applications. For example, the rapidly attaching adhesive of Cuvierian tubules, the releasable adhesive of tube feet or the powerful adhesive of asteroid larvae could each be useful to address particular bioadhesion problems. If the tube foot de-adhesive secretion proves effectively to be an enzyme, its incorporation into antifouling paints would be particularly relevant as many fouling larvae use temporary adhesion during settlement (Crisp 1984).

Work is currently in progress to identify, purify and characterize the different molecules involved in echinoderm adhesion. The complete elucidation of their structure and physico-chemical characteristics is an obligatory prerequisite before any application can be envisaged.

Acknowledgements. Thanks are extended to Dr. M. Jangoux and Dr. J.H. Waite for useful comments on the manuscript. This work was supported by the US Office of Naval Research (grant no.°N00014-99-1-0853). D.H. benefited from an FRIA doctoral grant (Belgium) and R.S. from a doctoral grant of the Foundation for Science and Technology of Portugal (no.°SFRH/BD/4832/2001). P.F. is a Research Associate for the National Fund for Scientific Research of Belgium (FNRS). This study is a contribution of the "Centre Interuniversitaire de Biologie Marine" (CIBIM).

References

Abarzua S, Jakubowski S (1995) Biotechnological investigation for the prevention of biofouling. I. Biological and biochemical principles for the prevention of biofouling. Mar Ecol Prog Ser 123:301–312

Albala DM (2003) Fibrin sealants in clinical practice. Cardiovascular Surg 11(S1):5–11

Barker MF (1978) Structure of the organs of attachment of brachiolaria larvae of *Stichaster australis* (Verrill) and *Coscinasterias calamaria* (Gray) (Echinodermata: Asteroidea). J Exp Mar Biol Ecol 33:1–36

Benedict CV, Picciano PT (1989) Adhesives from marine mussels. In: Hemingway RW, Conner AH, Branham SJ (eds) Proc ACS Symp Adhesives from Renewable Resources, vol 385. American Chemical Society, Washington, pp 465–483

Burzio LO, Burzio VA, Silva T, Burzio LA, Pardo J (1997) Environmental bioadhesion: themes and applications. Curr Opin Biotechnol 8:309–312

Callow ME, Callow JA (2002) Marine biofouling: a sticky problem. Biologist 49:1–5

Cameron JL, Fankboner PV (1984) Tentacle structure and feeding processes in life stages of the commercial sea cucumber *Parastichopus californicus* (Stimpson). J Exp Mar Biol Ecol 81:193–209

Chia FS, Burke RD, Koss R, Mladenov PV, Rumrill SS (1986) Fine structure of the doliolaria larva of the feather star *Florometra serratissima* (Echinodermata: Crinoidea), with special emphasis on the nervous system. J Morphol 189:99–120

Crisp DJ (1984) Overview of research on marine invertebrate larvae, 1940–1980. In: Costlow JD, Tipper RC (eds) Marine biodeterioration: an interdisciplinary study. E & FN Spon, London, pp 103–126

Dalsin JL, Hu B-H, Lee BP, Messersmith PB (2003) Mussel adhesive protein mimetic polymers for the preparation of nonfouling surfaces. J Am Chem Soc 125:4253–4258

De Moor S, Waite JH, Jangoux M, Flammang P (2003) Characterization of the adhesive from the Cuvierian tubules of the sea cucumber *Holothuria forskali* (Echinodermata, Holothuroidea). Mar Biotechnol 5:37–44

Deming TJ (1999) Mussel byssus and biomolecular materials. Curr Opin Biol Chem 3:100–105

Eeckhaut I, Flammang P, Lo Bue C, Jangoux M (1997) Functional morphology of the tentacles and tentilla of *Coeloplana bannworthi* (Ctenophora, Platyctenida), an ectosymbiont of *Diadema setosum* (Echinodermata, Echinoidea). Zoomorphology 117:165–174

Endean R (1957) The Cuvierian tubules of *Holothuria leucospilota*. Q J Microsc Sci 98:455–472

Flammang P (1996) Adhesion in echinoderms. In: Jangoux M, Lawrence JM (eds) Echinoderm studies, vol 5. AA Balkema, Rotterdam, pp 1–60

Flammang P (2003) The glue of sea cucumber Cuvierian tubules: a novel marine bioadhesive. In: Colliec-Jouault S, Bergé JP, Guézennec J, Fleurence J, Le Gal Y, Roy P (eds) Marine biotechnology: an overview of leading fields. Actes Colloq Ifremer 36:176–185

Flammang P, Walker G (1997) Measurement of the adhesion of the podia in the asteroid *Asterias rubens* (Echinodermata). J Mar Biol Assoc UK 77:1251–1254

Flammang P, Demeuleneare S, Jangoux M (1994) The role of podial secretions in adhesion in two species of sea stars (Echinodermata). Biol Bull 187:35–47

Flammang P, Michel A, Van Cauwenberge A, Alexandre H, Jangoux M (1998a) A study of the temporary adhesion of the podia in the sea star *Asterias rubens* (Echinodermata, Asteroidea) through their footprints. J Exp Biol 201:2383–2395

Flammang P, Gosselin P, Jangoux M (1998b) The podia, organs of adhesion and sensory perception in larvae and postmetamorphic stages of the echinoid *Paracentrotus lividus* (Echinodermata). Biofouling 12:161–171

Flammang P, Ribesse J, Jangoux M (2002) Biomechanics of adhesion in sea cucumber Cuvierian tubules (Echinodermata, Holothuroidea). Integr Comp Biol 42:1107–1115

Franc JM (1978) Organization and function of ctenophore colloblasts: an ultrastructural study. Biol Bull 155:527–541

Gosselin P, Jangoux M (1998) from competent larva to exotrophic juvenile: a morphofunctional study of the perimetamorphic period of *Paracentrotus lividus* (Echinodermata, Echinoidea). Zoomorphology 118:31–43

Haesaerts D, Jangoux M, Flammang P (2003) Study of the perimetamorphic period of the sea star *Asterias rubens* by scanning electron microscopy. In: Féral JP, David B (eds) Echinoderm research 2001. AA Balkema, Lisse, pp 155–159

Haeserts D, Jangoux M, Flammang P (2005) The attachment complex of brachioloaria larvae of the sea star *Asterias rubens* (Echinodermata): An unstructural and immunocytochemical study. Zoomorphology (in press)

Hamel J-F, Mercier A (2000) Cuvierian tubules in tropical holothurians: usefulness and efficiency as a defence mechanism. Mar Fresh Behav Physiol 33:115–139

Hermans CO (1983) The duo-gland adhesive system. Oceanogr Mar Biol Ann Rev 21:281–339

Jangoux M, Lahaye M-C (1990) The attachment complex of the dololaria larvae of *Antedon bifida* (Echinodermata, Crinoidea). In: De Ridder C, Dubois P, Lahaye M-C, Jangoux M (eds) Echinoderm research. AA Balkema, Rotterdam, pp 99–105

Kamino K, Inoue K, Maruyama T, Takamatsu N, Harayama S, Shizuri Y (2000) Barnacle cement proteins. Importance of disulfide bonds in their insolubility. J Biol Chem 275:27360–27365

Kinloch AJ (1987) Adhesion and adhesives, science and technology. Chapman and Hall, London

Koehl MAR, Hadfield MG (2004) Soluble settlement cue in slowly moving water within coral reefs induces larval adhesion to surfaces. J Mar Syst 49:75–88

Lahaye M-C (1987) Comportement larvaire et ontogénèse postembryonnaire chez la comatule *Antedon bifida* (Echinodermata, Crinoidea). PhD Thesis, Université Libre de Bruxelles, Brussels

Lahaye M-C, Jangoux M (1987) The skeleton of the stalked stages of the comatulid crinoid *Antedon bifida* (Echinodermata). Fine structure and changes during growth. Zoomorphology 107:58–65

Lahaye M-C, Jangoux M (1988) Morphologie externe et comportement des larves doliolaria d'*Antedon bifida* (Echinodermata, Crinoidea). Ann Soc R Zool Belg 118:183–189

Mladenov PV, Chia FS (1983) Development, settling behaviour, metamorphosis and pentacrinoid feeding and growth of the feather star *Florometra serratissima*. Mar Biol 73:309–323

Nachtigall V (1974) Biological mechanisms of attachment. The comparative morphology and bioengineering of organs for linkage, suction, and adhesion. Springer, Berlin Heidelberg New York

Nakano H, Hibino T, Oji T, Hara Y, Amemiya S (2003) Larval stages of a living sea lily (stalked crinoid echinoderm). Nature 421:158–160

Ninan L, Monahan J, Stroshine RL, Wilker JJ, Shi R (2003) Adhesive strength of marine mussel extracts on porcine skin. Biomaterials 24:4091–4099

Paine VL (1926) Adhesion of the tube feet in starfishes. J Exp Zool 45:361–366

Peppas NA, Langer R (1994) New challenges in biomaterials. Science 263:1715–1720

Ridgway SH, Lindner E, Mahoney KA, Newman WA (1997) Gray whale barnacles *Cryptolepas rhachianecti* infest white whales, *Delphinapterus leucas*, housed in San Diego Bay. Bull Mar Sci 61:377–385

Santos R (2003) Etude biomécanique et morphologique de l'adhérence chez les oursins (Echinodermata, echinoidea) et sa relation avec la distribution des espèces dans le milieu naturel. Master Thesis, Université de Mons-Hainaut, Belgium

Santos R, Haesaerts D, Jangoux M, Flammang P (2005) Comparative histological and immunohistochemical study of sea star tube feet (Echinodermata, Asteroidea). J Morphol 263:259–270

Singer AJ, Thode HC (2004) A review of the literature on octylcyanoacrylate tissue adhesive. Am J Surg 187:238–248

Smith BL, Schäffer TE, Viani M, Thompson JB, Frederick NA, Kindt J, Belcher A, Stucky GD, Morse DE, Hansma PK (1999) Molecular mechanistic origin of the toughness of natural adhesives, fibres and composites. Nature 399:761–763

Strathmann RR (1978) Larval settlement in echinoderms. In: Chia FS, Rice ME (eds) Settlement and metamorphosis of marine invertebrate larvae. Elsevier-North Holland, New York, pp 235–246

Strausberg RL, Link RP (1990) Protein based medical adhesives. Trends Biotechnol 8:53–57

Tatham AS, Shewry PR (2000) Elastomeric proteins: biological roles, structures and mechanisms. Trends Biochem Sci 25:567–571

Taylor CM, Weir CA (2000) Synthesis of the repeating decapeptide unit of Mefp1 in orthogonally protected form. J Org Chem 65:1414–1421

Taylor SW, Waite JH (1997) Marine adhesives: from molecular dissection to application. In: McGrath K, Kaplan D (eds) Protein-based materials. Birkhäuser, Boston, pp 217–248

Thomas LA, Hermans CO (1985) Adhesive interactions between the tube feet of a starfish, *Leptasterias hexactis*, and substrata. Biol Bull 169:675–688

Tyler S (1988) The role of function in determination of homology and convergence – examples from invertebrates adhesive organs. Fortsch Zool 36:331–347

Vaitilingon D, Eeckhaut I, Fourgon D, Jangoux M (2004) Population dynamics, infestation and host selection of *Vexilla vexillum* (Gmelin, 1791), an ectoparasitic muricid of echinoids, in Madagascar. Dis Aquat Org 61:241–255

VandenSpiegel D, Jangoux M (1987) Cuvierian tubules of the holothuroid *Holothuria forskali* (Echinodermata): a morphofunctional study. Mar Biol 96:263–275

VandenSpiegel D, Jangoux M, Flammang P (2000) Maintaining the line of defense: regeneration of Cuvierian tubules in the holothuroid *Holothuria forskali* (Echinodermata). Biol Bull 198:34–49

Waite JH (1987) Nature's underwater adhesive specialist. Int J Adhesion Adhesives 7:9–14

Waite JH (2002) Adhesion à la moule. Integr Comp Biol 42:1172–1180

Walker G (1987) Marine organisms and their adhesion. In: Wake WC (ed) Synthetic adhesives and sealants. Wiley, Chichester, pp 112–135

Whittington ID, Cribb BW (2001) Adhesive secretions in the Platyhelminthes. Adv Parasitol 48:101–224

Zahn RK, Müller WEG, Michaelis M (1973) Sticking mechanisms in adhesive organs from a *Holothuria*. Res Mol Biol 2:47–88

Mutable Collagenous Tissue: Overview and Biotechnological Perspective

I.C. Wilkie

Abstract. The mutable collagenous tissue (MCT) of echinoderms can undergo extreme changes in passive mechanical properties within a timescale of less than 1 s to a few minutes, involving a mechanism that is under direct neural control and coordinated with the activities of muscles. MCT occurs at a variety of anatomical locations in all echinoderm classes, is involved in every investigated echinoderm autotomy mechanism, and provides a mechanism for the energy-sparing maintenance of posture. It is therefore crucially important for the biology of extant echinoderms. This chapter summarises current knowledge of the physiology and organisation of MCT, with particular attention being given to its molecular organisation and the molecular mechanism of mutability. The biotechnological potential of MCT is discussed. It is argued that MCT could be a source of, or inspiration for, (1) new pharmacological agents and strategies designed to manipulate therapeutically connective tissue mechanical properties and (2) new composite materials with biomedical applications.

1 Introduction

1.1 Collagenous Tissue

The collagens are a family of at least 19 proteins characterised by the presence of triple helical regions composed of repeating Gly-X-Y triplets. The majority of collagens occur in the extracellular matrix (ECM) of connective tissue, the most prevalent of these being types I, II and III, which self-assemble into par-

I.C. Wilkie (e-mail: i.wilkie@gcal.ac.uk)
Department of Biological and Biomedical Sciences, Glasgow Caledonian University, 70 Cowcaddens Road, Glasgow G4 0BA, Scotland, UK

Progress in Molecular and Subcellular Biology
Subseries Marine Molecular Biotechnology
V. Matranga (Ed.), Echinodermata

allel aggregations known as fibrils, and type IV, which forms a network in the lamina densa of all basement membranes. In the ECM, collagen fibrils are accompanied by a variety of other components, which may include other fibrous structures, such as elastin fibres or microfibril aggregations, proteoglycans, glycoproteins and glycosaminoglycans. Some of these other components link together adjacent collagen fibrils and are thereby responsible for interfibrillar cohesion (Kannus 2000; Sarras and Deutzmann 2001).

Connective tissue in which the mechanically dominant component is collagen is the most widespread and evolutionarily ancient structural material in the animal kingdom. With the exception of possibly only the phyla Rotifera (Clément 1993) and Placozoa (Syed and Schierwater 2002), it is present in the bodies of all multicellular animals, including sponges, the internal 'mesohyl' of the latter being histologically and functionally a collagenous connective tissue that, unlike the situation in other animals, is not separated from its investing epithelia by basement membranes (Garronne 1978; Bonasoro et al. 2001). (It is possible that the absence of basement membranes from most sponges is due to secondary loss: see Maldonado 2004.)

1.2 Mechanical Adaptability of Collagenous Tissue

In its role as a structural material, the principal functions of collagenous tissue are to transmit, resist and dissipate mechanical forces and to store and release elastic strain energy. These functions depend on the tissue as a whole possessing certain 'fit for purpose' mechanical properties (tensile strength, stiffness, resilience, etc.), which are the result of a combination of the mechanical properties of its individual structural components, the micro-architectural organisation of these components and the nature of the interactions between them.

The net mechanical properties of most collagenous structures are relatively stable within a physiological timescale (seconds to hours). They may change during maturation and ageing, partly as a result of alterations in collagen fibril diameter and the composition of the interfibrillar phase mediated by adjustments in the synthetic activities of fibroblasts and, in the case of ageing, partly as a result of the non-adaptive biochemical modification of the ECM, e.g. by non-enzymatic glycation (Vogel 1980; Bruel and Oxlund 1996; Reddy et al. 2002; Silver et al. 2003). The mechanical properties of collagenous structures can also be modified in response to a long-term shift in the pattern of force to which they are subjected. For example, exercise increases the tensile strength and stiffness of tendons, again probably through fibroblast-mediated changes in fibril diameter and interfibrillar composition (Buchanan and Marsh 2002; see also Chiquet 1999 for a discussion of the regulatory pathways that might be involved). These age-related and adaptive changes occur over timescales of days to years and are quantitatively undramatic. However, some collagenous structures undergo more drastic changes in mechanical properties over a shorter timescale, which result in a qualitative modification of the mechanical functioning of the tissue and thus justify it being regarded as a mechano-effector. In *Homo sapiens* and other mammals, this phenomenon is demonstrated by various collagenous structures associated with the female reproductive tract. For example, at the end of pregnancy, the compliance of the uterine cervix increases temporarily by a factor of 12. It switches from being tough and inextensible, in which state it helps to prevent expulsion of the conceptus, to being soft and easily extensible, a condition that permits the dilatation and effacement required for the passage of the fetus through the birth canal. Cervical ‘relaxation’ depends on two separate processes: (1) changes in the synthetic activity of the stromal cells, which secrete less collagen types I and III and small proteoglycans and secrete more of the large proteoglycan versican; and (2) the enzymatic degradation of the ECM by matrix metalloproteinases (MMPs) including MMP-1, MMP-3 and MMP-8, together with increased expression of the tissue inhibitors of metalloproteinases TIMP-1 and TIMP-2 (these presumably acting as a brake on the degradative process). The resulting change in mechanical properties occurs over hours to days and is controlled by a combination of hormonal events and the direct effect of stretch on the stromal cells (Uldbjerg 1994; Westergren-Thorsson et al. 1998; Yoshida et al. 2002; Sennström et al. 2003). Similar mechanisms and controlling factors are involved in the rupture of the fetal membranes at term (Bryant-Greenwood 1998) and of the ovarian follicle at ovulation (Robker et al. 2000).

The above examples of connective tissue mechanical adaptability have been given in order to highlight the distinctiveness of the subject of this chapter – the mutable collagenous tissue (MCT) of echinoderms. MCT can undergo extreme alterations of mechanical properties within a timescale of less than 1 s to a few minutes, involving a mechanism that is under direct neural control and coordinated with the activities of other mechano-effectors, viz. muscles. The rest of this chapter provides a review of current knowledge of the physiology and organisation of MCT and discusses its biotechnological potential.

2 Mutable Collagenous Tissue: Physiology and Organisation

2.1 Overview

MCT is present at a variety of anatomical locations in all living echinoderm classes, is involved in every investigated autotomy (defensive self-detachment) mechanism and provides a mechanism for the energy-sparing maintenance of posture (Wilkie 1996, 2001, 2002; Wilkie et al. 2004a). MCT is therefore crucially important for the biology of extant echinoderms and, in view of its apparent absence from other animals, has been regarded as one of the distinguishing characteristics of the phylum (Byrne 2001).

MCT occurs in the form of dermal connective tissue, interossicular ligaments, tendons linking muscles to skeletal elements and the connective tissue layer in the walls of tubular structures. It performs the same mechanical functions as the collagenous connective tissue at analogous locations in the bodies of vertebrates. Although there have been few attempts to correlate the supramolecular organisation of mutable collagenous structures with their 'conventional' mechanical functions (O'Neill 1989; O'Neill and Withers 1995), there is no reason to doubt that, despite their additional property of variable tensility, all such structures are as exquisitely designed to meet the subtly different functional demands of each anatomical location as are their vertebrate equivalents (see, e.g., Bauer et al. 1989; Munns et al. 1994; Gupte et al. 2002).

2.2 Mechanical Adaptability of MCT

2.2.1 Passive Mechanical Properties

Each mutable collagenous structure exhibits one of three patterns of change in its passive tensile properties: (1) only *reversible* changes (e.g. in viscosity or stiffness); (2) *irreversible* destabilisation (always associated with autotomy) as well as reversible changes; or (3) only irreversible destabilisation (Wilkie 2002).

The most extreme manifestation of MCT mechanical adaptability is the rapid and irreversible loss of tensile strength undergone by collagenous structures that cross echinoderm autotomy planes. For example, when an arm of the ophiuroid *Ophiocomina nigra* is autotomised, the ultimate tensile

Fig. 1A–I. Passive mechanical behaviour of MCT. **A** Creep test (in which extension of sample under constant load is recorded): in response to 100 mM K^+ (*K*) syzygial ligament of crinoid *Antedon mediterranea* shows sudden decrease in viscosity culminating in rupture (*star*). *Horizontal scale bar* 1 min; *vertical scale bar* 1 mm (adapted from Wilkie et al. 1999). **B, C** Creep tests: viscosity of dental ligament of echinoid *Diadema setosum* is increased reversibly by 1 mM acetylcholine (*ACh*) and 100 mM K^+ (*K*). *ASW* Artificial seawater (adapted from Birenheide et al. 1996). **D** Stress relaxation tests (in which samples are subjected to constant deformation and force is recorded): average force relaxation curves of 20 dental ligaments of echinoid *Dendraster excentricus* treated with artificial seawater (*ASW*) and 15 ligaments treated with divalent cation-free seawater (*DCF*) (adapted from Ellers and Telford 1996). **E, F** Stress-strain tests (in which force is recorded while samples are stretched at fixed extension rate) conducted on spine ligament of echinoid *Anthocidaris crassispina*: Examples of stress-strain curves produced by samples treated with **E** 0.1 mM acetylcholine

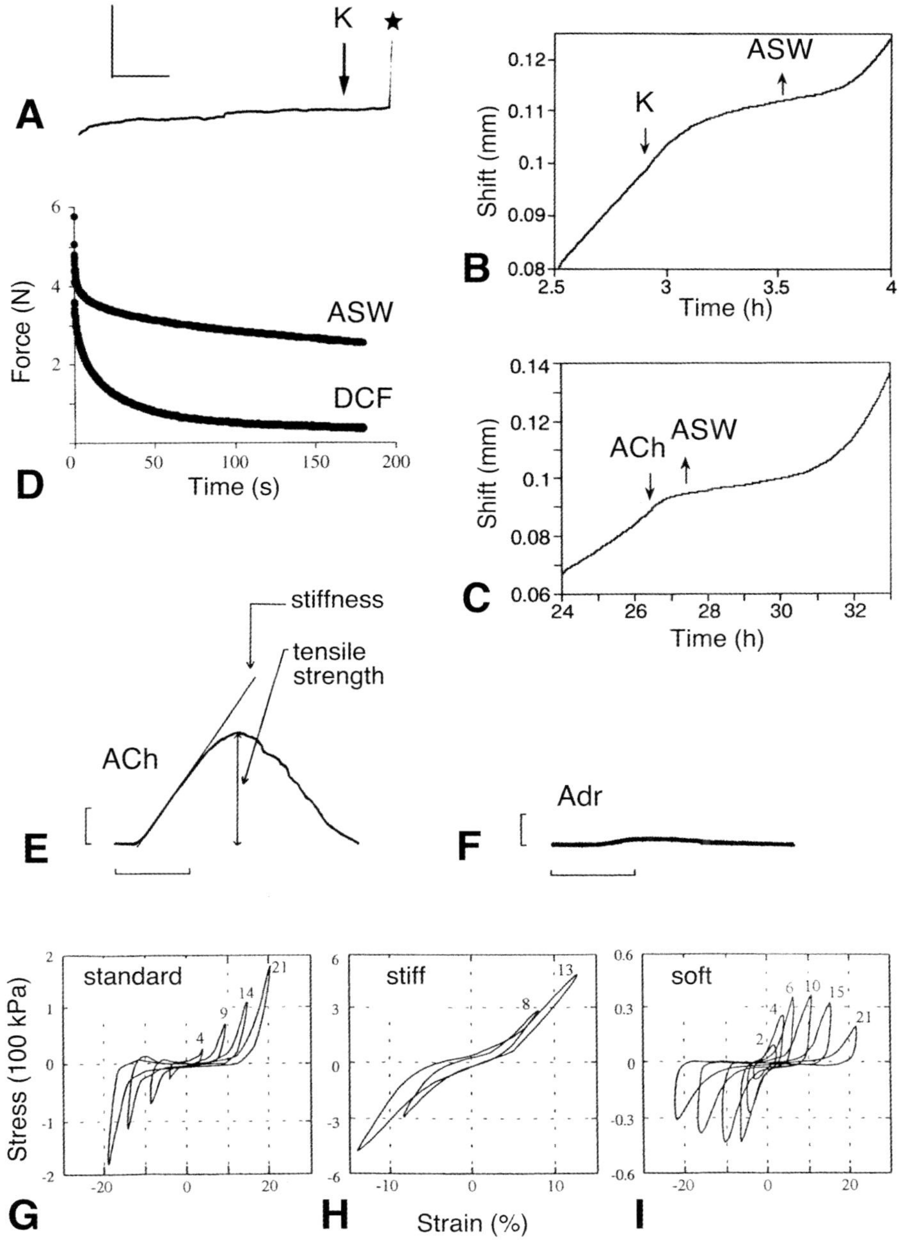

(*ACh*) and **F** 0.1 mM adrenaline (*Adr*). *Horizontal scale bar* Strain of 10 % in both cases; *vertical scale bar* 10 MPa in **E** and 1 MPa in **F** (adapted from Hidaka and Takahashi 1983). **G–I** Dynamic stress-strain tests (in which force is recorded while samples are subjected to oscillating strain): hysteresis loops produced by repetitive testing of dermis of holothurian *Actinopyga mauritiana* in three mechanical states. During these tests, maximum strain (indicated by *number above each curve*) was increased incrementally; note that sample in '*soft*' state showed strain-induced softening. (Adapted from Motokawa and Tsuchi 2003)

strength of the intervertebral ligament at the autotomising joint drops to less than 0.1 % of the normal value in a timescale of 0.4–5.4 s (Wilkie 1988). These drastic changes in mechanical properties can also be demonstrated experimentally using isolated tissue preparations undergoing creep tests (in which their extension under constant load is recorded). Treatment with neuro-active agents such as elevated [K^+] or appropriate neurotransmitter chemicals causes an abrupt decrease in viscosity (stress-strain rate) culminating in tissue rupture (Fig. 1A).

Reversible changes in mechanical properties have been quantified and analysed by means of various testing methods. Reversible changes in viscosity have been demonstrated in creep tests and stress relaxation tests (in which specimens are stretched to a particular length, which is fixed whilst force decay is recorded) (Fig. 1B–D). Reversible changes in tensile strength, tensile stiffness, dynamic shear stiffness, relative damping, etc. have been investigated by means of standard stress-strain tests (Fig. 1E,F) and by dynamic testing methods in which samples are subjected to oscillating strain (Fig. 1G–I). Evidence of mutability is often apparent in the wide variability in the mechanical properties of untreated tissues. Motokawa (1983), for example, found a 200-fold difference in the viscosity of untreated central spine ligaments of *Diadema setosum* taken from different joints of the same animal. In order to overcome this problem and compare tissues in predictable mechanical states, a common strategy has been to subject them to different treatments that induce either maximal or minimal values of stiffness, viscosity, etc. (see Fig. 1G–I). Although biologically relevant stimuli, such as mechanical compression, have been used in these investigations, the artificial nature of some treatments, especially those employed to bring about a compliant state, engender some uncertainty about the physiological relevance of the results thus obtained. Nevertheless, these methods no doubt give an indication of the magnitude of the reversible changes that MCT can accommodate in vivo. Hidaka and Takahashi (1983), for instance, found that at a low strain rate the ultimate tensile strength and elastic modulus of an echinoid spine ligament in the *compliant* condition (induced by 0.1 mM adrenaline) were around 1 % of the values measured in the *stiffened* condition (induced by 0.1 mM acetylcholine), the latter falling within the range reported for mammalian tendon (Redaelli et al. 2003; Fig. 1E,F).

Data from these biomechanical studies have been used to generate models with the ultimate intention of specifying the contribution of different extracellular components to net mechanical properties and determining which of them contribute to variable tensility. Due to the diversity of testing regimes employed, the conclusions from different investigations have been incompatible (see, e.g., Szulgit and Shadwick 2000; Motokawa and Tsuchi 2003) and are difficult to interpret in the light of the increasingly complex picture of the MCT extracellular matrix that is emerging (see Sect. 2.3 below). It seems likely that the full potential of this methodology will not be realised until it can be applied to tissues from which specific components have been eliminated chemically, enzymatically or, ideally, by genetic knockout (see Bornstein et al. 2000 and Chakravarti 2002 for examples of this last approach applied to mammalian connective tissues).

2.2.2 *Active Contractility*

Certain mutable collagenous structures can generate tensile force. In the case of the capsular ligament, or 'catch apparatus', of the echinoid spine-test joint, this is attributable to the presence of fine (diameter 0.1–1.0 μm) muscle fibres that are distributed between the bundles of collagen fibrils and constitute 1–3 % of the total cross-sectional area of the ligament. The neurotransmitter acetylcholine increases the tensile strength and stiffness of this ligament (Hidaka and Takahashi 1983; Morales et al. 1989), but also causes isolated preparations of it to shorten and develop mechanical force, which relaxes as soon as the acetylcholine is removed. Both the contraction and relaxation phases of this response can be very fast, the former in some cases reaching 90 % of the maximum in under 1 s (Fig. 2A; Vidal et al. 1993). The functional significance of the muscle fibres is at present unknown. Del Castillo et al. (1995), ignoring incontrovertible evidence that the spine ligament consists of MCT (see, e.g., Hidaka and Takahashi 1983; Szulgit and Shadwick 1994), hypothesised that they are responsible for varying its passive stiffness, which they achieve by adjusting the frictional forces between the ligament fibres and the skeletal ossicles into which they are inserted. Having provoked a frank exchange of views (see Del Castillo and Smith 1996; Wilkie 1996, 2002; Pérez-Acevedo et al. 1998; Elphick and Melarange 2001), this hypothesis was tested and disproved by Takemae and Motokawa (2002). The muscle fibres also do not seem to be involved in straightening out the wrinkles that form transiently in regions of the ligament that are compressed by contraction of the spine muscle (Pérez-Acevedo et al. 1998). This leaves the possibilities that they assist the reshortening of stretched ligament fibres or that they operate synergistically with the spine muscle, perhaps during specific manoeuvres such as re-erection of the spine.

The reputation of echinoderms for being an inexhaustible mine of biological novelty has been enhanced by the discovery that, as well as varying their passive mechanical properties, ligaments in the cirri and arms of crinoids have the capacity for active contractility, though they lack myocytes. Cirri are finger-like appendages supported by a single series of interarticulating ossicles. The only mechanically significant structures connecting adjacent ossicles are myocyte-free collagenous ligaments. Cirri attached to the stalk of sea lilies bend upwards, against gravity, when the stalk or the cirri themselves are stimulated mechanically. In stress relaxation tests, isolated cirri display slow force production in response to the cholinergic agonists muscarine and methacholine at concentrations as low as 0.1 μM (Fig. 2B). Since force produc-

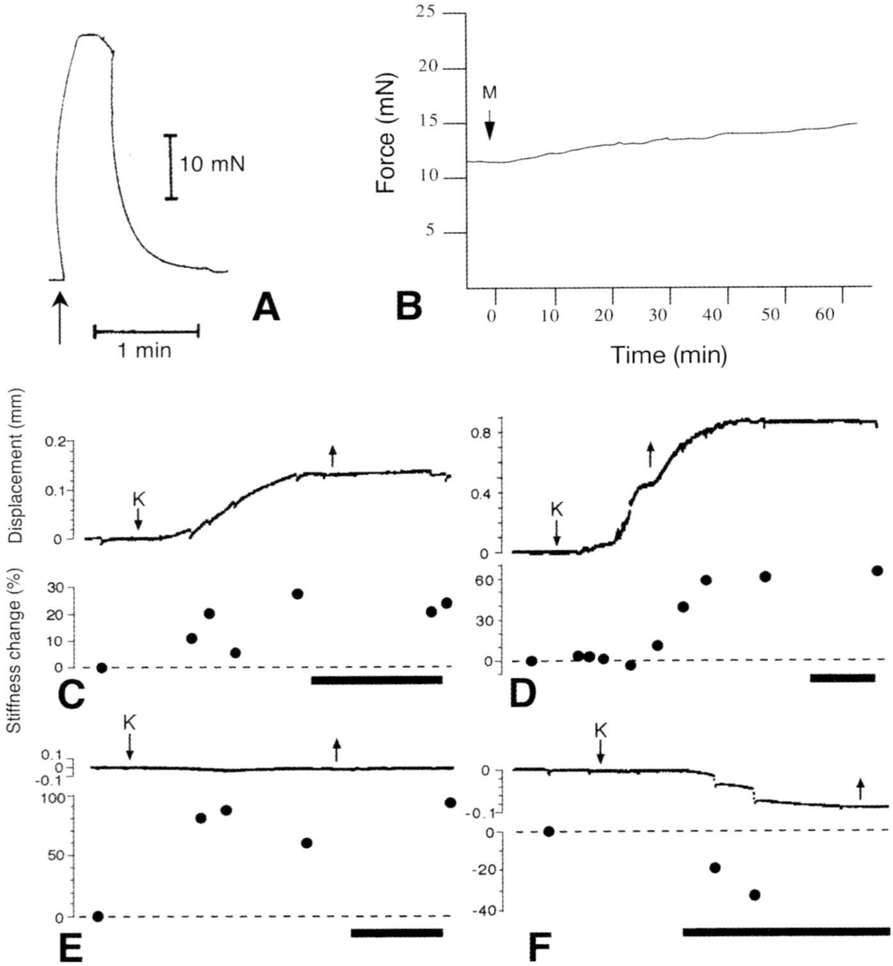

Fig. 2A–F. Force generation by MCT. **A** Contraction of spine ligament of echinoid *Eucidaris tribuloides* induced by 0.1 mM acetylcholine (*arrow*) which was removed as soon as force peaked (adapted from Vidal et al. 1993). **B** Contraction of cirral ligaments of crinoid *Metacrinus rotundus* induced by 0.1 μm methacholine (*M*) (adapted from Birenheide et al. 2000). **C–F** Responses to 100 mM K^+ (*K*) of arm ligaments of *M. rotundus*. In each case, *upper trace* shows upward displacement of arm tip (caused by shortening of ligaments) and *lower trace* shows stiffness changes in ligament. *Horizontal scale bar* 3 min in all cases. **C** Contraction associated with stiffening. **D** Contraction without stiffness change in 100 mM K^+ and contraction with stiffening when excess K^+ was removed. **E** No contraction, but with marked stiffening. **F** No contraction, but with destiffening. (Adapted from Motokawa et al. 2004)

tion can be induced in preparations that have undergone some stress relaxation prior to stimulation, it cannot be explained in terms of the passive recoil of a previously stretched elastic element (Birenheide and Motokawa 1995, 1998; Birenheide et al. 2000).

The mobility of crinoid arms depends on the presence of muscular articulations between adjacent arm ossicles. Below the fulcral ridge of each articulation is a single aboral ligament and above it are paired oral ligaments and paired muscles. Contraction of the muscles bends the arm orally (upwards), yet the power stroke for locomotion by swimming, crawling or climbing is generated by the aboral (downward) flexion of the arm and must be effected by the ligaments. It had long been assumed that this involved the purely passive elastic recoil of stretched aboral ligaments, and perhaps of compressed oral ligaments, these functioning like expanded and compressed springs respectively (see, e.g., Young and Emson 1995). However, Birenheide and Motokawa (1996, 1998) established that the aboral ligament can shorten slowly and generate a tension of up to ca. 5.6 kPa. Because this ligament is a mutable collagenous structure, it is feasible that its apparent contractility results from its becoming stiff whilst it is stretched (by muscle-mediated oral flexion); it would then store strain energy until appropriate stimulation induced its destiffening and thereby allowed it to recoil elastically and reshorten. This 'spring-with-a-lock' hypothesis was shown to be untenable by Motokawa et al. (2004) who, by recording simultaneously stiffness and shortening, demonstrated the independence of passive mechanical properties and contraction: contraction, for example, does not require the ligament to be in a destiffened condition (Fig. 2C–F). The fact that contracting ligaments are usually destiffened, however, indicates that there is coordination of the passive and active mechanical properties, both of which appear to be under cholinergic control.

2.3 Organisation of MCT

2.3.1 Cells

Myocytes, which were discussed in the preceding section, have been found in only a small minority of mutable collagenous structures. All of these structures, however, always contain two other types of cell. The first is characterised by the presence of heterogeneous vacuoles that appear to be lysosomal and may enclose cytoplasmic debris or collagen fibrils. These cells tend to have a roughly fusiform outline and long, sometimes branching, processes (see, e.g., Wilkie 1988; Wilkie et al. 1992). They occur also in echinoderm connective tissue that is not mutable (Wilkie et al. 2004b) and, given the absence of obvious fibroblasts, they are likely to be pluripotential cells that can adopt a fibrogenic phenotype.

The second cellular component that is invariably present in MCT are cell bodies and/or processes containing large, electron-dense, membrane-bounded granules. In ophiuroids, these cellular elements, known as 'juxtaligamental cells', form a complex system of ganglion-like clusters innervated by hyponeural (motor) nerves, with a separate juxtaligamental cluster serving each collagenous structure, including the autotomy tendons of the intervertebral muscles (Wilkie 1979). It is assumed that at least some of the granule-containing cells associated with MCT in the other echinoderm classes are homologous to the juxtaligamental cells, although the cell processes in holothurians differ from those of the other four classes in being separated from the extracellular matrix by a basal lamina (see, e.g., Koob et al. 1999), which presumably represents an ontogenetic rather than a phylogenetic or functional distinction. There is clear morphological evidence for functional contact between motor neurons and juxtaligamental cells in at least ophiuroids, echinoids and crinoids (Cobb 1985; Peters 1985; Welsch et al. 1995), and so, mainly on the grounds that they (1) provide a link between the motor nervous system and the extracellular matrix, (2) terminate in MCT and (3) have no possible cellular targets, it has been hypothesised that they are the effector cells that directly alter the tensile properties of MCT. In all classes it is usual for at least two types of process to co-occur in the same tissue, these being distinguishable by the size and shape of their granules (see, e.g., Welsch et al. 1995; Koob et al. 1999), and it has been proposed that these include separate 'stiffener' and 'plasticiser' cell types. The recent demonstration by immunological methods that juxtaligamental granules of holothurian dermis contain identified chemical factors that influence interfibrillar cohesion provides further evidence that these cells have a role in variable tensility (see below).

2.3.2 *Extracellular Matrix*

Overview

The extracellular matrix of almost all mutable collagenous structures is dominated by fibres consisting of parallel assemblages of collagen fibrils. These fibres exhibit a range of patterns, including parallel fibre arrays (e.g. echinoid spine ligament: Trotter and Koob 1989), crossed-fibre arrays (e.g. echinoid peristomial membrane: Wilkie et al. 1994) and three-dimensional meshworks (e.g. holothurian dermis: Motokawa 1982), which is comparable in diversity to that shown by vertebrate connective tissue structures. An important exception to this fibrous organisation is provided by the autotomy tendons of ophiuroids. These are extensions of muscle cell basal laminae and consist of non-fibrillar collagen related probably to vertebrate type IV (Wilkie and Emson 1987), although, since their biochemical composition is not known, this is a supposition based only on their histochemistry and ultrastructure. The fol-

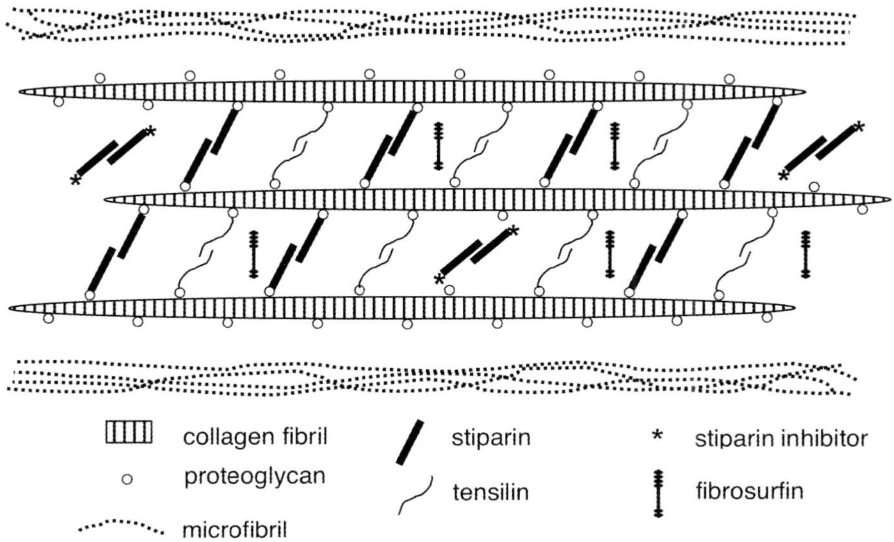

Fig. 3. Model of MCT molecular organisation, which is based on current evidence and assumes that the few mutable structures from which that evidence has been derived are representative. Most MCT consists of parallel aggregates of discontinuous, spindle-shaped collagen fibrils to which are attached PGs and other GAG-containing molecules whose functions include serving as binding sites for molecules responsible for interfibrillar cohesion. Amongst the latter are the proteins stiparin and tensilin, the fibril-aggregating activities of which are modified by a variety of specific inhibitors. The mechanisms by which stiparin and tensilin cause fibril aggregation are incompletely understood. For simplicity, the model assumes that they form dimers that act as interfibrillar crossbridges. Fibril bundles are delimited by loose networks of elastic fibrillin-containing microfibrils that return the tissue to its resting dimensions after it has undergone deformation when in a compliant condition. The functions of fibrosurfin are as yet unknown

lowing sections summarise current knowledge of the molecular constituents of fibrous MCT alone. A generalised model of the molecular organisation of MCT is illustrated in Fig. 3.

Collagen

The fibres of which most MCT is composed are parallel aggregations of cross-banded collagen fibrils that are discontinuous, i.e. shorter than the length of the fibres. The fibrils of two mutable structures with very different microarchitectures – the echinoid spine ligament and holothurian dermis – resemble those of mammalian connective tissue such as rat tail tendon in being composed of molecules with a triple helix length of 300 nm which are assembled in parallel arrays with a regular stagger of 67 nm between adjacent molecules, in having cross-striations in similar positions though varying in stain intensity (an indication of differences in the charge density associated with their constituent amino acids), and in being stabilised by high levels of trivalent hydroxypyridinium intermolecular crosslinks (Trotter and Koob 1989, 1994; Trotter et al. 1994, 1995). The chain compositions of the echinoid and holothurian collagen molecules were not the same, the former being a heterotrimer of two $\alpha 1$ and one $\alpha 2$ polypeptides, as in mammalian type I and most other echinoderm collagens (see, e.g., Omura et al. 1996; Robinson 1997), and the latter being a homotrimer of three $\alpha 1$ polypeptides. These two collagens also have different solubility characteristics and amino acid compositions (Trotter and Koob 1994; Trotter et al. 1995). The general conclusion from this research is that MCT collagens possess no consistent set of biochemical or structural features that distinguish them from the collagens of other echinoderms or other phyla, or that could be correlated with the mutability of their parent tissues.

Data on gene sequence and gene organisation indicate that at least some echinoderm collagen polypeptides are evolutionarily close to those of vertebrate fibrillar collagens (D'Alessio et al. 1989, 1990; Exposito et al. 1992; Tomita et al. 1994; Cluzel et al. 2000). So far, no primary sequence data on collagen extracted from any MCT have been published. However, an epitope of a fully characterised collagen polypeptide has been detected in two confirmed mutable structures. Cluzel et al. (2001) immunolocalised the amino propeptide of the 2α collagen chain, which, apart from the amino propeptide itself, is closely similar to that of vertebrate fibrillar collagen (D'Alessio et al. 1990), in the mutable peristomial membrane and spine ligament of the echinoid *Paracentrotus lividus*. Since the amino and carboxyl propeptide regions of collagen chains are usually removed by specific proteases during the extracellular maturation process, their retention in the tissues of adult echinoids may have functional significance (Lethias et al. 1997).

Because complete collagen fibrils can be isolated from MCT using mild, non-denaturing extraction methods, more is known about their supramolecular organisation than that of vertebrate fibrils. Fibrils from both echinoid spine ligament and holothurian dermis are spindle-shaped with paraboloidal tips. Despite varying greatly in length, they have a constant aspect (length:diameter) ratio in the order of 2000 and are molecularly bipolar, i.e. in both halves of each fibril the amino termini of the collagen molecules are orientated towards the nearer tip, and near the axial midpoint of each fibril there is a region of symmetrical transition from parallel to antiparallel molecular packing (Trotter and Koob 1989; Thurmond and Trotter 1994; Trotter et al. 1994). As in other fibrillar collagens, this organisation results from the self-assembly of the constituent molecules which occurs automatically after enzymatic removal of their N- and C-propeptides. More recent work by Trotter et al. (1998, 2000a), using digital scanning-transmission microscopy to determine mass per unit length (and therefore the number of molecules) along whole collagen fibrils from both echinoid and holothurian tissues, has provided evidence that the self-assembly mechanism is different from that of vertebrate fibrils.

Although it is not possible at present to isolate whole collagen fibrils from normal adult tissues of animals other than echinoderms, the limited data that are available on vertebrate fibrils indicate that, despite the different mechanism of fibrillogenesis, they are also spindle-shaped with paraboloidal tips (see Trotter et al. 1998). This is the ideal shape for fibrils that reinforce a discontinuous fibre composite, since it allows the full tensile strength and stiffness of the fibril to be exploited along its whole length and avoids shear-stress concentrations near its ends (Trotter and Koob 1989; Trotter et al. 2000b). The fusiform shape of the collagen fibrils in MCT is therefore unrelated to its variable tensility.

Proteoglycans

Proteoglycans (PGs) are present in the fibrous connective tissue of all animals and consist of a protein core to which are attached covalently side chains of polyanionic sulphated glycosaminoglycans (GAGs). The use of the polycationic dyes cuprolinic and cupromeronic blue has revealed that polyanions are localised to specific sites in each D-period on the surface of fibrils in crinoid, echinoid and holothurian MCT, as is the case in vertebrate collagenous tissue (Trotter and Koob 1989; Erlinger et al. 1993; Trotter et al. 1995). Biochemical methods have demonstrated that PGs are attached to the fibrils non-covalently or covalently. Non-covalently bound PGs in the chondroitin/dermatan sulphate class are attached to the collagen fibrils of the echinoid spine ligament (Trotter and Koob 1989). Collagen fibrils in holothurian dermis are associated covalently with three different GAG-containing macromolecules. The most abundant of these includes a fucose-containing GAG that is associated with the fibrils via a non-reducible covalent bond. The structure of the highly sulphated fucose branches of chondroitin sulphate E from *Astichopus japonicus* has been characterised fully by Kariya et al. (1997). The other two covalently bound GAG-containing molecules in holothurian dermis are high molecular weight PGs that are linked to collagen fibrils probably via disulphide bonds and at least one of which acts as a binding site for the glycoprotein stiparin (see below). In addition to these insoluble PGs, holothurian dermis contains at least two soluble PGs that bind stiparin, inhibit stiparin-fibril binding and may be involved in the regulation of stiparin-fibril binding (Trotter et al. 1995 and unpubl.).

Other Non-Collagenous Proteins

Stiparin is the most abundant soluble glycoprotein in the dermis of the holothurian *Cucumaria frondosa* and can be extracted from minced tissue by prolonged treatment with seawater alone (which also results in tissue disaggregation, an indication that the collagen fibrils are normally held together by weak bonds). Trotter et al. (1996) demonstrated that stiparin, which has a molecular weight of about 375 kDa, causes calcium-independent aggregation in vitro of collagen fibrils that have been treated with guanidine-HCl (which removes non-covalently bound PGs) but has no effect on the mechanical properties of samples of intact dermis (Koob et al. 1999). Whilst it seems likely that stiparin binds to collagen fibrils via a surface-bound PG, the molecular mechanism of stiparin-induced fibril aggregation has still to be determined.

The dermis of *C. frondosa* contains a 62-kDa sulphated glycoprotein that does not bind collagen fibrils but does bind stiparin and thereby inhibits stiparin's fibril-aggregating activity. This molecule has the highest negative charge density of all macromolecules extracted from the dermis, and all of its inhibitory activity is associated with the polygalactose sulphate moiety of the molecule rather than with its protein component. The relative concentration of stiparin inhibitor is 200 times greater in the loose outer dermis of *C. frondosa* than in the dense inner dermis (Trotter et al. 1999).

Tensilin (also known as 'stiffener') is a constituent of the inner dermis of *C. frondosa* that can be isolated only after treatments that cause cell lysis, such as repeated freeze–thaw cycles, indicating that it is present mainly in intracellular locations. Like stiparin, it causes aggregation of isolated collagen fibrils, but, unlike stiparin, it stiffens intact inner dermis, both effects being calcium-independent (Koob et al. 1999). The peptide sequence of tensilin deduced from a full-length cDNA clone (Fig. 4) suggests significant similarity to the tissue inhibitor of metalloproteinase (TIMP) proteins with 21–36 % identity between tensilin and the mammalian TIMPs (Tipper et al. 2003). It seems likely that tensilin interacts with collagen fibrils via surface GAGs. For example, it binds to isolated collagen fibrils that have surface GAGs, but does not bind to GAG-free molecular collagen (Trotter et al. 1995; Tipper et al. 2003). The binding activity of tensilin is unaffected by stiparin inhibitor (Trotter et al. 2000b).

'Plasticiser' is a cell-sequestered <15-kDa protein that is present in only the outer dermis of *C. frondosa* and destiffens samples of intact inner dermis. It appears to act directly on the extracellular matrix, since it is as effective on cell-lysed samples as it is on fresh samples (Koob et al. 1999). Nothing more is known about its mode of action.

Cluzel et al. (2001) characterised a sea-urchin gene that encodes a multidomain interfibrillar protein they called 'fibrosurfin'. This contains 17 epidermal

```
1     MEVAFLVLLI GALSLSSADA QCAGCSVKHP QHHFCDATFV MKVTIIDVIL
51    DRQGGDKLIN AEINRSWKKG PSSGDFQFYA PSSFCGATFD SGDTYVVTGT
101   KEETSDGRRY WLHGSCDYMI KWDDMSDQQK AGFKGGYKAR CGECQIAESL
151   TAASVKVEDI AANDYPLATT YWTPTGCYYN PLMTRQFVGR KGSSVVDCED
201   VYGLCKPNEA DKCQWTLTPD YERCLKERDD FVKADSSAFA ITRVEQCDVY
251   TNKRKRKNCR QRFRELQAEM GADEELIFYR
```

Fig. 4. Deduced peptide sequence of tensilin. Leu225 (*underlined*) represents last residue that was found to be in alignment with a known TIMP sequence (*Drosophila* TIMP). Protein database comparison revealed that beyond Leu225 tensilin shows no identity or similarity to any known TIMP sequence or any other sequence. (Tipper et al. 2003)

growth factor (EGF) motifs, 11 of which could potentially bind calcium, and was detected in protein extracts of mainly the mutable spine ligaments and peristomial membrane of *Paracentrotus lividus*. Immunogold labelling indicated that fibrosurfin occurs between, or close to, the collagen fibrils of the spine ligament. Since proteins that contain EGF domains are often involved in protein–protein interactions, it is possible that fibrosurfin contributes to interfibrillar cohesion. Its relevance to mutability is at present unknown.

Microfibrils

Hollow microfibrils 10–14 nm in diameter and sometimes beaded with a periodicity of 30–100 nm are ubiquitous in MCT. They can be aggregated into fibres or sheets (see, e.g., Wilkie et al. 1994, 1998, 2004b), but most often form loose sheaths that surround and separate bundles of collagen fibrils. Thurmond and Trotter (1996) and Thurmond et al. (1997) demonstrated that the microfibrils of *C. frondosa* dermis resemble the fibrillin-containing microfibrils of mammalian connective tissue in their morphology, biochemistry and immunological properties. Isolated microfibrillar networks from *C. frondosa* possess long-range elasticity (Thurmond and Trotter 1996). They thus may confer elasticity on MCT that is in a compliant state and provide it with a predetermined set of dimensions to which it returns when external forces are removed (Trotter et al. 2000b). Microfibrils are also present in echinoderm collagenous structures that are non-mutable (Del Castillo et al. 1995; Wilkie et al. 2004b). There is no evidence that they have a role in the variable tensility of MCT.

2.4
Molecular Mechanism of MCT Mutability

2.4.1
Are Collagen Fibrils Involved?

There is no evidence that the variable tensility of MCT involves changes in the mechanical properties of the collagen fibrils. This would be highly unlikely on a priori grounds, in view of the similarities between the collagen fibrils of MCT and those of vertebrate connective tissue in terms of (1) fibril shape, supramolecular organisation and intermolecular crosslink biochemistry, and (2) the structure of their constituent collagen molecules. Furthermore, numerous ultrastructural investigations have failed to provide evidence that alterations in mechanical properties are accompanied by modification of the shape or organisation of the collagen fibrils. Erlinger et al. (1993), however, interpreted 10–11 nm filaments in a crinoid ligament as being collagen 'protofibrils' produced by a reversible fibril disaggregation mechanism possibly associated with mutability. The ultrastructural observations of Birenheide and Motokawa (1994) indicate that these filaments are more likely to be homologous to the fibrillin-containing microfibrils that are ubiquitous in echinoderm connective tissue. It is therefore almost certain that mutability depends on changes not in the tensility of the collagen fibrils, but in the cohesive forces holding the fibrils together.

2.4.2
Are Calcium Ions Involved?

The mechanical properties of MCT are sensitive to changes in the extracellular calcium ion concentration. Increasing $[Ca^{2+}]_o$ stiffens and decreasing $[Ca^{2+}]_o$ destiffens almost all mutable structures that have been investigated. These and other findings led to the hypothesis that Ca^{2+} ions contribute directly to interfibrillar cohesion in MCT and that the juxtaligamental cells alter tissue stiffness by controlling the amount of extracellular Ca^{2+} available for such a role (reviewed by Wilkie 1996). This hypothesis was discredited by the demonstration that certain treatments stiffen MCT in the absence of Ca^{2+} ions and that agents that interfere with calcium-dependent cellular processes can change MCT tensility in the presence of a normal $[Ca^{2+}]_o$ (Szulgit and Shadwick 1994; Trotter and Koob 1995; Trotter and Chino 1997). The weight of evidence now favours the view that the influence of $[Ca^{2+}]_o$ manipulation on MCT tensility is due mainly to direct effects on cellular elements rather than on the extracellular matrix itself, and that, although Ca^{2+} ions contribute directly to interfibrillar cohesion in an unknown way (Szulgit and Shadwick 2000), variable tensility does not involve modulation of $[Ca^{2+}]_o$.

2.4.3
Tensilin–Tensilin Protease Hypothesis

The stiffness of MCT is changed dramatically by a range of treatments that cause cell membrane lysis, such as freeze–thawing or exposure to deionised water or detergents (Szulgit and Shadwick 1994, 2000; Trotter and Koob 1995; Trotter and Chino 1997; Wilkie et al. 1999). Extracts prepared from the dermis of *C. frondosa* after it has been subjected to freeze–thawing have the same effects on isolated tissue samples as freeze–thawing itself (Trotter and Koob 1995; Koob et al. 1999; Szulgit and Shadwick 2000), and the analysis of such extracts resulted in the isolation of the active agents tensilin ('stiffener') and 'plasticiser'. The observation that these proteins can be isolated from tissues only after cell lysis indicates that they are present mainly in intracellular reservoirs, and led to the hypothesis that they are regulatory molecules which are secreted from cells and bring about changes in MCT tensility (Koob et al. 1999; Trotter et al. 2000b). The case for tensilin being a secreted effector molecule has been strengthened by its recent immunolocalisation in granules of juxtaligamental cells in *C. frondosa* dermis (D.R. Keene and J.A. Trotter, unpubl.).

It was noted by Tipper et al. (2003) that tensilin tends to undergo proteolysis in vitro and that the degraded product neither binds collagen fibrils nor induces fibril aggregation. Since analysis of trypsin digests suggested that the C-terminus, which includes a putative fibril-binding site, is susceptible to proteolysis, these authors hypothesised that tensilin-induced stiffening is reversed in vivo by a specific protease. Such a protease could be expressed constitutively, resulting in 'automatic' decay back to the destiffened state, or it could be secreted or activated in response to specific signals (Fig. 5).

At present, the significance for variable tensility of other recently isolated molecules is not clear. Some may have a regulatory and others a constitutive role. Indirect immunofluorescence and immunogold labelling have revealed that stiparin is associated much more abundantly with collagen fibrils than is tensilin (D.R. Keene and J.A. Trotter, unpubl.). This observation and the fact that stiparin, unlike tensilin and 'plasticiser', has no effect on whole tissue samples, suggest it may be a constitutive factor that is not involved in short-term changes in mechanical properties, but functions to hold collagen fibrils in a weak association that facilitates the action of effector molecules such as tensilin (Trotter et al. 2000b). However, the demonstration by immunocytochemistry that stiparin, like tensilin, is present in the juxtaligamental granules

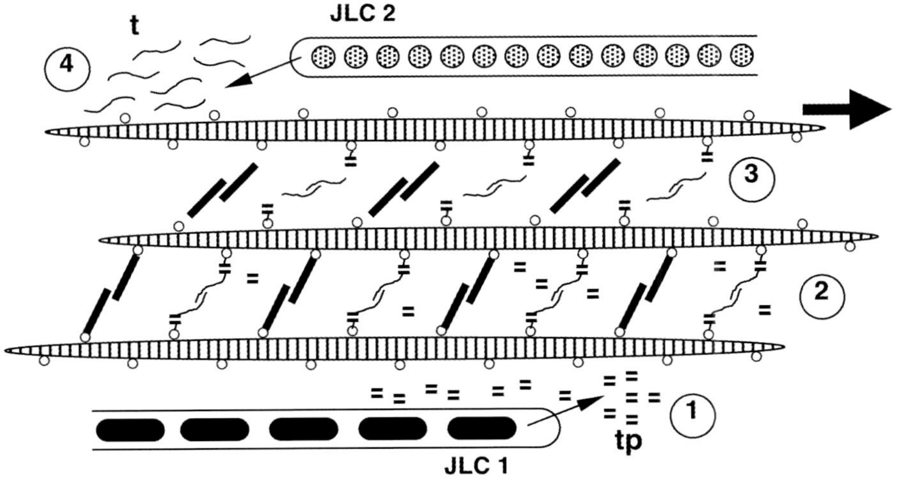

Fig. 5. Model of the tensilin–tensilin protease hypothesis. MCT plasticisation or *destiffening* results from (*1*) the release from, or activation by, a specific type of juxtaligamental cell (*JLC 1*) of tensilin protease (*tp*), which (*2*) cleaves tensilin near its GAG-binding site. This (*3*) allows fibrils to slide past each other, since they are held together only weakly by stiparin. *Restiffening* results from (*4*) the release of fresh tensilin (*t*) from a second type of juxtaligamental cell (*JLC 2*). MCT constituents represented as in Fig. 3

of *C. frondosa* (D.R. Keene and J.A. Trotter, unpubl.) raises the possibility that it also could be a regulatory molecule (or that juxtaligamental cells are a source of both constitutive and regulatory factors).

2.4.4 *Active Force Generation*

No information is available currently on the mechanism of active force generation in crinoid ligaments.

3 Mutable Collagenous Tissue: Biotechnological Perspective

3.1 Current Commercial Uses of MCT

At the present time, MCT is being exploited commercially for purposes that have nothing to do with its mechanical adaptability, though they may rely on biochemical properties that underpin mutability. The best-known example of this is the use of holothurian body wall as a food item ('trepang' or 'beche-de-mer') throughout SE Asia, China and Japan. The demand for trepang is met by the vigorous fishing of natural populations and by an expanding mariculture sector. It is estimated that in the period 1986–1996 more than 50,000 tonnes of trepang, valued at over US$ 40 million, was imported into Hong Kong, the main world market (Conand 2001; Jiaxin 2003).

For millennia, holothurian body wall has also been revered for its prophylactic and curative properties and it is still a component of many herbal medicines. It has become apparent that its reputed medicinal properties may have a scientific basis. For example, its efficacy as a remedy for joint pain is thought to result from the presence of chondroitin sulphate (which may be an important determinant of the mechanical behaviour of MCT: see above) and, because of its antiviral properties, holothurian chondroitin sulphate has been patented for HIV therapy (Jiaxin 2003). In addition, the body wall of certain holothurians possesses antibacterial activity (Villasin and Pomory 2000) and has a fatty acid profile that suggests it could be of clinical benefit in promoting wound repair (Fredalina et al. 1999).

Echinoids are fished and cultured throughout the world for their edible gonads (Andrew et al. 2002). It has been suggested that discarded echinoid tests, which include sutural ligaments (which may be mutable: Johnson et al. 2002), spine ligaments and peristomial membrane (both of which are mutable: Hidaka and Takahashi 1983; Wilkie et al. 1993), could be used as a source of collagen for the food, cosmetic and biomedical industries (Nagai and Suzuki 2002).

3.2 Biotechnological Potential of MCT

Consideration of the biotechnological potential of MCT must of necessity be highly speculative in view of our currently incomplete knowledge of the molecular organisation of MCT and the molecular mechanism underpinning its variable tensility. Theoretically MCT could be a source of, or an inspiration for, (1) new pharmacological agents or strategies and (2) new composite materials.

3.2.1 *Pharmacological Agents or Strategies*

It hardly needs to be reiterated that the outstanding property of MCT is its capacity for reversible changes in stiffness. There are certain clinical conditions that would benefit from the therapeutic manipulation of connective tissue mechanical properties, perhaps as an alternative to surgery or other interventions. Most of these conditions would require the temporary or permanent *plasticisation* or weakening of the connective tissue, rather than its strengthening. This applies to problems like joint contractures due to immobilisation, burn scar contractures, breast capsule contractures following enhancement procedures, Dupuytren's contracture and peritendinous lesions following tendon surgery. Previous suggestions for the pharmacological treatment of some of these conditions have focused on the suppression of collagen synthesis and deposition by, for example, the topical application of lathyrogens such as β-aminopropionitrile, which inhibits an enzyme – lysyl oxidase – involved in intermolecular cross-link formation (Chvapil 1988). On the other hand, structures that need to be *strengthened* include ligaments and tendons weakened by immobilisation (Nordin and Frankel 1980) and repair sites in traumatically or surgically transected tendons, which rarely regain full tensile strength (Koob 2002).

Is it possible that MCT contains molecules that could affect the mechanical properties of mammalian connective tissue? As noted above, holothurian chondroitin sulphate relieves joint pain, though there is at the moment no reason to believe that this is due to anything more than the anti-inflammatory or anti-oxidant effect exerted by other chondroitin sulphates and other glycosaminoglycans (GAGs) (Delehedde et al. 2002; Campo et al. 2003). It is intriguing, however, that GAGs contribute significantly to interfibrillar force transfer, and therefore the overall mechanical properties, of mammalian connective tissue (Redaelli et al. 2003), and that stiffness changes in the uterine cervix are accompanied by significant shifts in the expression of certain GAGs (Westergren-Thorsson et al. 1998). Whilst this implies that the best way to treat fibrotic lesions might be to engineer in them a GAG composition mimicking that of the compliant cervix, it is feasible that, since holothurian GAGs are components of much more mutable connective tissue, they possess features that might facilitate the 'loosening' of fibrotic tissue and that could be incorporated pharmacologically or genetically into the latter. This illustrates the need to determine both the chemical structure and the precise role in MCT mutability of GAGs, the proteoglycans of which they are constituents, and other, as yet incompletely characterised, interfibrillar components. The benefits that would accrue from a therapeutic strategy that treats successfully the fibrotic conditions referred to above, as well as other common pathophysiological processes such as pulmonary fibrosis and connective tissue-related stiffening of the walls of hypertensive blood vessels, cannot be exaggerated.

Regarding a completely separate feature of MCT, Szulgit and Shadwick (1998) discovered that the mutable dermis of the holothurian *Parastichopus parvimensis* has remarkable self-adhesive properties. Dermal autografts or allografts adhered to their implantation site without external pressure or assistance of any sort and shear stresses of 200–500 Pa were required to separate isolated samples after they had been in contact with each other for only 2 h. This property is not due to the entangling of collagen fibres, capillary adhesion or viscous shear forces, but seems to be based on weak chemical bonds. It is independent of the mechanical state of the tissue and is not cell-dependent, and so it seems to be unrelated to mutability, although Szulgit and Shadwick (1998) speculated that a collagen fibril-aggregating factor, such as stiparin, might be involved. This phenomenon merits further investigation, since its elucidation might lead to the identification of chemical factors or mechanisms that could be exploited to promote the adhesion of tissue grafts or artificial skin to wound areas or could be incorporated into MCT-derived artificial tissue (see below).

3.2.2 New Composite Materials

In a recent review, Langer and Tirrell (2004) have drawn attention to the outstanding impact that biomaterials have had on health care, particularly in the context of prosthetic and drug delivery devices, and they commented that the extracellular matrix "provides an important model for biomaterial design". With regard to the development of new structural materials that have medical applications, connective tissue has been employed in three different ways:

1. An entire connective tissue, either living (i.e. with cellular elements left in situ) or with cellular elements removed, may be used as a graft or prosthesis. Examples of this include the use of Achilles tendon allografts for the reconstruction of cruciate ligaments (DeFrate et al. 2004) and skin repair using dermis from different species prepared by various methods (Ramos-e-Silva and Ribeiro de Castro 2002). Obviously, it would be ideal if the mechanical properties of each implant could be adjusted precisely to match the needs of the respective implantation site. For example, skin replace-

ments need to be more extensible and elastic over the extension sides of joints, such as those of the finger, than over the flexion sides. It has to be admitted, however, that there is unlikely to be much scope for using whole MCT in this way, due to the immunological challenge it would present and the low probability that any echinoderm structure would have a micro-architecture more suitable than that of a mammalian alternative. Techniques would also have to be developed to fix the MCT xenograft in the optimal mechanical state. Furthermore, because of the labile nature of the bonds upon which the integrity of MCT depends, many mutable collagenous structures become unmanageably friable in the softened state. It is possible, though, that some of those that demonstrate a limited range of tensile changes and never become compliant to the point of disintegration could be exploited. The echinoid compass depressor ligament and peristomial membrane are examples of such structures (Wilkie et al. 1992, 1993).

2. Components may be isolated from the connective tissue and then reassembled with other biological or synthetic elements to form a novel composite. Examples of biomaterials produced in this way are *Integra* a combination of bovine collagen, shark chondroitin-6-sulphate and a silicone sheet (the last acting as an artificial epidermis), which is used as a skin substitute (Ramos-e-Silva and Ribeiro de Castro 2002), and acellular blood vessel grafts that consist of intestinal submucosa and bovine collagen (Huynh et al. 1999). It is as a source of such components that MCT could make the most direct contribution to biomaterial design, largely by virtue of the extractability of its collagen fibrils. It is notoriously difficult to extract intact collagen fibrils from the post-fetal connective tissue of vertebrates (Trotter et al. 1997). For this reason, the collagen used in existing reassembled biomaterials is in the form of disaggregated molecules. These have the advantageous property of aggregating spontaneously to form fibrils, but, due to the absence of covalent intermolecular bonds, the fibrils have a low tensile strength. Koob (2002) has developed a method for chemically crosslinking such reconstituted fibrils to make a product that could be used to bridge gaps in damaged tendons. In stark contrast to the vertebrate situation, intact collagen fibrils can be isolated easily from MCT by mild non-denaturing techniques. A solution containing 0.5 M NaCl, 0.05 M EDTA, 0.2 M β-mercaptoethanol and 0.1 M TRIS buffer (pH 8.0) disaggregates holothurian dermis, asteroid body wall and echinoid spine ligaments (Matsumura 1973; Matsumura et al. 1973; Trotter and Koob 1989). Even more remarkably, Trotter et al. (1996) discovered that fibrils could be isolated from holothurian dermis using sequential 24-h extractions in artificial seawater alone. This is evidently a consequence of the weak nature of the interactions that maintain interfibrillar cohesion in MCT, and it means that holothurian dermis and other mutable collagenous structures represent a cheap and easily accessible source of intact collagen fibrils that retain their tensile strength and stiffness and could be used for the manufacture of artificial tissues.

However, a more exciting way in which MCT could contribute to the design of new biomaterials, whether these incorporate MCT-derived components or not, is through the mimicking of the mechanisms responsible for its variable tensility and contractility. There is a clinical need for artificial tissues with site-specific micro-architectures and mechanical properties that are either pre-set or continuously adjustable and responsive to changing physiological parameters or to therapeutic manipulation (Langer and Tirrell 2004). Amongst possible applications for such 'smart' materials (i.e. which combine the functions of sensors and actuators) would be vascular implants or cuffs that controlled blood pressure or local blood flow, and whose stiffness and/or contractile state were directly sensitive to blood biochemistry or to precisely targeted pharmacological agents, thereby offering an alternative to systemic antihypertensive drugs and their spectrum of unwanted side effects.

Trotter et al. (2000b) have already examined the possibility of developing a simple 'hybrid' biomaterial assembled from collagen fibrils extracted from holothurian dermis and a synthetic interfibrillar matrix. The interaction between the fibrils and the matrix, and therefore the tensile properties of the whole system, would depend on a pair of synthetic molecules that have been shown to selectively and reversibly associate with each other in physiological conditions. These are a catechol and a phenylboronic acid, which complex to form a boronic ester. This interaction can be reversed by oxidizing the catechol to orthoquinone which does not bind to boronic acid (Fig. 6). The synthetic material would consist of collagen fibrils, to which catechol groups had been attached chemically, linked by a soluble polyacrylamide polymer complexed with phenylboronic acid groups. Trotter et al. envisaged that the association between fibrils and matrix could be repeatedly switched on and off by sequential oxidations and reductions controlled perhaps by optical or electrical signals that change the redox potential of the matrix (Fig. 7). The only MCT-derived elements in this device are collagen fibrils. It is possible that further investigation of MCT will reveal other components that could be built into new controllable biomaterials.

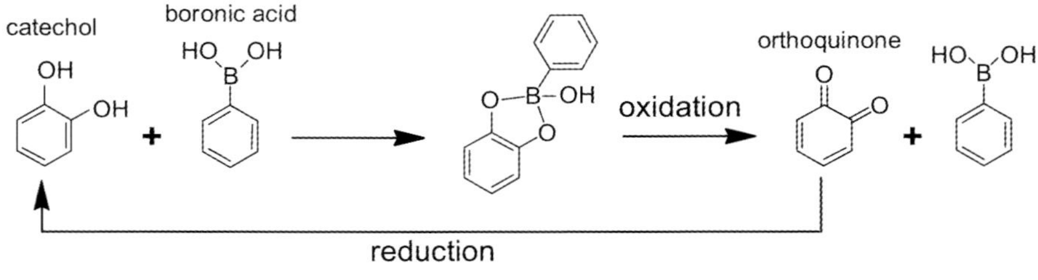

Fig. 6. Interaction between catechol and phenylboronic acid, which provides the reversible cross-links in proposed hybrid biomaterial. (Adapted from Trotter, unpubl.)

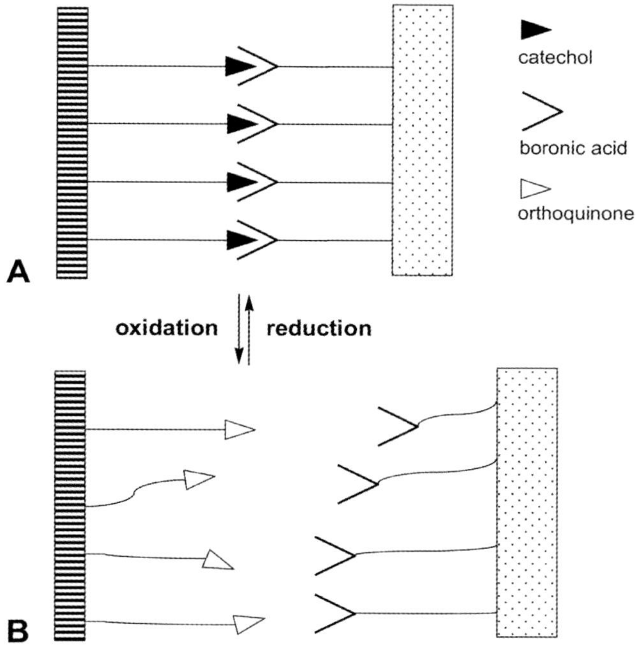

Fig. 7A, B. Model of proposed hybrid biomaterial. Collagen fibrils (*horizontally striated*; only one shown) with attached catechol groups are embedded in polyacrylamide polymer (*stippled*) complexed with phenylboronic acid. **A** Stiff condition, in which cross-links are formed by interaction of catechol and phenylboronic acid. **B** Compliant condition, in which cross-links are reversed by oxidation of catechol to orthoquinone. (Adapted from Trotter, unpubl.)

3. Connective tissue may provide inspiration for entirely synthetic materials. A simple example resulting from this approach is an artificial tendon which is manufactured from poly(ethylene terephthalate) fibres embedded in a swollen hydrogel matrix and can be designed to have mechanical properties that suit specific implantation sites (Kolarik 1995). Trotter (unpubl.) has speculated that MCT could serve as a model system for the construction of dynamically controllable ligaments with adjustable stiffness and adjustable damping. Trotter has also suggested that these might be incorporated into energy-efficient robots, and it is therefore relevant that there has been interest in the design of 'compliant' robots for specialist purposes such as pipeline inspection (Suzumori 1996). As for reassembled biomaterials (see above), both the variable passive mechanical properties and the contractility of MCT may yield design principles applicable to the development of fully synthetic devices.

4 Concluding Remarks

Recent research has done much to elucidate the molecular organisation and functioning of MCT. Whilst a number of interfibrillar molecules have been isolated, only tensilin ('stiffener') has been characterized fully, and the significance of none of these molecules for either interfibrillar cohesion or variable tensility is fully understood. It would be particularly interesting to find out more about the proteoglycans in view of their critical influence on the supramolecular organisation and mechanical properties of vertebrate collagenous tissue (Redaelli et al. 2003). Further investigation of the biochemistry and molecular biology of MCT could take advantage of the presence in some echinoderms of adjacent structures that differ significantly in their capacity for undergoing tensile changes. Examples include the autotomy and non-autotomy tendons of ophiuroid intervertebral muscles (only the former are mutable: Wilkie and Emson 1987), the distal and proximal oral arm plate ligaments of ophiuroids (the former show both reversible and irreversible changes in mechanical properties and the latter only reversible changes: Wilkie 1992), and the capsular and central ligaments of cidaroid echinoid spines (only the former are mutable: Del Castillo et al. 1995). Qualitative and quantitative comparison of these structures might help to identify the molecular correlates of variable tensility.

A fascinating recent discovery is the homology between tensilin and the tissue inhibitors of metalloproteinases (TIMPs) (Tipper et al. 2003). Matrix metalloproteinases (MMPs) are ubiquitous, connective tissue degrading enzymes and TIMPs are important modulators of MMP activity. One mammalian TIMP (TIMP-3) binds strongly to the extracellular matrix via a GAG-binding site (Yu et al. 2000), as does tensilin. This similarity and the involvement of MMPs and TIMPs in the mechanical changes undergone by female mammalian reproductive tract collagenous tissues suggest that the mechanism underpinning MCT mutability could have evolved from a MMP-TIMP system. There are also surprising similarities between MCT and the collagenous mesohyl of demosponges, which also shows reversible changes in stiffness (Wilkie et al. 2004 c). Mutability may therefore be an ancestral property of the extracellular matrix, which was lost during the evolution of most animals.

It has emerged that MCT is more versatile than previously suspected. As well as showing adaptable passive mechanical properties, some mutable collagenous structures are actively contractile, and another has been found to demonstrate non-cell-mediated self-adhesiveness. This chapter has provided some initial thoughts on the biotechnological potential of these properties, which would seem to be considerable in view of the currently expanding interest in biological systems as a source of molecules and mechanisms that could be developed for biomedical and other applications (Langer and Tirrell 2004). Improved knowledge of the basic biology of MCT should lead to better understanding of how its unique properties could be exploited for such purposes.

Acknowledgement. I am particularly indebted to Professor J.A. Trotter for providing me with unpublished information.

References

Andrew NL, Agatsuma Y, Ballesteros E, Bazhin AG, Creaser EP, Barnes DKA, Botsford LW, Bradbury A, Campbell A, Dixon J, Einarsson S, Gerring PK, Hebert K, Hunter M, Hur SB, Johnson CR, Juinio-Menez MA, Kalvass P, Miller RJ, Moreno CA, Palleiro JS, Rivas D, Robinson SML, Schroeter SC, Steneck RS, Vadas RL, Woodby DA, Xiaoqi Z (2002) Status and management of world sea urchin fisheries. Oceanogr Mar Biol Annu Rev 40:343–425

Bauer AM, Russell AP, Shadwick RE (1989) Mechanical properties and morphological correlates of fragile skin in gekkonid lizards. J Exp Biol 145:79–102

Birenheide R, Motokawa T (1994) Morphological basis and mechanics of arm movement in the stalked crinoid *Metacrinus rotundus* (Echinodermata, Crinoida). Mar Biol 121:273–283

Birenheide R, Motokawa T (1995) Motility and stiffness of cirri of the stalked crinoid *Metacrinus rotundus*. In: Emson RH, Smith AB, Campbell AC (eds) Echinoderm research 1995. AA Balkema, Rotterdam, pp 91–94

Birenheide R, Motokawa T (1996) Contractile connective tissue in crinoids. Biol Bull 191:1–4

Birenheide R, Motokawa T (1998) Crinoid ligaments: catch and contractility. In: Mooi R, Telford M (eds) Echinoderms: San Francisco. AA Balkema, Rotterdam, pp 139–144

Birenheide R, Tsuchi A, Motokawa T (1996) To be stiff or to be soft – the dilemma of the echinoid tooth ligament. II. Mechanical properties. Biol Bull 190:231–236

Birenheide R, Yokoyama K, Motokawa T (2000) Cirri of the stalked crinoid *Metacrinus rotundus*: neural elements and the effect of cholinergic agonists on mechanical properties. Proc R Soc Lond B 267:7–16

Bonasoro F, Wilkie IC, Bavestrello G, Cerrano C, Candia Carnevali MD (2001) Dynamic structure of the mesohyl in the sponge *Chondrosia reniformis* (Porifera, Demospongiae). Zoomorphology 121:109–121

Bornstein P, Armstrong LC, Hankenson KD, Kyriakides TR, Yang Z (2000) Thrombospondin 2, a matricellular protein with diverse functions. Matrix Biol 19:557–568

Bruel A, Oxlund H (1996) Changes in biomechanical properties, composition of collagen and elastin, and advanced glycation endproducts of the rat aorta in relation to age. Atherosclerosis 127:155–165

Bryant-Greenwood GD (1998) The extracellular matrix of the human fetal membranes: structure and function. Placenta 19:1–11

Buchanan CI, Marsh RL (2002) Effects of exercise on the biomechanical, biochemical and structural properties of tendons. Comp Biochem Physiol 133A:1101–1107

Byrne M (2001) The Echinodermata. In: Anderson DT (ed) Invertebrate zoology, 2nd edn. Oxford University Press, Sydney, pp 366–395

Campo GM, Avenoso A, Campo S, Ferlazza A, Altavilla D, Micali C, Calatroni A (2003) Aromatic trap analysis of free radicals production in experimental collagen-induced arthritis in the rat: protective effect of glycosaminoglycans treatment. Free Radical Res 37:257–268

Chakravarti S (2002) Functions of lumican and fibromodulin: lessons from knockout mice. Glycoconjugate J 19:287–293

Chiquet M (1999) Regulation of extracellular matrix gene expression by mechanical stress. Matrix Biol 18:417–426

Chvapil M (1988) Method of treatment of fibrotic lesions by topical administration of lathyrogenic drugs. In: Nimni ME (ed) Collagen, vol 2. CRC Press, Boca Raton, pp 161–175

Clément P (1993) The phylogeny of rotifers: molecular, ultrastructural and behavioural data. Hydrobiologia 255/256:527–544

Cluzel C, Lethias C, Garrone R, Exposito J-Y (2000) Sea urchin fibrillar collagen 2α chain participates in heterotrimeric molecules of (1 α)2 α stoichiometry. Matrix Biol 19:545–547

Cluzel C, Lethias C, Humbert F, Garrone R, Exposito J-Y (2001) Characterization of fibrosurfin, an interfibrillar component of sea urchin catch connective tissue. J Biol Chem 276:18108–18114

Cobb JLS (1985) Motor innervation of the oral plate ligament in the brittlestar *Ophiura ophiura* (L.). Cell Tissue Res 242:685–688

Conand C (2001) Overview of sea cucumbers fisheries over the last decade – what possibilities for a durable management? In: Barker MF (ed) Echinoderms 2000. Swets and Zeitlinger, Lisse, pp 339–344

D'Alessio M, Ramirez F, Suzuki HR, Solursh M, Gambino R (1989) Structure and developmental expression of a sea urchin fibrillar collagen gene. Proc Natl Acad Sci USA 86:9303–9307

D'Alessio M, Ramirez F, Suzuki HR, Solursh M, Gambino R (1990) Cloning of a fibrillar collagen gene expressed in the mesenchymal cells of the developing sea urchin embryo. J Biol Chem 265:7050–7054

DeFrate LE, van der Ven A, Gill TJ, Li G (2004) The effect of length on the structural properties of an Achilles tendon graft as used in posterior cruciate ligament reconstruction. Am J Sports Med 32:993–997

Del Castillo J, Smith DS (1996) We still invoke friction and Occam's razor to explain catch in the spines of *Eucidaris tribuloides*. Biol Bull 190:243–244

Del Castillo J, Smith DS, Vidal AM, Sierra C (1995) Catch in the primary spines of the sea urchin *Eucidaris tribuloides*: a brief review and a new interpretation. Biol Bull 188:120–127

Delehedde M, Allain F, Payne SJ, Borgo R, Vanpouille C, Fernig DG, Deudon E (2002) Proteoglycans in inflammation. Curr Med Chem Anti-inflam Anti-allergy Agents 1:89–102

Ellers O, Telford M (1996) Advancement mechanics of growing teeth in sand dollars (Echinodermata, Echinoidea): a role for mutable collagenous tissue. Proc R Soc Lond B 263:39–44

Elphick MR, Melarange M (2001) Neural control of muscle relaxation in echinoderms. J Exp Biol 204:875–885

Erlinger R, Welsch U, Scott JE (1993) Ultrastructural and biochemical observations on proteoglycans and collagen in the mutable connective tissue of the feather star *Antedon bifida* (Echinodermata, Crinoidea). J Anat 183:1–11

Exposito J-Y, D'Alessio M, Solursh M, Ramirez F (1992) Sea urchin collagen evolutionarily homologous to vertebrate pro-α2(I) collagen. J Biol Chem 267:15559–15562

Fredalina BD, Ridzwan BH, Zainal Abidin AA, Kaswandi MA, Zaiton H, Zali I, Kittakoop P, Mat Jais AM (1999) Fatty acid composition in local sea cucumber, *Stichopus chloronotus*, for wound healing. Gen Pharmacol 33:337–340

Garronne R (1978) Phylogenesis of connective tissue. Morphological aspects and biosynthesis of sponge intercellular matrix. In: Robert L (ed) Frontiers of matrix biology, vol 5. S Karger, Basel

Gupte CM, Smith A, Jamieson N, Bull AMJ, Thomas RD, Amis AA (2002) Meniscofemoral ligaments – structural and material properties. J Biomech 35:1623–1629

Hidaka M, Takahashi K (1983) Fine structure and mechanical properties of the catch apparatus of the sea-urchin spine, a collagenous connective tissue with muscle-like holding capacity. J Exp Biol 103:1–14

Huynh T, Abraham G, Murray J, Brockbank K, Hagen PO, Sullivan S (1999) Remodeling of an acellular collagen graft into a physiologically responsive neovessel. Nat Biotechnol 17:1083–1086

Jiaxin C (2003) Overview of sea cucumber farming and sea ranching practices in China. SPC Beche-de-mer Info Bull 18:18–23

Johnson AS, Ellers O, Lemire J, Minor M, Leddy HA (2002) Sutural loosening and skeletal flexibility during growth: determination of drop-like shapes in sea urchins. Proc R Soc Biol Sci 269:215–220

Kannus P (2000) Structure of tendon connective tissue. Scand J Med Sci Sports 10:312–320

Kariya Y, Watabe S, Kyogashima M, Ishihara M, Ishii T (1997) Structure of fucose branches in the glycosaminoglycans from the body wall of the sea cucumber *Stichopus japonicus*. Carbohydr Res 297:273–279

Kolarik J (1995) Composite artificial tendons with hydrogel matrix. Coll Czech Chem Commun 60:1995–2005

Koob TJ (2002) Biomimetic approaches to tendon repair. Comp Biochem Physiol 133A:1171–1192

Koob TJ, Koob-Emunds MM, Trotter JA (1999) Cell-derived stiffening and plasticizing factors in sea cucumber (*Cucumaria frondosa*) dermis. J Exp Biol 202:2291–2301

Langer R, Tirrell DA (2004) Designing materials for biology and medicine. Nature 428:487–492

Lethias C, Exposito JY, Garrone R (1997) Collagen fibrillogenesis during sea urchin development. Retention of SURF motifs from the N-propeptide of the 2α chain in mature fibrils. Eur J Biochem 245:434–440

Maldonado M (2004) Choanoflagellates, choanocytes, and animal multicellularity. Invertebr Biol 123:1–22

Matsumura T (1973) Shape, size and amino acid composition of collagen fibril of the starfish *Asterias amurensis*. Comp Biochem Physiol 44B:1197–1205

Matsumura T, Shinmei M, Nagai Y (1973) Disaggregation of connective tissue: preparation of fibrous components from sea cucumber body wall and calf skin. J Biochem 73:155–162

Morales M, del Castillo J, Smith DS (1989) Acetylcholine sensitivity of the spine-test articular capsule of the sea urchin *Eucidaris tribuloides*. Comp Biochem Physiol 94C:547–554

Motokawa T (1982) Fine structure of the dermis of the body wall of the sea cucumber, *Stichopus chloronotus*, a connective tissue which changes its mechanical properties. Galaxea 1:55–64

Motokawa T (1983) Mechanical properties and structure of the spine-joint central ligament of the sea urchin, *Diadema setosum* (Echinodermata, Echinoidea). J Zool Lond 201:223–235

Motokawa T, Tsuchi A (2003) Dynamic mechanical properties of body-wall dermis in various mechanical states and their implications for the behaviour of sea cucumbers. Biol Bull 205:261–275

Motokawa T, Shintani O, Birenheide R (2004) Contraction and stiffness changes in collagenous arm ligaments of the stalked crinoid *Metacrinus rotundus* (Echinodermata). Biol Bull 206:4–12

Munns SW, Jayaraman G, Luallin SR (1994) Effects of pretwist on biomechanical properties of canine patellar tendon. Arthroscopy 10:404–411

Nagai T, Suzuki N (2002) Collagen from the underutilized resources of aquatic organisms. Trends Comp Biochem Physiol 9:259–268

Nordin M, Frankel VH (1980) Biomechanics of collagenous tissues. In: Frankel VH, Nordin M (eds) Basic biomechanics of the skeletal system. Lea and Febiger, Philadelphia, pp 87–110

Omura Y, Urano N, Kimura S (1996) Occurrence of fibrillar collagen with structure of (α1)2 α2 in the test of sea urchin *Asthenosoma ijimai*. Comp Biochem Physiol 115B:63–68

O'Neill PO (1989) Structure and mechanics of starfish body wall. J Exp Biol 147:53–89

O'Neill PO, Withers PC (1995) An analysis of the load curve of the body wall of *Coscinasterias calamaria* (Echinodermata: Asteroidea). Mar Fresh Behav Physiol 25:245–260

Pérez-Acevedo NL, Marrero H, del Castillo J (1998) Transient wrinkles in a variable length tendon. In: Mooi R, Telford M (eds) Echinoderms: San Francisco. AA Balkema, Rotterdam, pp 783–790

Peters BH (1985) The innervation of spines in the sea-urchin *Echinus esculentus* L. Cell Tissue Res 239:219–228

Ramos-e-Silva M, Ribeiro de Castro MC (2002) New dressings, including tissue-engineered living skin. Dermatol Clin 20:715–723

Redaelli A, Vesentini S, Soncini M, Vena P, Mantero S, Montevecchi FM (2003) Possible role of decorin glycosaminoglycans in fibril to fibril force transfer in relative mature tendons – a computational study from molecular to microstructural level. J Biomech 36:1555–1569

Reddy GK, Stehno-Bittel L, Enwemeka CS (2002) Glycation-induced matrix stability in the rabbit Achilles tendon. Arch Biochem Biophys 399:174–180

Robinson JJ (1997) Comparative biochemical analysis of sea urchin peristome and rat tail tendon collagen. Comp Biochem Physiol 117B:307–313

Robker RL, Russell DL, Yoshioka S, Sharma CS, Lydon JP, O'Malley BW, Espey LL, Richards JS (2000) Ovulation: a multi-gene, multi-step process. Steroids 65:559–570

Sarras MP, Deutzmann R (2001) *Hydra* and Niccolo Paganini (1782–1840) – two peas in a pod? The molecular basis of extracellular matrix structure in the invertebrate, *Hydra*. BioEssays 23:716–724

Sennström MB, Brauner A, Byström B, Malmström A, Ekman G (2003) Matrix metalloproteinase-8 correlates with the cervical ripening process in humans. Acta Obstet Gynecol Scand 82:904–911

Silver FH, Freeman JW, Seehra GP (2003) Collagen self-assembly and the development of tendon mechanical properties. J Biomech 36:1529–1553

Suzumori K (1996) Elastic materials producing compliant robots. Robotics Auton Syst 18:135–140

Syed T, Schierwater B (2002) The evolution of the Placozoa: a new morphological model. Senckenbergiana Lethaea 82:315–324

Szulgit GK, Shadwick RE (1994) The effects of calcium chelation and cell perforation on the mechanical properties of sea urchin ligaments. In: David B, Guille A, Féral JP, Roux M (eds) Echinoderms through time. AA Balkema, Rotterdam, pp 887–892

Szulgit GK, Shadwick RE (1998) Novel non-cellular adhesion and tissue grafting in the mutable collagenous tissue of the sea cucumber *Parastichopus parvimensis*. J Exp Biol 201:3003–3013

Szulgit GK, Shadwick RE (2000) Dynamic mechanical characterization of a mutable collagenous tissue: response of sea cucumber dermis to cell lysis and dermal extracts. J Exp Biol 203:1539–1550

Takemae N, Motokawa T (2002) Is muscular contraction necessary for the stiffness changes of catch apparatus? Zool Sci 19:1460

Thurmond FA, Trotter JA (1994) Native collagen fibrils from echinoderms are molecularly bipolar. J Mol Biol 235:73–79

Thurmond FA, Trotter JA (1996) Morphology and biomechanics of the microfibrillar network of sea cucumber dermis. J Exp Biol 199:1817–1828

Thurmond FA, Koob TJ, Bowness JM, Trotter JA (1997) Partial biochemical and immunological characterization of fibrillin microfibrils from sea cucumber dermis. Conn Tissue Res 36:211–222

Tipper JP, Lyons-Levy G, Atkinson MAL, Trotter JA (2003) Purification, characterization and cloning of tensilin, the collagen-fibril binding and tissue stiffening factor from *Cucumaria frondosa* dermis. Matrix Biol 21:625–635

Tomita M, Kinoshita T, Izumi S, Tomino S, Yoshizato K (1994) Characterizations of sea urchin fibrillar collagen and its cDNA clone. Biochim Biophys Acta 1217:131–140

Trotter JA, Chino K (1997) Regulation of cell-dependent viscosity in the dermis of the sea cucumber *Actinopyga agassizi*. Comp Biochem Physiol 118A:805–811

Trotter JA, Koob TJ (1989) Collagen and proteoglycan in a sea urchin ligament with mutable collagenous properties. Cell Tissue Res 258:527–539

Trotter JA, Koob TJ (1994) Biochemical characterization of fibrillar collagen from the mutable spine ligament of the sea-urchin *Eucidaris tribuloides*. Comp Biochem Physiol 107B:125–134

Trotter JA, Koob TJ (1995) Evidence that calcium-dependent cellular processes are involved in the stiffening response of holothurian dermis and that dermal cells contain an organic stiffening factor. J Exp Biol 198:1951–1961

Trotter JA, Thurmond FA, Koob TJ (1994) Molecular structure and functional morphology of echinoderm collagen fibrils. Cell Tiss Res 275:451–458

Trotter JA, Lyons-Levy G, Thurmond FA, Koob TJ (1995) Covalent composition of collagen fibrils from the dermis of the sea cucumber, *Cucumaria frondosa*, a tissue with mutable mechanical properties. Comp Biochem Physiol 112A:463–478

Trotter JA, Lyons-Levy G, Luna D, Koob TJ, Keene D, Atkinson MAL (1996) Stiparin: a glycoprotein from sea cucumber dermis that aggregates collagen fibrils. Matrix Biol 15:99–110

Trotter JA, Dahners L, de Vente J, Lester G (1997) Isolation of intact collagen fibrils from healing ligament. J Electron Microsc 46:353–356

Trotter JA, Chapman JA, Kadler KE, Holmes DF (1998) Growth of sea cucumber collagen fibrils occurs at the tips and centers in a coordinated manner. J Mol Biol 284:1417–1424

Trotter JA, Lyons-Levy G, Chino K, Koob TJ, Keene DR, Atkinson MAL (1999) Collagen fibril aggregation inhibitor from sea cucumber dermis. Matrix Biol 18:569–578

Trotter JA, Kadler KE, Holmes DF (2000a) Echinoderm collagen fibrils grow by surface-nucleation-and-propagation from both centers and ends. J Mol Biol 300:531–540

Trotter JA, Tipper J, Lyons-Levy G, Chino K, Heuer AH, Liu Z, Mrksich M, Hodneland C, Dillmore WS, Koob TJ, Koob-Emunds MM, Kadler K, Holmes D (2000b) Towards a fibrous composite with dynamically controlled stiffness: lessons from echinoderms. Biochem Soc Trans 28:357–362

Uldbjerg N (1994) Connective tissue changes related to cervical maturation. Eur J Obstet Gynecol Reprod Biol 55:8

Vidal AM, del Castillo J, Smith DS (1993) contractile properties of the articular capsule or ligament, in the primary spines of the sea-urchin *Eucidaris tribuloides*. Comp Biochem Physiol 106C:643–647

Villasin J, Pomory CM (2000) Antibacterial activity of extracts from the body wall of *Parastichopus parvimensis* (Echinodermata: Holothuroidea). Fish Shellfish Immunol 10:465–467

Vogel HG (1980) Influence of maturation and aging on mechanical and biochemical properties of connective tissue in rats. Mech Ageing Dev 14:283–292

Welsch U, Lange A, Bals R, Heinzeller T (1995) Juxtaligamental cells in feather stars and isocrinids. In: Emson RH, Smith AB, Campbell AC (eds) Echinoderm research 1995. AA Balkema, Rotterdam, pp 129–135

Westergren-Thorsson G, Norman M, Bjorrnsson S, Endresen U, Stjernholm Y, Ekman G, Malmstrom A (1998) Differential expressions of mRNA for proteoglycans, collagens and transforming growth factor-β in the human cervix during pregnancy and involution. Biochim Biophys Acta 1406:203–213

Wilkie IC (1979) The juxtaligamental cells of *Ophiocomina nigra* (Abildgaard) (Echinodermata: Ophiuroidea) and their possible role in mechano-effector function of collagenous tissue. Cell Tissue Res 197:515–530

Wilkie IC (1988) Design for disaster: the ophiuroid intervertebral ligament as a typical mutable collagenous structure. In: Burke RD, Mladenov PV, Lambert P, Parsley RL (eds) Echinoderm biology. AA Balkema, Rotterdam, pp 25–38

Wilkie IC (1992) Variable tensility of the oral arm plate ligaments of the brittlestar *Ophiura ophiura* L. (Echinodermata: Ophiuroidea). J Zool Lond 228:5–26

Wilkie IC (1996) Mutable collagenous tissues: extracellular matrix as mechano-effector. Echinoderm Stud 5:61–102
Wilkie IC (2001) Autotomy as a prelude to regeneration in echinoderms. Microsc Res Tech 55:369–396
Wilkie IC (2002) Is muscle involved in the mechanical adaptability of echinoderm mutable collagenous tissue? J Exp Biol 205:159–165
Wilkie IC, Emson RH (1987) The tendons of *Ophiocomina nigra* and their role in autotomy (Echinodermata, Ophiuroida). Zoomorphology 107:33–44
Wilkie IC, Candia Carnevali MD, Bonasoro F (1992) The compass depressors of *Paracentrotus lividus* (Lamarck) (Echinodermata, Echinoida): ultrastructural and mechanical aspects of their variable tensility and contractility. Zoomorphology 112:143–153
Wilkie IC, Candia Carnevali MD, Andrietti F (1993) Variable tensility of the peristomial membrane of the sea-urchin *Paracentrotus lividus* (Lamarck). Comp Biochem Physiol 105A:493–501
Wilkie IC, Candia Carnevali MD, Andrietti F (1994) Microarchitecture and mechanics of the sea-urchin peristomial membrane. Boll Zool 61:39–51
Wilkie IC, Candia Carnevali MD, Bonasoro F (1998) Organization and mechanical behaviour of myocyte-ligament composites in a sea-urchin lantern: the compass depressors of *Stylocidaris affinis*. Zoomorphology 118:87–101
Wilkie IC, Candia Carnevali MD, Bonasoro F (1999) Evidence for the 'cellular calcium regulation hypothesis' from 'simple' mutable collagenous structures: the brachial and cirral syzygial ligaments of *Antedon mediterranea* (Lam). In: Candia Carnevali MD, Bonasoro F (eds) Echinoderm research 1998. AA Balkema, Rotterdam, pp 119–125
Wilkie IC, Candia Carnevali MD, Trotter JA (2004a) Mutable collagenous tissue: recent progress and an evolutionary perspective. In: Heinzeller T, Nebelsick J (eds) Proc 11th Int Echinoderm Conf, Munich 2003 (in press)
Wilkie IC, McKew M, Candia Carnevali MD (2004b) Functional morphology of the compass-rotular ligament of *Echinus esculentus* (Echinodermata: Echinoida): a non-mutable collagenous component of Aristotle's lantern. Zoomorphology (in press)
Wilkie IC, Bonasoro F, Bavestrello G, Cerrano C, Candia Carnevali MD (2004 c) Mechanical properties of the collagenous mesohyl of *Chondrosia reniformis*: evidence for physiological control. Boll Mus Istit Univ Genova (in press)
Yoshida M, Sagawa N, Itoh H, Yura S, Takemura M, Wada Y, Sato T, Ito A, Fujii S (2002) Prostaglandin $F_{2\alpha}$, cytokines and cyclic mechanical stretch augment matrix metalloproteinase-1 secretion from cultured human uterine cervical fibroblast cells. Mol Human Reprod 8:681–687
Young CM, Emson RH (1995) Rapid arm movements in stalked crinoids. Biol Bull 188:89–97
Yu WH, Yu SC, Meng Q, Brew K, Woessner JF (2000) TIMP-3 binds to sulfated glycosaminoglycans of the extracellular matrix. J Biol Chem 275:31226–31232

Bioresources from Echinoderms

Y. Yokota

Abstract. More than 6,500 species have been recorded in the phylum Echinodermata. A variety of biologically active substances have been isolated from the echinoderm species: saponins, glycolipids, carotenoids, porphyrins, naphthoquinones, venoms and others. Several substances unique to the echinoderm have also been reported and some of them showed high potentiality as a new medicament. This chapter gives an overview of the history of the exploitation of echinoderm species in the Orient, presents studies on the biologically active substance obtained from them, and discusses questions related to the exploitation of the echinoderm and prospects of development of new medications.

1 Introduction

The sea occupies 71 % of the terrestrial globe. Its average depth is approximately 3,800 m. It is an entire complex of environmental conditions rather than a huge uniform hydrosphere. Marine faunas and floras resulted from adaptation to a variety of its environmental conditions. There is almost no light in the sea at depths below 150–200 m, as the seawater absorbs visible radiation. The dark environment, where visual communication is more disadvantageous than under terrestrial conditions, has forced organisms to evolve other means of communication, e.g., chemical communication. Marine organisms may contain substances that are not found in terrestrial organisms.

Marine organisms are anticipated as new resources of biologically active substances, e.g., as pharmacological reagents. The Echinodermata phylum

Y. Yokota (e-mail: yokota@ist.aichi-pu.ac.jp)
Department of Applied Information Technology, Aichi Prefectural University, Aichi 480-1198, Japan

Progress in Molecular and Subcellular Biology
Subseries Marine Molecular Biotechnology
V. Matranga (Ed.), Echinodermata

consists of five major classes: Ophiuroidea, 1,900 species; Crinoidea, 600 species; Echinoidea, 1,000 species; Asteroidea, 1,500 species; and Holothuroidea, 500 species. Less pharmaceutical exploitation of marine organisms has taken place than that of terrestrial animals and plants, because we are not as familiar with marine life as with terrestrial life. Research and development of novel chemicals from unexplored echinoderms are current concerns, e.g., chemicals to fight tumors and microbes, depressors (hypotensives), pesticides and agrichemicals with low environmental impacts.

When we exploit echinoderm species as a new biological resource, we have to consider the economic and ecological standpoints. Historically and currently, the most fished animal among the echinoderms is the sea urchin. Some *Regularia* species have long been consumed and exploited as a gastronomic or an exotic food. In addition, the sea cucumber is fished and consumed in Japan and China. Therefore, market prices of new products from the edible sea urchin and sea cucumber must be higher than those of raw sea urchin or raw sea cucumber as a food. The price of sea urchin roes of the highest grade is 2,000–3,000 Japanese yen per 100 g. Judging only from the current situation of the sea urchin market, development of new products from sea urchins looks unprofitable, although most echinoderm species are non-edible and unexplored. The economic bottleneck described above will be solved by making use of non-edible species.

From an ecological viewpoint, we must not duplicate the error of overexploitation of sea urchins which caused a decrease in sea urchin resources in Japan and the Mediterranean Sea. However, the use of unexplored echinoderm species such as starfish may contribute to the management of fishery resources. Most asteroid species are carnivorous and final predators in the benthos population. In the 1960s, outbreaks of *Acanthaster planci* damaged the Great Barrier Reef; similarly in the 1970s, the coral reef in Okinawa, Japan, was also preyed upon seriously by *A. planci*. Blooms of asteroids in the Frigid and Temperate zones and the consequent impacts on shell fishery are also often reported. Exploitation of asteroids as a biological resource will contribute to the management of the benthic ecosystem. Although aquacultural farms of scallops, clams, and other shellfish remove starfish as a measure of damage control, the countermeasure is not intended to eradicate the starfish population. The landed starfish usually generates a new problem of industrial waste, and its disposal is an expense for aquacultural farms. When useful substances are prepared from collected starfish, two problems – how to decrease both predation by asteroids and the cost of waste disposal – are simultaneously solved.

Man has used a variety of organisms for medicines from time immemorial. Furthermore, marine organisms, in addition to terrestrial organisms, have also been employed as medicine in Japan and China. In this chapter, biologically active substances from echinoderms are described and their potential application is discussed.

2
Oriental Medicine and Historical Background

In Oriental medicine, traditional ideology has been inherited, expressing that the meal equals the medication. According to this ideology, every food is considered to be physiologically effective. For instance, in the Orient, foods abundant in collagen and chondroitin sulfate have long been recognized to contribute to maintaining smooth skin and preventing senescence and arteriosclerosis. The dried body wall of sea cucumber has been used as a corroborant and nutritious diet.

The sea cucumber is called "haishen" in Chinese, which literally means marine ginseng. Ginseng, *Panax ginseng*, is the most famous plant in Oriental medicine, best known for its multipotent pharmaceutical activities. It contains saponins which show depression effects on the central nervous system, and vasodilator effects. Ancient Japanese names "umiko" and "nameriko" for the sea cucumber were found as medications in an antique medical book, "Daidoruijuho" (808 A.D.). In the 17th century, a Chinese scholar, Xie Qi (1567–1624), reported that sea cucumbers were comparable to ginseng with respect to pharmaceutical activities in a book named "Wuzazhu". In Japan, dried and roasted sea cucumbers have been employed as a hemostatic for bleeding such as in hemophilia, a relaxant for convulsions and other remedies since the 11th century. It is recognized in Japan and China that sea cucumbers contain medicinal substances. The historical background described above seems to support pharmaceutical investigations into the Holothuroidea.

3
Saponins

It has long been known that some holothurian species contain toxins. In the southern Pacific and Tokara Islands (islands of southern Japan), a peculiar fishing method using sea cucumbers is employed. The autochthones catch paralyzed fish after they throw extracts or fragments of sea cucumbers into tidal pools. This toxic effect results from saponins. Cooper (1880) reported that some holothurians discharged white filamentous structures, called Cuvierian tubules. Contact with these tubules also caused skin irritation. Saponins concentrate in Cuvierian tubules. On the other hand, in holothurian species without these structures, saponins are distributed in the body wall. Dried sea cucumbers were used as a home remedy for skin diseases, and powdered starfish was employed as a pesticide in Japan. The pharmacological activity described is assumed to be due to saponins. In the animal kingdom, the occurrence of saponins is limited to only two echinoderm classes, namely holothurians and asteroids; in contrast, saponins are widely distributed in the plant kingdom. Saponins have been used as an expectorant, a cure for congestive heart failure and as a component of contraceptives and antirheumatic agents. The surfactant activity of saponins has been known and utilized for a long time in medicine. Studies on toxic and pharmaceutical activities of holothurians were initiated by Yamanouchi in the 1940s (Yamanouchi 1942, 1943a, b). Later, Nigrelli and Zahl (1952) named the toxic substance from Cuvierian tubules of *Actinpya agassizi* holothurin, and initiated a series of further investigations into the chemical and pharmacological properties of the substance. They demonstrated its lethal activities towards various organisms in e.g., Protozoa, Cnidaria, Nematoda, Mollusca, Annelida, and Amphibia. Antifungal activity of the holothurian saponin, holotoxin, was reported by Shimada (1969). He patented holotoxin as a cure for athlete's foot disease and commercialized it. Kitagawa et al. (1976a, b) showed that holotoxin is in fact composed of three molecules, holotoxin A, B and C. The inhibitory activities on fungal growth were much higher than those of various plant saponins (Table 1). Aglycons of holothurian saponins are terpenoids (Fig. 1), whereas those of asteroid saponins are sterols (Fig. 2). The molecular difference in saponins between the asteroid and holothurian seems to reflect phylogenic differences. Recently, novel cytotoxic triterpene glycosides have been isolated from sea cucumbers *Pentamera calcigera* (Avilov et al. 2000), *Staurocucumis liouvillei* (Maier et al. 2001), *Hemoiedema spectabilis* (Chludil et al. 2002) and *Mensamaria intercedens* (Zou et al. 2003). The glycosides from *P. calcigera*, *S. liouvillei* and *M. intercedens* showed antineoplastic activity against mammalian cancer cells. Terpene glycosides deserve to be investigated as anticancer pharmaceuticals. Two triterpene glycosides from *H. spectabilis* also showed antifungal activity against the phytopathogenic fungus. Antifungal activities of substances from sea cucumbers are thought to be due to their terpenoid structures.

Table 1. Antifungal activities of holotoxins. Minimum inhibitory concentration for growth of microorganisms (μg/ml)

Species	A	B	C
Trichophyton rubrum	0.78	0.78	6.25
Trichophyton mentagrphyte	1.56	1.56	12.5
Mictosporum gypseum	3.12	1.56	12.5
Candida albicans	6.25	6.25	25.0
Candida utilis	3.12	3.12	12.5
Tomla utilis	3.12	3.12	12.5
Aspergillus oryzae	6.25	12.52	5.0
Penicillium chrysogenum	3.12	6.25	12.5
Trishomonas vaginalis	3.12	1.56	3.12

Fig. 1. Structural formulae of holotoxin A and B

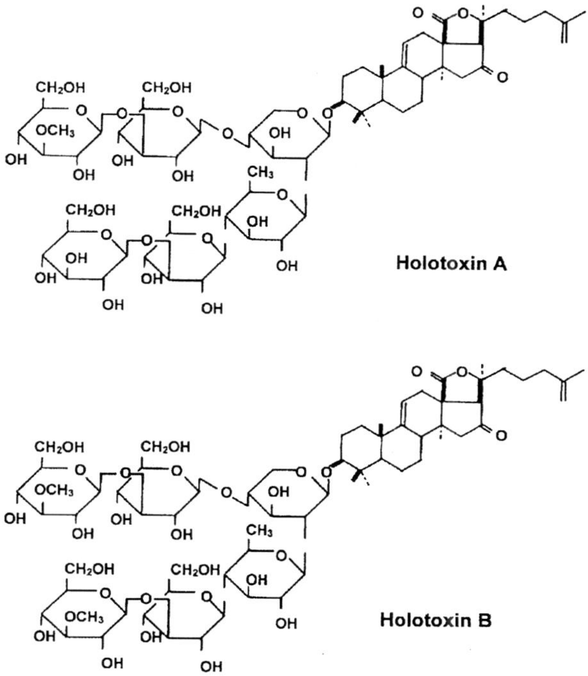

Fig. 2. Structural formula of asterosaponin

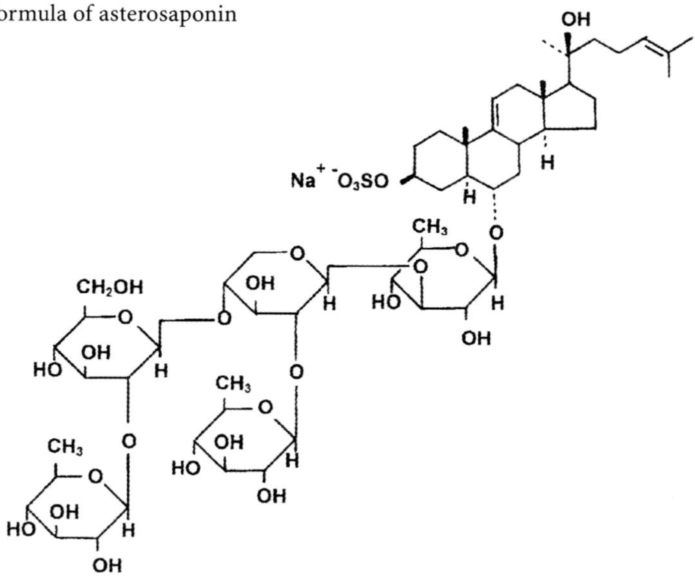

4 Glycolipids

Glycosphingolipids, which are important constituents of a variety of cell membranes of vertebrates, are also reported to be contained abundantly in echinoderms (Kubo et al. 1990; Higuchi et al. 1991, 1994). Glycosphingolipids have various physiological functions, e.g. as receptors of cytotoxins, cells, viruses, transmitters and hormones, in the regulation of cell proliferation and differentiation, cell recognition regulatory function of the cell membranes, signal transduction, etc. Complex carbohydrates have been revealed to play important roles in cancer and neuropathy. Some glycosphingolipids such as cerebrosides and gangliosides are believed to be useful in new medications.

Recently, the bioactive glycolipids of echinoderms have been intensively investigated, focusing on the development of new medicinal materials (Yamada 2002). Higuchi and coworkers examined neuritogenic activity of gangliosides isolated from the echinoderms *Stichopus japonicus*, *Holothuria leucospilota* (Yamada et al. 2001), *Asterias amurensis* (Higuchi et al. 1993) and *Holothuria pervicas* (Yamada et al. 2000), on cultured rat pheochromocytoma cells. They also demonstrated antitumoral activity of gangliosides from *Asteropecten latespinosus* (Higuchi et al. 1995). It was revealed that neuritogenic activity of echinoderm gangliosides is dependent on the amount of sialic acids in the carbohydrate moieties. Holothurians contain sphingoglycolipids with molecular structures different from those of mammalian sphingoglycolipids. A ganglioside, which shows a pharmaceutical activity higher than that of a mammalian ganglioside employed for the therapy of neuropathy, was found among holothurian gangliosides. The holothurian ganglioside appears capable of being employed in a new cure for neuropathy, since the chemical synthesis of sphingoglycolipids is less advantageous than isolation from echinoderms from an economic standpoint.

5 Pigments

Marine animals, including echinoderms, attract our attention because of their fantastic and bizarre colors. The coloration of echinoderms, in most cases, is due to chemical pigments. Carotenoids, melanins, porphyrins and naphthoquinones contribute to the pigmentation of the integument. In addition, the appetizing color of sea urchin roes is precisely that of carotenes. Carotenoids are known as antioxidants and widely used as a food supplement. As echinoderms are not capable of synthesizing carotenoids de novo, they obtain carotenoids from algae or animals that take them from algae. The occurrence of major carotenoids in echinoderms was documented by Fox and Hopkins (1966). The well-known antitumor promoter β-carotene and its derivatives are the most abundant carotenoids in echinoids and

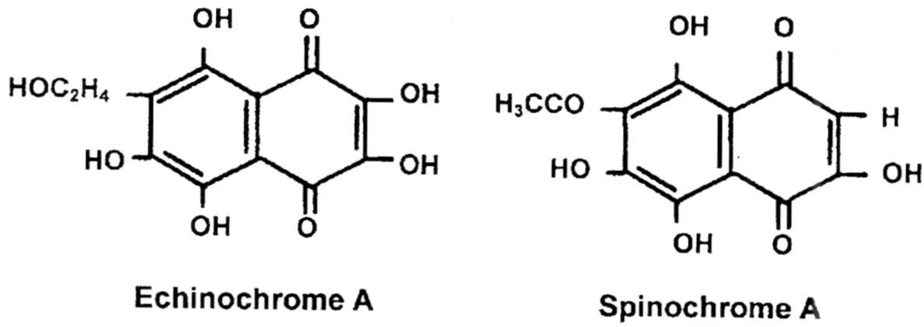

Fig. 3. Structures of echinochrome A and spinochrome A

asteroids. Preparations of major carotenoids from plants seem more economical than from echinoderms. Pigments other than major carotenoids, which are specific to echinoderm species, will be a focus of pharmaceutical development. Tsushima et al. (1995) carried out extensive studies on echinoderm carotenoids from the viewpoints of comparative biochemistry and pharmacology. They examined 51 carotenoids with different structures for the inhibitory effects on Epstein-Barr virus activation. A novel marine carotenoid from *Cucumaria japonica*, cucumariaxanthin C, showed such an effect (Tsushima et al. 1996). Quinone sulfates isolated from the crinoids *Tropiometra afra macrodisucus* and *Oxycomanthus japonicus* have shown antifeedant activity on fish (Takahashi et al. 2002). Spinochrome and echinochrome are assumed to show hypotensive activity (Kuzuya et al. 1973; Fig. 3). Pharmaceutical investigation into echinodermal quinones is required for the development of new anticancer reagents.

6 Venoms

Venoms occur in two echinoderm classes, namely echinoids and asteroids. Their distribution is limited to three echinoid families, Echinothuriidae, Diadematidae and Toxopneustidae. *Acanthaster planci* is the only species that has been reported to be venomous among starfish. A few studies on sea urchin venoms have been reported. Peditoxin, purified from the pedicellariae of *Toxopneustes pileolus*, is composed of a protein called pedin and a prosthetic group called pedoxin (Kuwabara 1994). Venomous activity results from pedoxin, which has a molecular mass of 206 Da. It causes sedation and anesthetic coma accompanied by muscular relaxation at sublethal doses. Another toxin, which was named contractin A, from the venom of *T. pileolus* pedicellaria, has been purified and characterized (Nakagawa et al. 1991). Contractin A, having an apparent molecular weight of 18,000 Da for a total of 138 amino-acid residues, caused contraction of the tracheal smooth muscle. Spine venom from *A. planci* has been extensively investigated by Japanese researchers. They demonstrated that *A. planci* venom has various toxic activities. The lethal factor was shown to be a potent hepatotoxic basic glycoprotein with a molecular weight of 20,000–25,000 Da (Shiomi et al. 1988, 1990). A new anticoagulant peptide with a native molecular mass of 7,500 Da from the spine venom of *A. planci*, plancinin, inhibits factor X activation in the human blood coagulation cascade (Koyama et al. 1998). A fraction of venom from the crown-of-thorns starfish causes smooth muscle contraction mediated by prostaglandins (Karasudani et al. 1996). In addition, the venom showed vasorelaxing and hypotensive effects, which were assumed to be due to the release of a platelet-activating factor or a factor-like substance (Yara et al. 1992). The fact that the molecular mass of pedoxin from *T. pileolus* is 206 Da may suggest that the development of its synthesis is possible. Further, pedoxin seems to show low antigenicity because of its low molecular mass when it is subcutaneously or intramuscularly injected. There are still many sea urchin species to be investigated, the venoms of which are not yet characterized.

7 Hemagglutinins

In echinoderms, hemagglutinins were reported in a starfish and sea urchins. Hemagglutinin occurs in the coelomic fluid of *Anthocidaris crassispina* (Giga et al. 1987) and *Hemicentrotus pulcherrimus* (Yamada and Aketa. 1982), and in eggs of *Anthocidaris crassispina* (Ozeki et al. 1991). Hemagglutinins of *A. crassispina* have been extensively investigated and their complete amino-acid sequence has been reported. Also in holothurians, hemagglutinin was reported (Hatakeyama et al. 1994; Matsui et al. 1994).

In general, hemagglutinins from echinoderms do not seem advantageous for clinical use, as most of them have not shown any blood group specificity.

8 Prospects

Organisms are capable of synthesizing many kinds of biologically active substances with diverse molecular structures. Various antibiotics have been discovered and incorporated in the treatment of infection in the last century. Development of biologically active chemicals from the half million species of marine organisms is one interest. In order to efficiently fulfill biological and pharmaceutical surveys of substances, a globally standardized system for biological screening must be established. Biologically active substances are generally divided into two categories: those that are antigenic and those that are non-antigenic. When an antigenic substance is subcutaneously or intramuscularly administered as a medication, there is a conflict with the host immune system. It is not to say that oral administration of peptides and proteins is impossible. This is why antitumoral proteins found in the sea urchins *Strongylocentrotus purpuratus*, *S. intermedius* and *S. nudus* (Pettit et al. 1979; Sasaki and Endo 1987) are not yet employed in therapy. However, analyses of the action mechanism and the notion of active sites in biologically active peptides may contribute to the development of novel pharmaceuticals without the side effect of antigenicity.

A large variety of toxic substances have been used for a long time as drugs. In addition, a number of useful biologically active substances such as anodynes have been synthesized based on the molecular structure of toxins. The fact that saponins and terpenoids occur only in Echinodermata in the animal kingdom suggests that echinoderms are capable of synthesizing unknown substances. Saponins from asteroids and holothurians display a variety of pharmaceutical activities. The utilization of synthetic activities specific to asteroids and holothurians is one way of producing new medications. Novel biologically active substances may act as potential leading compounds even though their isolation for such use is not commercially practical.

Finally, the blooming of starfish and sea cucumber, *Cucumaria echinata*, is a serious obstacle to fishery and the only countermeasure against it is their extirpation. The 2000 annual catch of starfish in Hokkaido, Japan, was reported to be 16,000 t. The disposal of extirpated animals is a serious problem for the fishery sectors. Exploitation of chemicals from these animals would be beneficial regarding both disposal of industrial waste and production of useful substances.

9 Appendix: Antique Illustrations of Echinoderms in Japan

We can go back to an encyclopedia, *Wakansansaizue*, edited by the physician Ryoan Terashima in 1712, to the scholarly description and illustration of the echinoderm. A sea urchin, a supposed starfish, a sand dollar and a sea cucumber were described and illustrated (Figs. 4–6). The encyclopedia is composed of 105 volumes. It was edited on the basis of an antique idea that human diseases were generated by three components: the heavens, earth (including animals and plants) and humans. *Wakansansaizue* literally means Japanese–Chinese encyclopedia on the three components.

In 1762, Yoritaka Matsudaira completed a series of picture books on natural history entitled *Shurinshukan* in order to enhance the local activity of fishery and related industries. Four of them, called *Shurinzu* and dealing with fishes, are highly valued for the quality of their drawings. Even echinoderms, sea cucumbers, starfish and brittle stars are illustrated in the third issue of *Shurinzu* (Figs. 7 and 8).

Senchufu (1811) by Tanshu Kurimoto (1756–1834) describes 27 species of echinoderms. This number seems incorrect because some species appeared

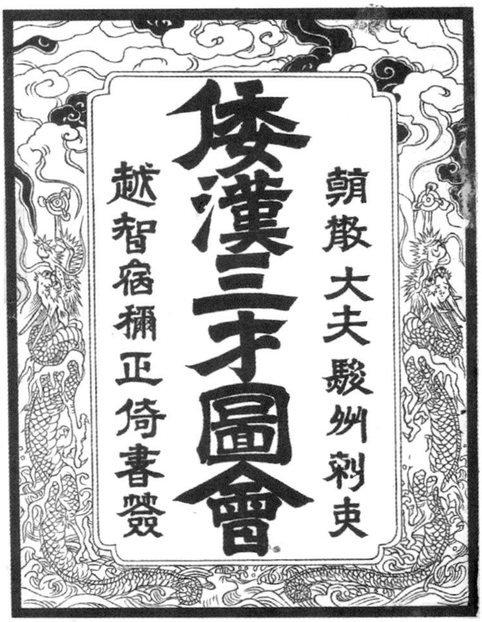

Fig. 4. Cover of *Wakansansaizue* (from the reprint published in 1902 by Chugai Shuppan, Tokyo)

もちかひ　海燕　ハアイエン

海燕　俗云餅貝　又云海星

陽遂足　俗云鮹枕

本綱海燕出東海大一寸狀扁面圓背上青黑腹下白脆似海鰾蛸有紋如蕈菌口在腹下食細沙口旁有五路正勾即其足也

陽遂足　本綱生海中色青黑腹白有五足不知頭尾生時躰耎死則乾脆（時珍以陽遂足出於海燕之下即以二物一）

△按海燕陽遂足二種時珍以爲一物混註之者非也

海燕圓薄扁如馬錢子而大一二寸灰白色有結梗花文其裏正中有一小孔即是爲口其旁有五路正勾文而似彫成勝之具山人見之疑貝石蕈菓器之間

Fig. 5. Illustration and description of starfish and sand dollar in *Wakansansaizue* (from the reprint published in 1902 by Chugai Shuppan, Tokyo). The author classified these two animals into different animal groups, although a famous Chinese scholar, Li Shizhen (1518–1594), described them as belonging to the same category in his pharmaceutical book *Bencaogangmu*. A starfish, named Takonomakura, is a blue-grey animal with five podia. A sand dollar, called Mochikai, has a thin circular body, on which a penta-radiated pattern is seen. A starfish is assumed *Asteropecten* sp. judging from the explanation given. An animal called Takonomakura is today *Clypeaster japonicus*

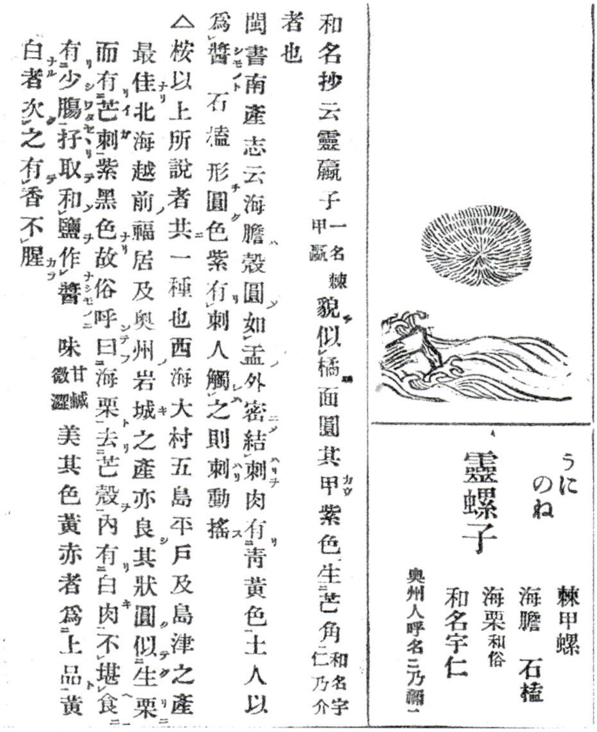
和名抄云霊蠃子一名棘甲蠃貌似橘而圓其甲紫色生芒角和名宇仁乃介
者也
閩書南產志云海膽殼圓如盂外密結刺肉有青黃色土人以
爲醬 石榼形圓色紫有刺人觸之則刺動搖
△按以上所說者共一種也西海大村五島平戶及島津之產
最佳北海越前福居及奧州岩城之產亦良其狀圓似生栗
而有芒刺紫黑色故俗呼曰海栗去芒殼內有白肉不堪食
有少腸抒取和鹽作醬 味甘鹹微澀 美其色黃赤者爲上品黃
白者次之有香不腥

Fig. 6. Description of sea urchin in *Wakansansaizue* (from the reprint published in 1902 by Chugai Shuppan, Tokyo). The body is round and similar to a bur. The animal moves its spines in response to contact stimuli. Localities of sea urchin production and quality of sea urchin roes were explained. The description of the animal in the text suggests that the species is *Anthocidaris crassispina*

Fig. 7. Sea cucumber in *Shurinzu* drawn by Bunryu Miki (1762) (from a publication of the Kagawa History Museum, 2003). The animal appears to be *Apostichopus japonicus*. Original drawing is in color

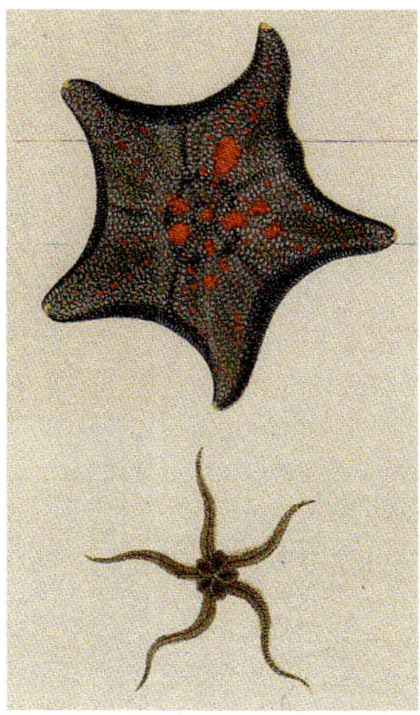

Fig. 8. Starfish and brittle star in *Shurinzu* (from a publication of the Kagawa History Museum, 2003). The starfish is recognized as *Asterina pectinifera*. Original in color

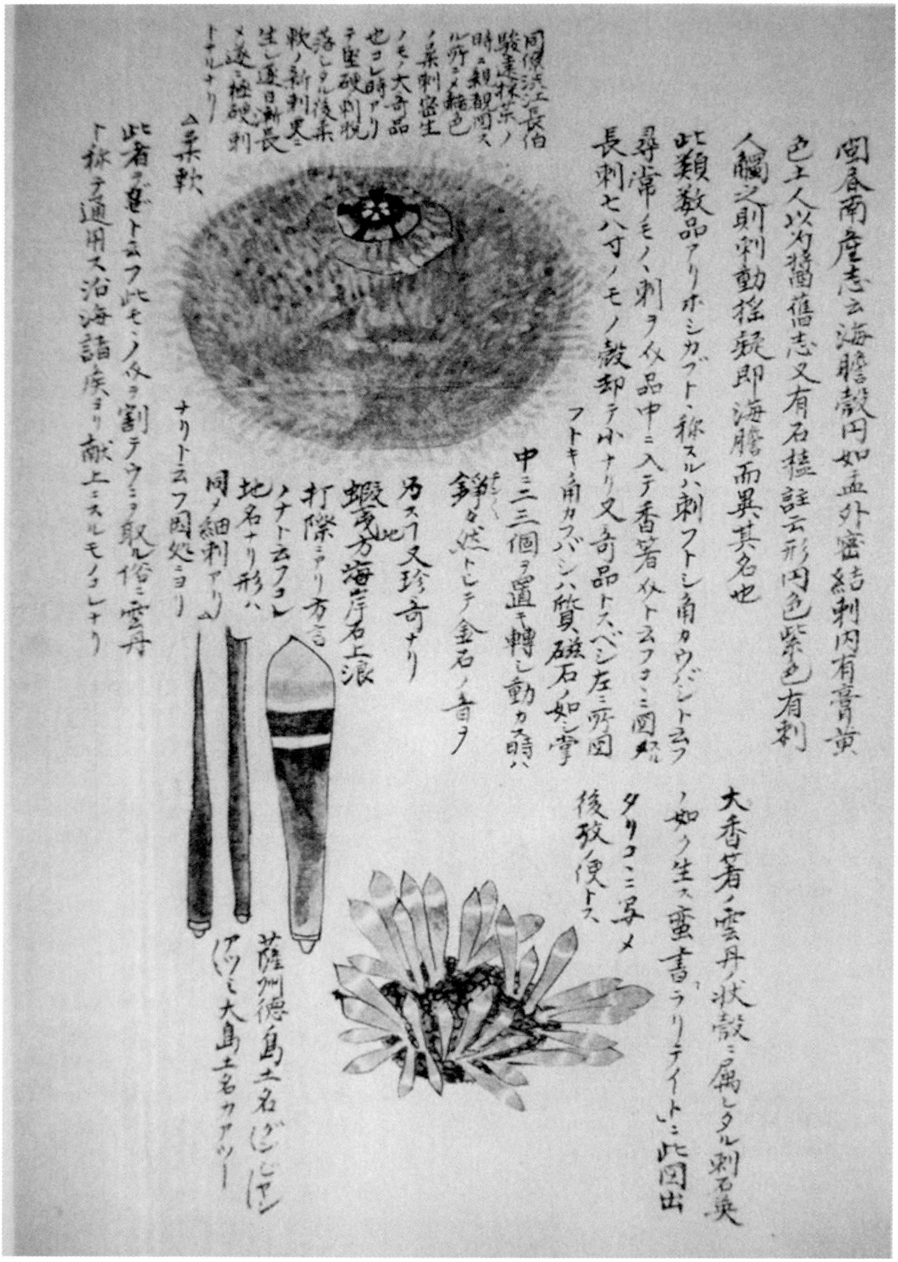

Fig. 9. Description of *Heterocentrotus mamillatus* in *Senchufu* (1811) (from the reprint published by Kowa Shuppan, 1982). It was described that this species occurred in southern Japan

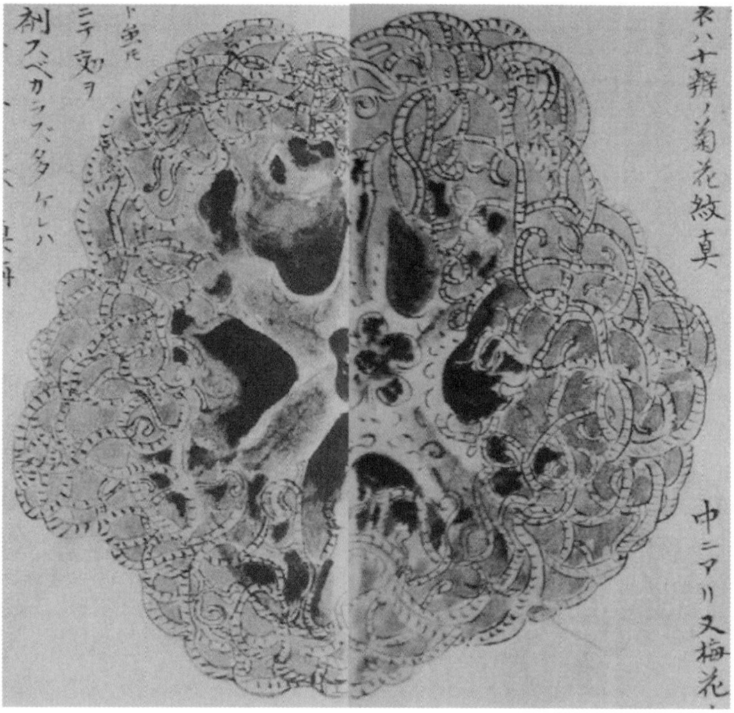

Fig. 10. Illustration of *Gorgonocephalus eucnemus* from *Senchufu* (1811) (from the reprint published by Kowa Shuppan, 1982). This animal was cited as a peculiar species of starfish. It was written that powdered *Gorgonocephalus* was used as a medication for bruising. The animal was reported to have fundamentally five large tentacles separating into a large number of small tentacles. A penta-radiated pattern like a cherry blossom was described in the center of the body. The animal is illustrated on two facing pages

two or more times in the different sections and the identification of species was unclear. *Stichopus japonicus*, *Asteropecten*, *Diadema setosum*, *Asterina pectinifera*, *Strongylocentrotus intermedius*, *Heterocentrotus mamillatus*, *Astriclypeus manni*, *Asterias amurensis*, *Gorgonocephalus eucnemus*, etc. were illustrated (Figs. 9 and 10). The original book of *Senchufu* was lost due to fire and only some reproductions remain.

References

Avilov SA, Antonov AS, Drozdova OA, Kalinin VI, Kalinovsky AI, Stonik VA, Riguera R, Lenis LA, Jimenez C (2000) Triterpene glycosides from the Far-Eastern sea cucumber *Pentamera calcigera*. 1. Monosulfated glycosides and cytotoxicity of their unsulfated derivatives. J Nat Prod 63:65–71

Chludil HD, Muniain CC, Seldes AM, Maier MS (2002) Cytotoxic and antifungal triterpene glycosides from the Patagonian sea cucumber *Hemoiedema spectabilis*. J Nat Prod 65:860–865

Cooper HS (1880) Coral sands, vol 2. Bentley and Sons, London

Fox DL, Hopkins TS (1966) The comparative biochemistry of pigments. In: Boolootian RA (ed) Physiology of Echinodermata. Interscience, New York, pp 277–300

Giga Y, Ikai A, Takahashi K (1987) The complete amino acid sequence of echinoidin, a lectin from the coelomic fluid of the sea urchin *Anthocidaris crassispina*. Homologies with mammalian and insect lectins. J Biol Chem 262:6197-6203

Gurst JE, Sheikh YM, Djerassi C (1973) Synthesis of corticosteroids from marine sources. J Am Chem Soc 95:628–629

Hatakeyama T, Kohzaki H, Nagatomo H, Yamasaki N (1994) Purification and characterization of four $Ca^{(2+)}$-dependent lectins from the marine invertebrate, *Cucumaria echinata*. J Biochem 116:209–214

Higuchi R, Inagaki M, Natori T, Komori T, Kawajiri S (1991) Biologically active glycosides from Asteroidea. XXV Glycosphingolipids from the starfish *Asterina pectinifera*. 2 Structure of three ganglioside molecular species and a homogeneous ganglioside, and biological activity of the ganglioside. Liebigs Ann Chem 1991:1–10

Higuchi R, Inukai K, Jhou JX, Honda M, Komori T, Tsuji S, Nagai Y (1993) Biologically active glycosides from Asteroidea. XXXI Glycosphingolipids from the starfish *Asterias amurensis versicolor* Slade. 2 Structure and biological activity of ganglioside molecular species. Liebigs Ann Chem 1993:359–366

Higuchi R, Inagaki M, Togawa K, Miyamoto T, Komori T (1994) Isolation and structure of three new cerebrosides, CE-2b, CE-2 c and CE-2d, from the sea cucumber *Cucumaria echinata*. Liebigs Ann Chem 1994:79–81

Higuchi R, Matsumoto S, Fujita M, Komori T, Sasaki T (1995) Biologically active glycosides from Asteroidea. XXXII Glycosphingolipids from the starfish *Asteropecten latespinosus*. Structure of two new ganglioside molecular species and biological activity of the ganglioside. Liebigs Ann 1995:545–550

Ikegami S, Kamiya Y, Tamura S (1972) A new sterol from asterosaponins A and B. Tetrahedron Lett 13:1601–1604

Karasudani I, Koyama T, Nakandakari S, Aniya Y (1996) Purification of anticoagulant factor from the spine venom of the crown-of-thorns starfish, *Acanthaster planci*. Toxicon 34:871–879

Kitagawa I, Sugawara T, Yosioka I, Kuriyama K (1976a) Saponin and sapogenol. XIV. Antifungal glycosides from the sea cucumber *Stichopus japonicus* Selenka. (1) Structure of Stichopogenin A_4, the genuine aglycon of holotoxin A. Chem Pharm Bull 24:266–274

Kitagawa I, Sugawara T, Yosioka I (1976b) Saponin and sapogenol. XV. Antifungal glycosides from sea cucumber *Stichopus japonicus* Selenka. (2) Structure of holotoxin A and holotoxin B. Chem Pharm Bull 24:275–284

Koyama T, Noguchi K, Aniya Y, Sakanashi M (1998) Analysis for sites of anticoagulant action of plancinin, a new anticoagulant peptide isolated from the starfish *Acanthaster planci*, in the blood coagulation cascade. Gen Pharmacol 31:277-282

Kubo H, Irie A, Inagaki F, Hoshi M (1990) Gangliosides from the eggs of the sea urchin, *Anthocidaris crassispina*. J Biochem 108:185–192

Kuwabara S (1994) Purification and properties of peditoxin and the structure of its prosthetic group, pedoxin, from the sea urchin *Toxopneustes pileolus* (Lamarck). J Biol Chem 269:26734–26738

Kuzuya H, Ikuta K, Nagatsu T (1973) Inhibition of dopamine-beta-hydroxylase by spinochrome A and echinochrome A, naphthoquinone pigments of echinoids. Biochem Pharmacol 22:2722–2724

Maier MS, Roccatagliata AJ, Kuriss A, Chludil H, Seldes AM, Pujol CA, Damonte EB (2001) Two new cytotoxic and virucidal trisulfated triterpene glycosides from the Antarctic sea cucumber *Staurocucumis liouvillei*. J Nat Prod 64:732–736

Matsui T, Ozeki Y, Suzuki M, Hino A, Titani K (1994) Purification and characterization of two $Ca^{(2+)}$-dependent lectins from coelomic plasma of sea cucumber, *Stichopus japonicus*. J Biochem 116:1127–1133

Nakagawa H, Tu AT, Kimura A (1991) Purification and characterization of Contractin A from the pedicellarial venom of sea urchin, *Toxopneustes pileolus*. Arch Biochem Biophys 284:279–284

Nigrelli R, Zahl PA (1952) Some biological characteristics of holothurin. Proc Soc Exp Biol Med 81:379–380

Ozeki Y, Matsui T, Titani K (1991) Amino acid sequence and molecular characterization of a D-galactoside-specific lectin purified from sea urchin (*Anthocidaris crassispina*) eggs. Biochemistry. 30:2391–2394

Pettit GR, Herald CL, Vanell LD (1979) Isolation and characterization of strogylostatin 1. J Nat Prod 42:407–409

Sasaki T, Endo Y (1987) Antineoplastic substances from marine animals (in Japanese). In: Yasumoto T, Kamiya H (eds) Bioactive marine natural products. Koseishakoseikaku, Tokyo, pp 21–31

Shimada S (1969) Antifungal steroid glycoside from sea cucumber. Science 163:1462

Shiomi K, Yamamoto S, Yamanaka H, Kikuchi T (1988) Purification and characterization of a lethal factor in venom from the crown-of-thorns starfish (*Acanthaster planci*). Toxicon 26:1077–1083

Shiomi K, Yamamoto S, Yamanaka H, Kikuchi T, Konno K (1990) Liver damage by the crown-of-thorns starfish (*Acanthaster planci*) lethal factor. Toxicon 28:469–475

Takahashi D, Maoka T, Tsushima M, Fujitani K, Kozuka M, Matsuno T, Shingu T (2002) New quinone sulfates from the crinoids *Tropionetra afra macrodiscus* and *Oxycomathus japonicus*. Chem Pharm Bull 50:1609–1612

Tsushima M, Maoka T, Katsuyama M, Kozuka M, Matsuno T, Tokuda H, Nishino H, Iwashima A (1995) Inhibitory effect of natural carotenoids on Epstein-Barr virus activation activity of a tumor promoter in Raji cells. A screening study for anti-tumor promoters. Biol Pharm Bull 18:227–233

Tsushima M, Fujiwara Y, Matsuno T (1996) Novel marine di-Z-carotenoids: cucumariaxanthins A, B, and C from the sea cucumber *Cucumaria japonica*. J Nat Prod 59:30–34

Yamada K (2002) Chemo-pharmaceutical studies on the glycosphingolipid constituents from echinoderm, sea cucumbers, as the medicinal materials (in Japanese) Yakugaku Zasshi 122:1133–1143

Yamada Y, Aketa K (1982) Purification and partial characterization of hemagglutinins in seminal plasma of the sea urchin, *Hemicentrotus pulcherrimus*. Biochim Biophys Acta 709:220–226

Yamada K, Harada S, Miyamoto T, Isobe R, Higuchi R (2000) Constituents of Holothuroidea. 9. Isolation and structure of a new ganglioside molecular species from the sea cucumber *Holothuria pervicax*. Chem Pharm Bull 48:157–159

Yamada K, Matsubara R, Kaneko M, Miyamoto T, Higuchi R (2001) Constituents of Holothuroidea. 10. Isolation and structure of a biologically active ganglioside molecular species from the sea cucumber *Holothuria leucospilota*. Chem Pharm Bull 49:447–452

Yamanouchi T (1942) On the poisonous extracts from holothurians (in Japanese). Teikoku Gakushiin Hokoku 73

Yamanouchi T (1943a) On the distribution of poison in the body of holothurians (in Japanese). Zool Mag Tokyo 55:87

Yamanouchi T (1943b) On the poison contained in holothurians (in Japanese). Zool Mag Tokyo 55:115

Yara A, Noguchi K, Nakasone J, Kinjo N, Hirayama K, Sakanashi M (1992) Cardiovascular effects of *Acanthaster planci* venom in the rat: possible involvement of PAF in its hypotensive effect. Toxicon 30:1281–1289

Zou ZR, Yi YH, Wu HM, Wu JH, Liaw CC, Lee KH (2003) Intercedensides A–C, three new cytotoxic triterpene glycosides from the sea cucumber *Mensamaria intercedens* Lampert. J Nat Prod 66:1055–1060

Subject Index

Taxonomic Index

Authors Index

Printing: Mercedes-Druck, Berlin
Binding: Stein + Lehmann, Berlin